RENEWALS 458-4574
DATE DUE

WITHDRAWN
UTSA Libraries

Ecological Studies, Vol. 151

Analysis and Synthesis

Edited by

I.T. Baldwin, Jena, Germany
M.M. Caldwell, Logan, USA
G. Heldmaier, Marburg, Germany
O.L. Lange, Würzburg, Germany
H.A. Mooney, Stanford, USA
E.-D. Schulze, Jena, Germany
U. Sommer, Kiel, Germany

Ecological Studies

Volumes published since 1995 are listed at the end of this book.

Springer

Berlin
Heidelberg
New York
Barcelona
Hong Kong
London
Milan
Paris
Singapore
Tokyo

K. Reise (Ed.)

Ecological Comparisons of Sedimentary Shores

With 82 Figures and 27 Tables

 Springer

Prof. Dr. Karsten Reise
Alfred-Wegener-Institut
für Polar- und Meeresforschung
Wattenmeerstation Sylt
25992 List
Germany

Cover illustration: Part of the picture "The Awakened World" by René Magritte.
© VG Bild-Kunst, Bonn 2000

ISSN 0070-8356
ISBN 3-540-41254-9 Springer-Verlag Berlin Heidelberg New York

Library of Congress Cataloging-in-Publication Data.

Ecological comparisons of sedimentary shores / K. Reise (ed.).
   p. cm. – (Ecological studies ; v. 151)
  Includes bibliographical references.
  ISBN 3540412549 (alk. paper)
   1. Coastal ecology. I. Reise, Karsten. II. Ecological studies ; v. 151.

QH541.5.C65 E24 2001
577.5'1–dc21   2001020043

This work is subject to copyright. All rights are reserved, whether the whole or part of the material is concerned, specifically the rights of translation, reprinting, reuse of illustrations, recitation, broadcasting, reproduction on microfilm or in any other way, and storage in data banks. Duplication of this publication or parts thereof is permitted only under the provisions of the German Copyright Law of September 9, 1965, in its current version, and permissions for use must always be obtained from Springer-Verlag. Violations are liable for prosecution under the German Copyright Law.

Springer-Verlag Berlin Heidelberg New York
a company of BertelsmannSpringer Science+Business Media GmbH

http://www.springer.de

© Springer-Verlag Berlin Heidelberg 2001
Printed in Germany

The use of general descriptive names, registered names, trademarks, etc. in this publication does not imply, even in the absence of a specific statement, that such names are exempt from the relevant protective laws and regulations and therefore free for general use.

Cover design: *design & production* GmbH, Heidelberg
Typesetting: Bader · Damm · Kröner, Heidelberg

SPIN 10737926   31/3130 YK – 5 4 3 2 1 0 – Printed on acid free paper

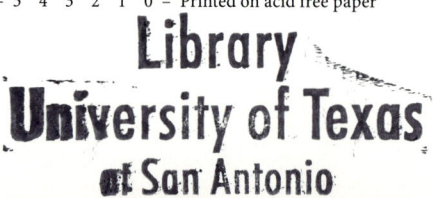

Preface

In the autumn of 1999 a group of marine ecologists gathered in a lecture hall amidst the sand dunes on the island of Sylt in the North Sea. Seaward of the dunes we walked along an exposed sandy beach and leeward of this slender island across extensive mud flats. In this environment we brought together and discussed our insights from ecological studies on sediment shores from coasts all around the world. We aimed at pairing analyses of local processes with global comparisons between distant coastal biota to find generalities and to identify singularities for these diverse and changing habitats.

Sedimentary coasts with their unique forms of life and highly productive ecosystems belong to the most threatened part of the biosphere. Nevertheless, sound knowledge of crucial processes and patterns in the coastal habitats is scant on a worldwide scale. This may severely limit the application of strategies by coastal managers from one coast to another. This book is not intended to be a synthesis of results but should help to enlarge the domain of competence. It is presented as an example of the kinds of basic research needed to gain a global perspective on the ecology of sedimentary shores. There is still much work to be done.

We wish to proceed from studies devoted to the analysis of contingent details on a local scale to an integration of knowledge across biogeographic regions. With this intention in mind about 80 participants from many coasts of the world came to the island of Sylt to discuss the comparative ecology of sedimentary shores. Thirty-three scientists have contributed to this volume and many others have carefully refereed individual chapters. Elisabeth Herre diligently compiled the species and subject index. Ulrich Sommer from the Editorial Board of Ecological Studies kindly approved the inclusion of this volume in the Series. Dieter Czeschlik, Andrea Schlitzberger and their staff of Springer-Verlag gave a remarkable support to this coastal project. The production editor Friedmut Kröner circumspectly managed the final step in the evolution of Volume 151. As the editor, I thank all these persons. We acknowledge the German Research Foundation (DFG) for funding the travel expenses to bring participants from as far as Chile, New Zealand and Japan to the shores of the North Sea; and the Alfred-

Wegener-Institut (AWI) for supporting the meeting. Our warmest thanks go to our hosts, the Akademie am Meer, Klappholttal auf Sylt, and to the helpful staff and students of the Wattenmeerstation Sylt, a field station for coastal research which celebrated its 75th anniversary in 1999.

Spring 2001   *Karsten Reise*, Sylt

Contents

Introduction 1
K. Reise

Part I Suspension Feeders in Coastal Mud and Sand

1   Benthic Suspension Feeders as Determinants of Ecosystem
    Structure and Function in Shallow Coastal Waters 11
    R. Dame, D. Bushek, T. Prins

1.1   Introduction 11
1.2   Reefs and Beds 12
1.2.1   Bivalve Molluscs (Bivalvia) 12
1.2.2   Vermetids (Gastropoda) 21
1.2.3   Sabellids (Polychaeta) 22
1.2.4   Serpulids (Polychaeta) 23
1.3   Encrustations 24
1.4   Species Groups 24
1.5   Discussion 26
References 31

2   Dynamics of Spatial and Temporal Complexity in European
    and North American Soft-Bottom Mussel Beds 39
    J.A. Commito, N.M.J.A. Dankers

2.1   Introduction 39
2.2   The Dynamic Nature of Mussel-Bed Structure 40
2.3   Quantifying Mussel-Bed Structure
        Using Fractal Geometry 44

2.4	Effects of Mussel Beds on Soft-Bottom Community Structure	46
2.5	Mechanisms of Mussel-Bed Impacts on Soft-Bottom Community Structure	50
2.6	Top-Down vs. Bottom-Up Control of Soft-Bottom Mussel-Bed Community Structure	52
2.7	Conclusions	54
References		54

3	**Suspension Feeders on Sandy Beaches**	61
	E. Jaramillo, M. Lastra	
3.1	Introduction	61
3.2	Beach Morphodynamic Types vs. Community Structure of the Macroinfauna	64
3.3	Beach Morphodynamic Types and Abundances and Population Biology of *E. analoga*	67
3.4	Tidal Movements and Burrowing Behaviour	68
3.5	Across- and Along-Shore Zonation	68
3.6	Conclusions	70
References		70

4	**Switching Between Deposit and Suspension Feeding in Coastal Zoobenthos**	73
	H.U. Riisgård, P. Kamermans	
4.1	Overview	73
4.1.1	Polychaetes	75
4.1.2	Echinoderms	80
4.1.3	Bivalves	81
4.1.4	Amphipods	82
4.1.5	Soft Corals	83
4.1.6	Most Examples Among Passive Suspension Feeders	84
4.1.7	Adaptation to Suspension Feeding	84
4.2	Example I: Switching to Suspension Feeding in *Nereis diversicolor*	85
4.2.1	Suspension-Feeding Behaviour	86
4.2.2	Mucus-Net and Particle-Retention Efficiency	86
4.2.3	Filtration Rates	87
4.2.4	Energy Cost of Pumping	87

4.2.5	Adaptation to Suspension Feeding	88
4.2.6	Time Spent on Suspension Feeding	88
4.2.7	Phytoplankton Reduction in Near-Bottom Water	89
4.3	Example II: Switching to Suspension Feeding in *Macoma balthica*	90
4.3.1	Switching to Suspension Feeding	91
4.3.2	Current Velocity	92
4.3.3	Food Availability	92
4.3.4	Feeding on Siphon Tips	94
4.3.5	Protection Against Lethal Predation	94
4.4	Conclusions	95
References		95

Part II Biogenic Stabilization and Disturbances in Coastal Sediments

5	**Microphytobenthos in Contrasting Coastal Ecosystems: Biology and Dynamics**	105
	D.M. Paterson, S.E. Hagerthey	
5.1	Contrasting Shores	105
5.2	The Microphytobenthos	105
5.3	Physical and Biological Sediment Properties	107
5.3.1	Sediment Types and Stability	107
5.3.2	Physical Dynamics	109
5.4	Redefining Intertidal Sediments – The Five Phases of Depositional Environments	112
5.5	Comparative Biodiversity	113
5.5.1	Non-cohesive Sediments	114
5.5.2	Cohesive Sediments	115
5.5.3	Niche Diversity	116
5.6	Sediment Stability	118
5.7	Conceptual Model	119
5.8	Conclusions	121
References		121

6	**Sediment Dynamics by Bioturbating Organisms**	127
	G.C. Cadée	
6.1	Introduction	127
6.2	History of Bioturbation Research	130

6.3	Types of Bioturbation	132
6.3.1	Crawling and Dwelling Traces	132
6.3.2	Deposit Feeders	133
6.3.3	Larger Predators and Grazers	134
6.4	Seasonal Variation	138
6.5	Latitudinal Variation	139
6.6	Changes in Historical Times	141
6.7	Conclusions	142
References		143

7 Competitive Bioturbators on Intertidal Sand Flats in the European Wadden Sea and Ariake Sound in Japan 149
E. Flach, A. Tamaki

7.1	Introduction	149
7.2	Large Bioturbators	150
7.3	Lugworms in the Wadden Sea	151
7.4	Effects of Lugworms on the Benthic Community	153
7.5	Ghost Shrimps in the Ariake Sound Estuarine System in Japan	158
7.6	Effects of the Ghost Shrimp Expansion and Decline	163
7.6.1	Effects on Sediment Properties	163
7.6.2	Effects on Invertebrates	164
7.7	Comparisons Between Biogeographic Regions	165
References		168

8 Biological and Physical Processes That Affect Saltmarsh Erosion and Saltmarsh Restoration: Development of Hypotheses 173
R.G. Hughes

8.1	Introduction	173
8.2	Managed Realignment	178
8.2.1	Physical Factors	178
8.2.2	Biological Factors	179
8.2.2.1	Effects of the Flora	179
8.2.2.2	Effects of Invertebrates	180
8.3	Managing Sediment Accretion and Development of Saltmarsh Vegetation in Managed Realignment Sites	182
8.3.1	Physical Factors	182

8.3.2	Biological Factors	183
8.4	Loss of Saltmarsh Vegetation by Lateral Erosion of Creeks	185
8.5	Managing Reduction of Lateral Creek Erosion	187
8.6	Conclusions	189
References		190

Part III Seagrasses and Benthic Fauna of Sediment Shores

9 Common Structures and Properties of Seagrass Beds Fringing the Coasts of the World ... 195
C. den Hartog, R.C. Phillips

9.1	Introduction	195
9.2	Distribution and Zonation	195
9.3	Structure of the Seagrass Community	199
9.4	Seagrass Production	202
9.5	Seagrass Dynamics	204
9.6	Worldwide Decline of Seagrass Beds	205
9.7	Conclusions	208
References		208

10 The Leaf Canopy of Seagrass Beds: Faunal Community Structure and Function in a Salinity Gradient Along the Swedish Coast ... 213
S.P. Baden, C. Boström

10.1	Introduction	213
10.1.1	*Zostera marina* (L.)	213
10.1.2	Aims of the Study	214
10.2	Features of the Study Area	215
10.2.1	Effects of Salinity	215
10.2.2	Physical Settings and Substrate Characteristics	217
10.3	Methods	219
10.3.1	Vegetation and Leaf Canopy Fauna	219
10.3.2	Predators	220
10.4	Results	220
10.4.1	*Zostera marina*: Standing Stock and Leaf Area	220
10.4.2	Leaf Fauna	222

10.4.3	Couplings Between Leaf Fauna and Infauna	227
10.4.4	Predators	228
10.5	Concluding Remarks	230
References		231

11   Energy Flow in Benthic Assemblages of Tidal Basins: Ria Formosa (Portugal) and Sylt-Rømø Bay (North Sea) Compared 237
M. Sprung, H. Asmus, R. Asmus

11.1	Introduction	237
11.2	Description of the Sites	239
11.3	Material and Methods	241
11.3.1	Primary Production	241
11.3.2	Secondary Production	242
11.4	Production	243
11.4.1	Primary Production	243
11.4.2	Secondary Production	245
11.5	Energy Flow and Nutrient Cycle	248
References		251

12   Soft-Bottom Fauna of a Tropical (Banc d'Arguin, Mauritania) and a Temperate (Juist Area, German North Sea Coast) Intertidal Area .. 255
H. Michaelis, W.J. Wolff

12.1	Introduction	255
12.2	Areas, Materials and Methods	256
12.3	Results	260
12.3.1	Habitat Division	260
12.3.2	Faunal Inventories	261
12.3.3	Structure and Distribution of Macrozoobenthos Assemblages	261
12.4	Discussion	268
References		272

13	Tropical Tidal Flat Benthos Compared Between Australia and Central America	275
	S. Dittmann, J.A. Vargas	

13.1	Introduction	275
13.2	Tropical Tidal Flats in Australia and Central America	277
13.3	Species Diversity and Abundance	279
13.3.1	Species Richness in Tropical Tidal Flats	279
13.3.2	Similarity in Taxonomic Compositions	281
13.3.3	Individual Abundances	282
13.4	Community Structure and Distribution	284
13.4.1	Spatial Zonation Along Environmental Gradients	284
13.4.2	Trophic Groups	284
13.4.3	Species Interactions	286
13.4.3.1	Promotive Interactions	286
13.4.3.2	Repressive Interactions	288
13.5	Conclusions	289
References		290

Part IV Structural Dynamics and Trophic Supplies to Sedimentary Shores

14	Recovery Dynamics in Benthic Communities: Balancing Detail with Simplification	297
	S.F. Thrush, R.B. Whitlatch	

14.1	Introduction	297
14.2	Searching for Generality Part I	299
14.3	Some General Mechanisms Influencing Recovery	301
14.3.1	Seasonality	301
14.3.2	Hydrodynamics	301
14.3.3	Mobility	302
14.3.4	Opportunistic Responses	302
14.3.5	Biotic Interactions	303
14.4	Searching for Generality Part II	304
14.5	Critical Scales of Disturbance and Recovery Dynamics	306
14.6.	Recovery: a Useful Tool for Assessing Broad-Scale and Cumulative Effects?	309
14.7	Searching for Generality Part III: the Need to Improve the Information Base	310

14.8	Conclusions	310
References		311

15 Population Dynamics of Benthic Species on Tidal Flats: the Possible Roles of Shorebird Predation 317
J. van der Meer, T. Piersma, J.J. Beukema

15.1	Introduction	317
15.2	Production-Consumption Comparisons	317
15.3	The Balgzand Area: a Long-Term Study	319
15.4	Long-Term Variability in Production and Consumption at the Balgzand	320
15.5	Density-Dependent Survival?	324
15.6	Recruitment and the Regulation of Populations	326
15.7	The Scale of Population Studies	329
15.8	Exclosure Experiments	329
15.9	Conclusions	330
References		332

16 Experimental Approaches to Integrating Production, Structure and Dynamics in Sediment Communities ... 337
D. Raffaelli, M. Emmerson

16.1	Introduction	337
16.2	Effects of Production Subsidies on Food Chain Dynamics	340
16.2.1	Effects of Subsidised Primary Producers	341
16.2.2	Effects of Subsidised Predators	343
16.3	Production and Body-Size Distributions	343
16.3.1	Constraint Space Plots	344
16.3.2	Biomass Size Spectra	345
16.4	Production and Biodiversity	347
16.5	Conclusions	350
References		350

Synthesis: Comparative Ecology of Sedimentary Shores 357
K. Reise

Suspension Feeders 357
Biogenic Stabilization and Disturbances 361
Seagrasses and the Benthic Fauna 361
Dynamic Structures and Trophic Supplies 367
General Conclusions 369
References 371

Species Index 373

Subject Index 379

Contributors

H. Asmus

   Alfred-Wegener-Institut für Polar- und Meeresforschung,
   Wattenmeerstation Sylt, 25992 List, Germany

R. Asmus

   Alfred-Wegener-Institut für Polar- und Meeresforschung,
   Wattenmeerstation Sylt, 25992 List, Germany

S.P. Baden

   Göteborg University, Department of Marine Ecology,
   Kristineberg Marine Research Station, 450 34 Fiskebäckskil, Sweden

J.J. Beukema

   Netherlands Institute for Sea Research (NIOZ),
   P.O. Box 59, 1790 AB Den Burg (Texel), The Netherlands

C. Boström

   Åbo Akademi University, Department of Biology, Environmental and
   Marine Biology, Akademigatan 1, 20520 Åbo, Finland

D. Bushek

   Belle W. Baruch Marine Field Laboratory,
   P.O. Box 1630, Georgetown, South Carolina 29442, USA

G.C. Cadée

   Netherlands Institute for Sea Research (NIOZ),
   P.O. Box 59, 1790 AB Den Burg, Texel, The Netherlands

J.A. Commito

Environmental Studies Department and Biology Department, Gettysburg College, Gettysburg, Pennsylvania 17325, USA

R.F. Dame

Coastal Carolina University, P.O. Box 261954, Conway, South Carolina 29528, USA

N. Dankers

Alterra, P.O. Box 167, NL 1790 AD Den Burg, Texel, The Netherlands

C. Den Hartog

Laboratory of Aquatic Ecology, Catholic University, Toernooiveld, 6225 ED Nijmegen, The Netherlands

S. Dittmann

Zentrum für Marine Tropenökologie, Fahrenheitstr. 1, 28359 Bremen, Germany

M. Emmerson

Culterty Field Station, University of Aberdeen, Newburgh, Ellon, Aberdeen, AB41 0HA, UK

E. Flach

Stockholm University, Department of System Ecology, 10691 Stockhom, Sweden

S.E. Hagerthey

Sediment Ecology Research Group, Gatty Marine Laboratory, University of St Andrews, St Andrews, Fife, KY16 8LB, UK

R.G. Hughes

School of Biological Sciences, Queen Mary and Westfield College, University of London, London E1 4NS, UK

E. Jaramillo

Instituto de Zoología, Universidad Austral de Chile, Valdivia, Chile

P. Kamermans

> Netherlands Institute for Fisheries Research (RIVO), Centre for Shellfish Research (CSO), P.O.Box 77, 4400 AB Yerseke, The Netherlands

M. Lastra

> Departamento de Ecología y Biología Animal, Universidad de Vigo, Spain

J. van der Meer

> Netherlands Institute for Sea Research (NIOZ), P.O. Box 59, 1790 AB Den Burg (Texel), The Netherlands

H. Michaelis

> NLÖ-Forschungsstelle Küste, An der Mühle 5, 26548 Norderney, Germany

D.M. Paterson

> Sediment Ecology Research Group, Gatty Marine Laboratory, University of St Andrews, St Andrews, Fife, KY16 8LB, UK

R.C. Phillips

> 1797 Meadow View Drive, Hermiston, Oregon 97838, USA

T. Piersma

> Netherlands Institute for Sea Research (NIOZ), P.O. Box 59, 1790 AB Den Burg (Texel), The Netherlands

T.C. Prins

> National Institute for Coastal and Marine Management/RIKZ, P.O. Box 8039, 4330 EA Middelburg, The Netherlands

D. Raffaelli

> Culterty Field Station, University of Aberdeen, Newburgh, Ellon, Aberdeen, AB41 0HA, UK

K. Reise

Alfred-Wegener-Institut für Polar- und Meeresforschung, Wattenmeerstation Sylt, 25992 List, Germany

H.U. Riisgaard

Research Centre for Aquatic Biology, Odense University, Hindsholmsvej 11, 5300 Kerteminde, Denmark

M. Sprung

CCMAR, UCTRA, Universidade do Algarve, Campus de Gambelas, 8000810, Faro, Portugal

A. Tamaki

Nagasaki University, Marine Research Institute, Faculty of Fisheries, Taira-Machi 1551-7, Nagasaki 851-2213, Japan

S.F. Thrush

National Institute of Water and Atmospheric Research, P.O. Box 11-115, Hamilton, New Zealand

J.A. Vargas

Centro de Investigacíon en Ciencias del Mar y Limnología (CIMAR), Universidad de Costa Rica, 2060 San José, Costa Rica

R.B. Whitlatch

Department of Marine Sciences, University of Connecticut, Groton, Connecticut 06340-6097, USA

W.J. Wolff

Rijksuniversiteit Groningen, Postbus 14, 9750 AA Haren, The Netherlands

Introduction

K. REISE

This book is an attempt to encourage comparative studies on the ecology of sedimentary shores. Most research on muddy and sandy shores has been local or regional, analyzing small to meso-scale patterns and the underlying processes. Comparisons between distant shores, across latitudinal gradients or between separate biogeographic provinces, have rarely been carried out, and such comparisons are often hampered by methodological differences in quantifying coastal biota. Field experiments may not provide comparable results when the spatio-temporal scales are different and, usually, little is known about how representative an analyzed process or pattern is across an entire region. Between different coasts, several important variables are likely to differ at the same time. When particular aspects are compared, either data from a large number of sites or a choice of carefully selected sites is required and researchers must be cautious in the interpretation of such data.

Zonation of organisms perpendicular to the shoreline is less distinct in sediments than on rock (Peterson 1991), and the ability of many species to transform habitat structure and the mobility of species in soft sediments contribute to dynamic and heterogeneous patterns over many spatial scales. Not many soft-sediment comparisons have been conducted within regions along gradients from exposed to sheltered, from atidal to macrotidal, from onshore to offshore currents, from brackish to marine, from oligotrophic to eutrophic, from small to large organic supplies originating from the sea or the land, and with respect to other similar variables. Even fewer comparisons are available between latitudes and between biogeographical provinces. Most of these are confined to surf beaches (Dahl 1952; McLachlan et al. 1993).

Nevertheless, the authors of this volume feel that the comparative ecology of sedimentary shores is a ripe area for a global view of patterns and processes studied at selected sites. This is necessary for both the development of conceptual theory, and for providing insights for effective coastal management. It is acknowledged that this volume cannot provide definitive answers to all these issues; its aim is to raise awareness of the need for a comparative perspective on the sedimentary shores of the world. Finding similarities and convergent biotic traits among biogeographic provinces in response to physical forcing constitutes the necessary basis for general models. Conversely, similar species with small differences in their life-history traits may ultimately lead to divergent flows of energy and matter in coastal ecosystems. In such cases, a similar physical regime and environment may generate quite different biotic effects. The global comparisons advocated here are required to examine and apply the generalities which are claimed to date and to separate these from peculiarities and unique phenomena, some of which may in themselves be noteworthy and deserve special appreciation.

Biotic assemblages of sedimentary shores are strongly dominated by a small set of species. This may ease comparative analyses. However, the question remains: what is the ecological role (if any) of the remaining species? Species richness recorded from intertidal surveys generally yield <100 species of benthic macrofauna in temperate zones and >100 in the tropics, often two to three times more than in temperate regions (i.e., Reise 1991; Pepping et al. 1999; Wijnsma et al. 1999). The biodiversity of the sedimentary shores of the world in terms of forms of life, species, ecological assemblages and habitats has never been compiled and brought into perspective.

Sedimentary shores may be evaluated across latitudes from the perspectives of migrant birds looking for food and roosting sites (Piersma et al. 1993). Global patterns of zoomass in marine intertidal communities not only show divergent trends of trophic groups (Ricciardi and Bourget 1999), but also reveal that studies comparable in terms of methodology and scale are few, and that the global distribution of such studies is very uneven. Blackburn and Gaston (1998) raised a number of methodological issues that need to be taken into account when individual data sets are assembled to detect trends over large geographical distances. In this volume, primarily pairwise comparisons are conducted with contrasting aspects between geographically distant shores.

Roughly two-thirds of the world's coastline consists of sedimentary shores, in long continuous stretches or in bays interspersed between rocky headlands (Fig. 1). Yet ecological perspectives and theory derived from sandy beaches and mudflats have remained in the shadow of the ecology of rocky shores. The reason is obvious. Most of the organisms of the rocky shore are showy and conspicuous, and are readily accessible to non-destructive quantification and experimentation. Patterns of zonation have been compared worldwide (Stephenson and Stephenson 1972) and the rocky shore served as a model system for the development of community theory (Connell 1975; Dayton 1975; Paine 1994).

Sedimentary shores, though an early focus of geologists (see Davis 1985), are more resistant to progress in ecological research. Most organisms are small, cryptic and difficult to extract from the sediment, and field experiments in this environment may suffer more than on rocky shores from

Introduction

Fig. 1. Exposed sandy beaches, sheltered tidal flats and sand dunes on the barrier island of Sylt in the North Sea (*top*), and sedimentary coves and an estuary between rocky headlands of the Seno de Reloncavi, southern Chile (*bottom*)

experimental artefacts. As a consequence, knowledge has advanced much more slowly. However, apart from their spatial extent and being subjected to many human impacts, sedimentary shores have a number of distinct features that deserve study, and it should not be regarded as the Cinderella to a rocky sister, the rugged beauty. Both shore types rely on food supplies from the sea, and suspension feeders are often prominent. Light allows benthic plants to develop in both shore types: kelp forms extensive stands on rocky coasts and seagrasses are common on sedimentary shores. Rocks basically provide a two-dimensional substrate, while sediments offer a third dimension in which to live. In this third dimension organic material can be stored and is available to deposit feeders. The third dimension also offers a buffer against the physical harshness of the intertidal environment. The slope of sedimentary shores tends to be gentler than that of rocky shores, so that the former has a much wider areal extent, and thus provides more food to visiting shrimps, fish and birds.

There are two key variables for the broad-scale comparison of biotic richness of sedimentary shores. Firstly, the external supplies of nutrients and organic material, mostly from the sea, but in estuaries also from the land. Secondly, the morphodynamics of the shore, determined by wave energy, the tidal regime and the sediment particles available. The broad-scale composition of shore biota further depends on salinity, the temperature regime and the biogeographic province.

Except for salinity, intermediate states of the above factors and their combinations seem to be the most favorable for the biota of the sedimentary shores. Too great an external food supply may result in anoxia, and too little will exclude the suspension feeders. Strong hydrodynamic forces and a coarse sediment grain render a sedimentary environment very unstable and only a few species are capable of coping with this combination. In embayments and the shallow subtidal these forces are often appreciably ameliorated by biogenic sediment stabilization. Shallow stagnant waters also require specialized adaptations. Without flow there may be little buffer against the high physical fluctuations of the surrounding terrestrial environment. Species richness of the shore increases from high to low latitudes, and is highest in the Indo-West Pacific province. On the other hand, benthic zoomass seems to be highest at mid-latitudes, presumably because under conditions of pronounced seasonality more phytoplankton escapes the pelagic algivores and is left as food for the zoobenthos (Ricciardi and Bourget 1999).

The ecological patterns and processes of almost all shores have been modified by humans. The past and present imprint of human activities on the shore biota not only has to be kept in mind when global comparisons are conducted, but in itself deserves comparative evaluation. On a global scale, the physical changes resulting from land-use practices, planned and unplanned coastal engineering have the most important effects. Often there is a habitat loss or truncation at the upper or inner shore because of embankments for agriculture and aquaculture, industry, airports, residents, and as coastal protection measures, whilst dredging and dumping activities, particularly in estuaries, hard coastal defence structures and replenished beach sand on open coasts and fishery operations, may all create strong physical disturbance.

The widespread introduction of species across biogeographic barriers from coast to coast is irreversible. This trans-oceanic exchange of species occurs unintentionally with shipping, either fouling or in ballast water, and with aquaculture (Carlton and Geller 1993). Only a few of the introduced species have turned into massive invaders transforming the dynamics and structure of the recipient ecosystem. The cumulative effect is to bring about a sameness between distant coasts which previously had no species in common, except for humans.

The exploitation of living resources on sedimentary shores affected primarily the upper trophic levels: mammals, birds and fish (Wolff 2000a,b). Gray whales have been exterminated from the coastal lagoons of the North Atlantic, and a re-introduction would require whales from the Pacific population. Dugongs and manatees have likewise become rare in parts of their former range and hunting has reduced populations of seals and birds. The latter suffered particularly from harvesting of eggs, a similar problem for sea turtles. More recently, nature conservation has allowed for a partial recovery of coastal bird populations. Overfishing and partly habitat degradations at coasts and rivers lead to declines in a variety of large anadromic fish. Some species of rays with a late and low reproductive output ended as by-catch in the fisheries and became rare and functionally extinct on the coast. Oysters became overexploited (i.e., Rothschild et al. 1994). It seems likely that on the intensively harvested tidal flats of the East-Asian coasts, the entire size-spectrum of benthic invertebrates became truncated at its upper range.

Particularly in densely populated areas around embayments and semi-enclosed seas, eutrophication and toxic pollutants have changed coastal ecosystems. Eutrophication primarily increased primary production (Schramm and Nienhuis 1996), but seagrasses have adapted to oligotrophic waters. These rooted plants give way to fast-growing algae when nutrient levels increase. Mats of green algae cover intertidal sediments enriched with nitrogen, with detrimental effects on the benthic fauna underneath. There is also evidence that eutrophication has increased benthic zoomass in estuaries (Beukema 1991). Toxic pollutants often occur at the same sites where nutrient levels have increased, and the opposite effects of these two components may mask one another. Global warming and the concomitant sea-level rise will interact with most of the factors mentioned above. Whilst anthropogenic effects on the biota of sedimentary shores is not a main focus of this Volume, the widespread occurrence of such effects means that they become an integral part of the comparisons; nowhere can the footprint of human activity be ignored.

Physical factors structuring the biota on exposed shores give way to a prevalence of biogenic structuring and biotic interactions on sheltered shores. On the latter, complex zonations and habitat mosaics develop. Here, the uppermost zone should be occupied by salt marshes or mangroves, except at the bare sepkhas of arid regions. These habitats of the upper fringe are merely touched upon in the present volume. In the lower zone of the shore at lower latitudes, coral reefs may grow and then transform the sedimentary environment into biogenic rock. Also this habitat has not been included. Instead we focus on the marine organisms of mud and sand in the nearshore zone. This book is divided into four thematic sections.

The first four chapters deal with suspension feeders. These often dominate in zoomass and form conspicuous reefs and dense beds. They rely on a supply of organic material from the sea, and mediate a particularly intricate bentho-pelagic coupling. On exposed beaches, this feeding mode is also combined with high mobility. The variability of food resources on sedimentary shores has selected for organisms able to alternate between suspension and deposit feeding.

The ability of organisms to modify the sediment structure and dynamics is considered in the next four chapters. In particular, benthic microalgae consolidate sediments, while grazers, burrowing and digging animals disturb these efforts. The interplay of biogenic stabilizing and destabilizing processes has cascading effects on the composition of the sedimentary coastal biota.

Next seagrasses and zoobenthos are compared between shores with respect to nutrient supplies, salinity, latitude and biogeographic provinces in five chapters. These large-scale comparisons reveal striking similarities as well as unique regional features. However, it is apparent that the global coverage of studied sites is still too uneven to arrive at sound generalities.

In the last three chapters attention is given to processes determining structural dynamics in communities and populations, and how these need to be studied in order to incorporate them into comparisons between distant sedimentary shores. The roles of disturbance, predators and subsidies from adjacent ecosystems are considered.

Finally, a synthesis provides an overview to major conclusions from all of the 16 chapters.

References

Beukema JJ (1991) Changes in composition of bottom fauna of a tidal-flat area during a period of eutrophication. Mar Biol 111:293–301

Blackburn TM, Gaston KJ (1998) Some methodological issues in macroecology. Am Nat 151:68–83

Carlton JT, Geller JB (1993) Ecological roulette: the global transport of nonindigenous marine organisms. Science 261:78–82

Connell JH (1975) Some mechanisms producing structure in natural communities. In: Cody ML, Diamond JM (eds) Ecology and evolution of communities. Harvard Univ Press, Cambridge, pp 460–490

Dahl E (1952) Some aspects of the ecology and zonation of the fauna on sandy beaches. Oikos 4:1–27

Davis RA Jr (ed) (1985) Coastal sedimentary environments. Springer, Berlin Heidelberg New York, 716 pp

Dayton PK (1975) Experimental evaluation of ecological dominance in a rocky intertidal algal community. Ecol Monogr 45:137–159

McLachlan A, Jaramillo E, Donn TE, Wessels F (1993) Sandy beach macrofauna communities and their control by the physical environment: a geographical comparison. J Coastal Res 15:27–38

Paine RT (1994) Marine rocky shores and community ecology: an experimentalist's perspective. In: Kinne O (ed) Excellence in ecology. Ecology Institute, Oldendorf/Luhe, Germany, 152 pp

Pepping M, Piersma T, Pearson G, Lavaleye M (eds) (1999) Intertidal sediments and benthic animals of Roebuck Bay, Western Australia. Neth Inst Sea Res NIOZ Report 1999-1993, 212 pp

Peterson CH (1991) Intertidal zonation of marine invertebrates in sand and mud. Am Sci 79:236–249

Piersma T, De Goeij P, Tulp I (1993) An evaluation of intertidal feeding habits from a shorebird perspective: towards relevant comparisons between temperate and tropical mudflats. Neth J Sea Res 31:503–512

Reise K (1991) Macrofauna in mud and sand of tropical and temperate tidal flats. In: Elliott M, Ducrotoy J-P (eds) Estuaries and coasts: spatial and temporal intercomparisons. Olsen and Olsen, Fredensborg, pp 211–216

Ricciardi A, Bourget E (1999) Global patterns of macroinvertebrate biomass in marine intertidal communities. Mar Ecol Prog Ser 185:21–35

Rothschild BJ, Ault J, Goulletquer P, Heral M (1994) Decline of the Chesapeake Bay oyster population: a century of habitat destruction and overfishing. Mar Ecol Prog Ser 111:29–39

Schramm W, Nienhuis PH (eds) (1996) Marine benthic vegetation. Ecological studies, vol 123. Springer, Berlin Heidelberg New York, 470 pp

Stephenson TA, Stephenson A (1972) Life between tide marks on rocky shores. Freeman, San Francisco, 425 pp

Wijnsma G, Wolff WJ, Meijboom A, Duiven P, De Vlas J (1999) Species richness and distribution of benthic tidal flat fauna of the Banc d'Arguin, Mauritania. Oceanol Acta 22:233–243

Wolff WJ (2000a) Causes of extirpations in the Wadden Sea, an estuarine area in The Netherlands. Conserv Biol 14:876–885

Wolff WJ (2000b) The south-eastern North Sea: losses of vertebrate fauna during the past 2000 years. Biol Conserv 95:209–217

Part I

Suspension Feeders in Coastal Mud and Sand

1 Benthic Suspension Feeders as Determinants of Ecosystem Structure and Function in Shallow Coastal Waters

R.F. DAME, D. BUSHEK, and T.C. PRINS

Publication No. 1259 of the Belle W. Baruch Institute for Marine Biology and Coastal Research.

1.1 Introduction

Suspension-feeding animals are common macroscopic inhabitants of hard- and soft-bottom habitats in shallow coastal waters. Many forms aggregate on specific bottom types where they take advantage of the free-energy subsidies provided by waves, tides and wind-driven currents that transport oxygen and particulate food to them, carry away waste, disperse larvae, and exclude predators. Because these organisms can move and process large amounts of material between the water column and the bottom, they are often major agents of benthic-pelagic coupling and nutrient cycling (Dame 1996; Wildish and Kristmanson 1997). As a group, they have been promoted as system filters or cleaners because of their suspension-feeding abilities (Newell 1988). Some reef builders are advocated as erosion-control mechanisms (Kirtley and Tanner 1968). All of these functional roles imply that benthic suspension feeders play a significant part in the overall functioning of many shallow coastal systems.

To be certain that we are clearly understood regarding suspension-feeding terminology, there are a few terms that should be defined. The more general term "suspension feeder" refers to animals that retain particles from the water passing their feeding organs. The term "filter feeder" is often used synonymously with the term suspension feeder; however, filter feeding specifically describes feeding by passing water through structures that retain particles according to size and shape (Jørgensen 1966). For example, vermetids (Gastropoda) are suspension feeders that use mucus nets to ensnare particles, but, in that mode, they are not filter feeders because the net is not very selective and the animal does not control water flow by the net. In comparison, many bivalves are suspension feeders and specifically filter feeders. At the ecosystem level, suspension feeders form a functional group that may be composed of a number of similar species.

When suspension feeders are abundant in a particular ecosystem they have the potential to directly modulate the availability of resources to other organisms and groups by causing physical state changes in biotic and abiotic materials (Carpenter and Kitchell 1988; Jones et al. 1994; Dame 1996). In this process, suspension feeders can modify, maintain and create habitats, and, as such, these animals are often referred to as ecosystem engineers (Jones et al. 1994). If they transform the environment via the structure of their bodies or skeletons we view these operations as autogenic (Jones et al. 1994). If they transform living or non-living materials from one physical state to another, Jones et al. (1994) calls these processes allogenic. Some suspension feeders, e.g., oysters in reefs, can exhibit both autogenic and allogenic characteristics (Dame 1996).

The goals of this chapter are to (1) describe the main types of benthic suspension feeders that can potentially control certain processes in shallow coastal ecosystems; (2) compare the system-influencing properties of the various types of benthic suspension feeders; (3) examine reports of these suspension feeders transforming specific systems; and (4) through synthesis suggest why suspension feeders are successful in specific ecosystems.

1.2 Reefs and Beds

1.2.1 Bivalve Molluscs (Bivalvia)

Bivalve suspension feeders can configure dense assemblages in the form of beds or reefs. Epifaunal species like mussels and oysters may reach densities with a biomass over 1000 g dry wt m^{-2} (Table 1.1; Nixon et al. 1971; Craeymeersch et al. 1986; Asmus 1987; Jørgensen 1990). Infaunal species are less conspicuous than epifaunal species, but they can also reach very high densities (Table 1.1). Seed and Suchanek (1992) identified three important components characterizing mussel beds, and probably also other bivalve communities: (1) a physical matrix of living and dead shells, (2) a bottom layer of sediments and biodeposits, and (3) a taxonomically diverse assemblage of associated flora and fauna.

The blue mussel, *Mytilus edulis*, forms beds on intertidal flats and rocky shores, with a foundation of empty shells and biodeposited material and living mussels attached to each other by byssus threads. The three-dimensional structure creates a special community, with a distinct faunal composition (e.g., Asmus 1987). These beds can persist for many years (Dankers and Koelemaij 1989) as complex communities of several year classes of mussels and their associated fauna (Asmus 1987). Their stability decreases when large amounts of biodeposits accumulate, causing the mussel beds to rise high above the surrounding tidal flat and making them vulnerable to storms and ice scouring. Suffocation from excessive biodeposition and predation (especially by humans, i.e., fishing) can be a significant mortality factor as well. A comprehensive description of these processes is given in Seed and Suchanek (1992).

Probably the most extensive bivalve structures are reefs built by oysters. For example, the American oyster, *Crassostrea virginica*, forms reefs that can be thousands of meters in length and several meters high (Fig. 1.1, see p. 16). In addition, these reefs may grow large enough to form dams (Fig. 1.1) that influence the hydrography of marsh creeks (Dame 1996). Like mussel beds, oyster reefs provide microhabitats for many motile and sessile species. The number of macrofaunal species found on intertidal reefs ranges from 37 (only macroinvertebrates, Dame 1979) to 303 (Wells 1961).

A few infaunal species may surreptitiously play similar roles. For example, estimates of the abundance of the cockle *Cerastoderma edule* (Table 1.1) show maximum abundance values up to 2000 m^{-2}, and maximum biomass levels up to 400 g dry wt m^{-2} (Verwey 1952). However, cockle beds cannot be characterized as communities with a distinct faunal composition (Dankers 1993). Recently, Gutierrez and Iribarne (1999) described the role of stout razor clams (*Tagelus plebeius*) in structuring benthic communities as a result of burrows and shell matrix accumulations.

Several factors affect biodiversity on bivalve reefs and beds. Studies have shown that the diversity of species increases with increasing patch size and with age of the mussel bed. Tidal exposure and mussel density may also be significant factors causing spatial variation in the associated fauna of the mussel bed (see Dame 1996 for review). Dame (1996) argued that it is important to recognize the interaction between the physical environment and the system structure in order to understand the temporal and spatial development of bivalve communities and their associated fauna.

A number of benthic suspension-feeding bivalve species are dominant grazers in shallow coastal areas and they can exert a strong control on particle concentrations in the water column. Based on estimates of the amount of suspended particulate matter processed by *Cerastoderma edule* and *Mytilus edulis* in the Dutch Wadden Sea, Verwey (1952) concluded that the bivalves play an important role in the sedimentation processes in this system. Studies of other systems also indicated that sedimentation of material might be significantly enhanced by bivalve filtration activity (Haven and Morales-Alamo 1966, 1972). In later studies, it was realized that bivalve filtration can have an impact on phytoplankton biomass. Data supporting this idea came

Table 1.1. Clearance rates of selected benthic suspension feeders. Clearance rates computed using ash-free dry body weight (AFDW)

Group	Species	Habitat	Clearance rate (l g^{-1} h^{-1})	Maximum density (No. m^{-2})	Source(s)
Ascidians	*Ascidiella aspersa*	Subtidal/epifauna	2.7	0.5	Hily (1991)
	Ciona intestinalis	Subtidal/epifauna	7.1[a]	277	Randløv and Riisgård (1979); Riisgård et al. (1995)
	Phallusia mammillata	Subtidal/epifauna	2.1	0.8	Fiala-Médoni (1978); Hily (1991)
	Styela clava	Subtidal/epifauna	7.0[a]		Riisgård (1988)
Barnacles	*Balanus crenatus*	Encrusting	0.1		Hily (1991)
	Balanus perforatus	Encrusting	8.8 ml/ind.	2000	Crisp and Southward (1961)
Bryozoans	General	Encrusting	0.42/zooid		Riisgård and Manriquez (1997)
Echinoderms	*Ophiothrix fragilis*	Subtidal/solitary	10.4	500	Hily (1991)
Mollusca					
Bivalvia	*Arctica islandica*	Subtidal/infauna	5.6[a]		Møhlenberg and Riisgård (1979)
	Argopectin irradians	Subtidal/epifauna	8.0		Kirby-Smith (1972)
	Cardium echinatum	Subtidal	4.2[a]		Møhlenberg and Riisgård (1979)
	Cerastoderma edule	Intertidal/infauna	1.5		Smaal et al. (1997)
			7.9	2000	Verwey (1952;) Vahl (1973)
	Chlamys hastata	Subtidal/infauna	8.6	5	Hily (1991)
	Crassostrea gigas	Inter/subtidal beds	6.7		Walne (1972)
	Crassostrea virginica	Inter/subtidal reefs	6.8[a]	4400	Dame (1975); Riisgård (1988)
	Geukensia demissa	Intertidal/epifauna	6.2[a]	2000	Bertness (1984); Riisgård (1988)
	Mercenaria mercenaria	Inter/subtidal beds/infauna	2.6[a]		Coughlan and Ansell (1964)
	Modiolus modiolus	Subtidal/epifauna	6.0[a]		Møhlenberg and Riisgård 1979

Group	Species	Habitat	Size	Number	Reference
	Mytilus edulis	Inter/subtidal beds	1.6		Smaal et al. (1997)
			7.5[a]	10,000	Møhlenberg and Riisgård (1979)
	Ostrea edulis	Inter/subtidal beds	8.8	350	Walne (1972)
Gastropoda	*Crepidula fornicata*	Encrusting	8.8	200	Hily (1991)
Polychaeta	*Chaetopterus variopedatas*	Intertidal/infauna	10[a]		Riisgård (1988)
	Lanice conchilega	Infauna	0.8[a]		Buhr (1976); Riisgård and Ivarsson (1990)
Sabellidae	*Myxicola infundibulum*	Subtidal/muddy	0.5[a]		Dales (1957)
	Sabella pencillus	Subtidal/epifauna	109[a]		Riisgård and Ivarsson (1990)
	Sabella spallanzanii	Subtidal/reefs	1	100?	Lemmens et al. (1996)
	Schizobranchia insignis		1.8[a]		Dales (1961)
Serpulidae	*Ficopomatus enigmaticus*	Inter/subtidal/reefs	9[a]		Davies et al. (1989)
	Hydroides norvegica	Encrusting	4.7[a]		Dales (1957)
	Pomatoceros triqueter	Subtidal/encrusting	5[a]		Klöckner (1976)
	Salmancina dysterii	Encrusting	10.4[a]		Dales (1957)
	Spiorbis borealis	Encrusting	4.8[a]		Dales (1957)
Porifera	*Ficulina ficus*	Subtidal/epifauna	1.2	40	Hily (1991)
	Halichondria panicea	Subtidal/epifauna	3.6[a]		Riisgård et al. (1993)

[a] Dry body weight (DW) is measure.

a

b

Fig. 1.1. a Natural intertidal oyster reefs in North Inlet, South Carolina, USA.
b Intertidal oyster reef forming a dam across a tidal creek in North Inlet

from field observations of reduced algal biomass in areas of extensive mussel culture (Cadée and Hegeman 1974) and depletion of phytoplankton biomass in waters passing bivalve beds (Wright et al. 1982; Carlson et al. 1984). More quantitative estimates of the potential effects of bivalve grazing on a system level, viz. South San Francisco Bay, came from mathematical models that included bivalve filtration rates and phytoplankton growth rates (Cloern 1982; Officer et al. 1982).

Studies over the last two decades indicate that bivalve grazing can be a major factor controlling phytoplankton biomass in numerous estuarine and coastal systems (see Smaal and Prins 1993; Dame 1996 for reviews). An elaborate computer model simulation has shown that bivalve filtration can effectively control phytoplankton biomass irrespective of nutrient loading (Herman and Scholten 1990). In the latter case, the open nature of these shallow coastal systems is essential, otherwise the accumulation of nutrients will inevitably result in eutrophication symptoms despite high grazing pressure (Herman 1993).

Predictions of the system level effects of bivalve grazing are based on the up-scaling of bivalve filtration rates from laboratory observations to the scale of an entire estuary. A few experimental studies have attempted to measure filtration rates under in situ conditions. Filtration rates calculated from chlorophyll depletion by mussel beds generally showed agreement between in situ values and filtration rate estimates from laboratory incubations with food in natural seawater (Prins et al. 1996; however, see Asmus et al. 1998). Other studies indicated that laboratory estimates of filtration rates using algal diets could significantly overestimate the natural filtration activity of bivalves (Doering and Oviatt 1986).

In addition to estimates of water column turnover rates by filtration, physical factors have to be included for the assessment of the grazing impact of bivalves on the pelagic system. The upper limit to the biomass of benthic suspension feeders in an estuarine system depends, to a large extent, on the residence time of the water as a measure of food exchange with the sea, and benthic biomass is ultimately limited by system productivity (Heip et al. 1995). Similarly, the strength of benthic grazing impact on the pelagic community depends on the water renewal rate (Smaal and Prins 1993). This point was elaborated by Dame (1996), who argued that systems with a short residence time need a high bivalve biomass to water volume ratio, resulting in a short clearance time, in order to control phytoplankton biomass via bivalve grazing. A prerequisite for benthic grazing control is a well-mixed water column. In poorly mixed systems, vertical stratification uncouples the water column from the benthos and creates phytoplankton bloom conditions (Koseff et al. 1993; Lucas et al. 1998).

Dame et al. (1980, 1984, 1985) recognized that, in addition to the processes leading to removal of material from the water column, bivalve grazing can act as an important link in a feedback loop with grazers inducing a flux of particulate nutrients from the water column to the sediment. That flux, in turn, feeds mineralization and a reverse flux of inorganic nutrients from the sediment to the water column. Several in situ studies, using flumes or benthic tunnels on mussel beds (Dame and Dankers 1988; Prins and Smaal 1990, 1994; Asmus and Asmus 1991, 1993; Dame et al. 1991a,b; Asmus et al. 1995) and natural oyster reefs (Dame et al. 1989; Dame and Libes 1993; Zurburg et al. 1994a,b), have confirmed that bivalve communities can be intense sites of mineralization and sources of inorganic nutrients. Generally, direct excretion by the bivalves is considered to be a less important nutrient source than geochemical processes. In addition to this positive feedback on inorganic nutrient pools through regeneration processes, the reduced storage of nutrients in phytoplankton biomass as a consequence of grazing forms another positive feedback that influences inorganic nutrient availability (Prins et al. 1995b, 1997). This process is important as it may stimulate phytoplankton primary production in the case of nutrient limitation. It may also influence phytoplankton development as differences in nutrient regeneration rates can change nutrient stoichiometry.

Evidence supporting the hypothesis that bivalve grazing exerts a significant control on plankton biomass comes from studies on experimental ecosystems, i.e., mesocosms (Table 1.2). In the MERL mesocosms at the University of Rhode Island, grazing by the clam *Mercenaria mercenaria* did not reduce phytoplankton biomass, but resulted in higher algal growth rates (Doering et al. 1986, 1987). However, other experiments did show reduction in phytoplankton biomass by bivalve grazing (Riemann et al. 1988; Olsson et al. 1992; Granéli et al. 1993; Prins et al. 1995b). Laboratory experiments have shown that bivalves can filter microzooplankton species like ciliates (Le Gall et al. 1997) as well as larger zooplankton like copepods (Kimmerer et al. 1994). Mesocosm studies show that bivalve grazing can reduce biomass of microzooplankton (Horsted et al. 1988; Prins et al. 1995a,b, 2000). Some enclosure studies showed no effects of mussels on copepods (Horsted et al. 1988), probably as a consequence of the short duration of the studies (2–3 weeks) relative to the generation time of the copepods. However, recent mesocosm experiments showed a strong inhibition of copepod development (Prins et al. 2000). Newell (1988) argued that mesozooplankton could have increased in abundance in Chesapeake Bay as a result of the decline in oyster stock. This increase in copepods may be responsible for the increase in jellyfish as well as a change in phytoplankton composition, because copepods have a more size-selective feeding mode than oysters. Declines in copepod biomass in the San Francisco Bay were ascribed to predation by the introduced clam *Potamocorbula amurensis* on nauplii (Kimmerer et al. 1994). Consequently, bivalve grazing may affect the structure of the pelagic food web, and control significant ecological processes within the estuary.

Table 1.2. Systems that exhibit potential control by suspension feeding ecosystem transformers

System	Transformers (engineers)	Morphological form	Environment	Advection	Source(s)
Bivalve dominated					
Königshafen	*Mytilus edulis* *Cerastoderma edule*	Beds	North temperate, intertidal and shallow estuarine	Tides	Asmus et al. (1990)
North Inlet	*Crassostrea virginica*	Reefs	Temperate, intertidal and shallow estuarine	Tides	Dame et al. (1980)
Carlingford Lough	*Crassostrea gigas* *Tapes semidiscussata* *Mytilus edulis*	Farming, beds	North temperate, fjord estuary	Tides River	Ball et al. (1997); Ferreira et al. 1997
Marennes-Oléron	*Crassostrea gigas* *Mytilus edulis*	Farming, beds	Temperate, intertidal and shallow estuarine	Tides Wind	Héral et al. (1988); Bacher (1989)
South San Francisco Bay	*Potamocorbula amurensis*	Beds	Temperate, estuarine	Rivers	Cloern (1982)
Narragansett Bay	*Mercenaria mercenaria*	Beds	Temperate, estuarine	Rivers, tides	Pilson (1985)
Oosterschelde	*Mytilus edulis* *Cerastoderma edule*	Farming, beds	North temperate, intertidal and shallow estuarine	Rivers, tides	Smaal et al. (1986)
Western Wadden Sea	*Mytilus edulis*, *Cerastoderma edule*	Beds	North temperate, intertidal and shallow estuarine	Runoff tides	Dame et al. (1991a); Van Stralen (1995)
Ria de Arosa	*Mytilus edulis*	Farming beds	Temperate	Upwelling	Tenore et al. (1982)
Delaware Bay	*Crassostrea virginica*		Temperate, estuarine	Rivers	Biggs and Howell (1984)
Chesapeake Bay	*Crassostrea virginica*	Beds/farming	Temperate, estuarine	Rivers	Newell (1988)
Gastropod dominated					
Bermuda	*Dendropoma irregulare*	Mini-atoll reefs	Subtropical, oceanic	Waves	Safriel (1974)
Shikmona	*Dendropoma petraeum* *Vermetus triquetrus*	Mini-atoll reefs	Subtropical, oceanic	Waves	Safriel (1974)

Table 1.2. (*continued*)

System	Transformers (engineers)	Morphological form	Environment	Advection	Source(s)
Polychaete dominated					
Marina da Gama	*Ficopomatus enigmaticus*	Reefs	Subtropical, lagoon	Low flow rivers	Davies et al. (1989)
Lac de Tunis	*Ficopomatus enigmaticus*	Reefs	Subtropical, lagoon	Low flow rivers	Keene (1980)
Ardbear Lough	*Serpula vermicularis*	Subtidal reefs	North temperate, fjord estuary	Rivers	Bosence (1973)
Loch Creran	*Serpula vermicularis*	Subtidal reefs	North temperate, fjord estuary	Rivers	Moore et al. (1998)
Ellis Fjord	*Serpula narconensis*	Subtidal reefs	Antarctic, fjord estuary	Wind-driven currents	Kirkwood and Burton (1988)
Southeast Florida	*Phragmatopoma lapidosa*	Surf reefs	Subtropical, ocean shore	Waves, long shore currents	Kirtley and Tanner (1968)
Mont Saint-Michel Bay	*Sabellaria alveolata*	Intertidal reefs	Temperate	Tides, waves	Gruet (1986)
Ascidian dominated					
Kertinge Nor	*Ciona intestinalis*	Epifauna on *Zostera*	North temperate, fjord estuary	Rivers	Petersen and Riisgård (1992); Riisgård et al. (1995)
Multiple Species					
Bay of Brest	Multiple	Subtidal	North temperate, estuary	Tides	Hily (1991)
Cockburn Sound	Multiple	Subtidal	Subtropical	Tides	Lemmens et al. (1996)

1.2.2 Vermetids (Gastropoda)

The Vermetidae are a small family of warm-water sessile gastropods capable of building upright feeding tubes to take advantage of the water flow (Schiparelli and Cattaneo-Vietti 1999). Most, if not all, vermetids are able to filter suspended particles out of the inhalant current with their ctenidial cilia, but for the majority their main source of food is particles trapped on a mucous net secreted by the pedal gland and released into the water by the pedal tentacles (Hadfield et al. 1972; Hughes 1979). The close proximity of individual vermetids in a colony or aggregation causes individual mucous nets to overlap. These communal nets are thought to be more efficient at catching particles in conditions of rapidly and unpredictably varying currents (Hughes 1979). Furthermore, Schiparelli and Cattaneo-Vietti (1999) report that, in the Mediterranean Sea, the reef builder *Dendropoma petraeum* does not produce erect feeding tubes in wave-swept high-energy environments. However, in adjacent subtidal and less energetic environments, vertical feeding tubes are common. These investigators argue that the ability of vermetids to modify the direction and shape of their calcareous feeding tubes gives them the plasticity to win the competition for substrate space over other sessile organisms and qualifies them as keystone species whose absence would dramatically change the functional character of the system. Hughes (1979) also contends that by forming reefs with individual densities up to 60,000/m^2 (Table 1.1), *Dendropoma* facilitates fertilization and provides a smooth profile less susceptible to wave damage. Safriel (1974) reports that *Dendropoma* reefs in the Mediterranean and near Bermuda are small intertidal structures 2–5 m in diameter resembling miniature atolls (Fig. 1.2). Later observations by Thomas and Stevens (1991) show that the coralline alga *Herposiphonia secunda* is also an essential component of this system and that the micro-atolls are always found at the low tide level where turbulence is maximal. We were unable to find any rate function data, i.e., feeding, excretion, etc., for these suspension-feeding gastropods. However, vermetids that form dense aggregations or reefs transform local environments because

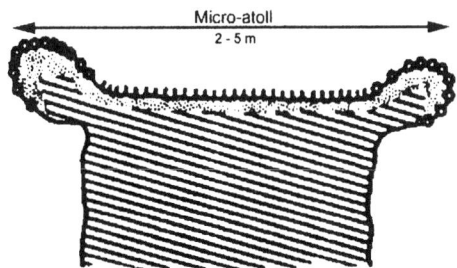

Fig. 1.2. A vermetid micro-atoll (cross section). (After Safriel 1974)

the structure of their bodies creates and alters a habitat based on their modulation of water currents.

1.2.3 Sabellids (Polychaeta)

Segmented worms in the Sabellariidae build tubes of agglutinated sand particles composed of quartz and heavy mineral grains, shell fragments, tests, fecal pellets and sponge spicules (Kirtley and Tanner 1968). These animals typically construct their tubes on top of one other, forming honey-comb-like aggregations. With densities of 25,000 individuals m^{-2} (Table 1.1), these structures are sometimes categorized as reefs (Herdman 1920; Multer and Milliman 1967) but, because the glue that holds the particles of the tube together loses its binding capabilities after the animal dies, geologists prefer not to use the term reef (Wood 1999).

Sabellid worm aggregations are generally found on sandy beaches in the intertidal and near subtidal zones where turbulence generated by waves or fast currents resuspends sand-sized particles that are needed for tube construction (Multer and Milliman 1967). Structures built by *Sabellaria alveolata*, sometimes more than 1 m across, are known from the English Channel, eastern Atlantic coast and western Mediterranean coast (Wilson 1971; Gruet 1986). Even larger reef-like structures, up to 100 m wide, are built in the surf zone by *Phargmatopoma lapidos* along the southeastern coast of Florida (Multer and Milliman 1967; Kirtley and Tanner 1968). In both types of aggregations, some type of hard substrate, beach rock, boulders, etc., is thought to be necessary for reef initiation by the settlement of planktonic larvae. As sabellid larvae are highly gregarious, this attribute usually establishes a succession of generations that ensures community persistence over long periods of time (Gruet 1986). There is some evidence that with time each of these aggregation types develops its own specific community of species (Gore et al. 1978; Porras et al. 1996).

Sabellids are filter feeders that have a specialized feeding structure, the funnel-shaped fan or crown-filament pump (Riisgård and Ivarsson 1990). This structure is covered with ciliated tracts that move water through the filter and convey particles of detritus and plankton to the mouth. Before entering the mouth, the particles are sorted with fine particles progressing to the mouth and large particles being rejected. In many sabellids there are three particle size categories and the medium-size grains are stored for tube construction. Sabellids efficiently remove particles between 3–8 µm (Jørgensen et al. 1984). Laboratory estimates of filtration rates by sabellids, none of which are reef builders, range from about 0.5 to 109 $l\ g^{-1}\ h^{-1}$ (Table 1.1; Dales 1957, 1961; Shumway et al. 1988; Riisgård and Ivarsson 1990). However, Riisgård and Ivarsson (1990) argue that the majority of these estimates are low because many of the studies used unrealistically high concentrations of suspended particles that probably saturated the digestive system of the worms. This idea is further supported by Riisgård and Ivarsson (1990) who determined that an average *Sabella penicillus* filtered an estimated 354 l of water per ml of oxygen consumed and that the worms are very efficient, only expending about 3.1% of the worm's metabolized energy for water processing.

Sabellid associations are physical structures that influence water flow and, in the case of *Phargmatopoma lapidos* systems, have been touted as erosion-control devices for Florida beaches (Kirtley and Tanner 1968). The structure of these systems also provides a special habitat in the high-energy flow zones of temperate and subtropical environments. We have found no published evidence of the influence of sabellid aggregations on the biogeochemical character of their ecosystems.

1.2.4 Serpulids (Polychaeta)

Serpulids are benthic suspension-feeding segmented worms that secrete their own calcareous tube. Serpulids have planktonic larvae that exhibit gregarious behavior by preferring settlement on or near adult worms of the same species at densities of 1 mm^{-2} or higher (ten Hove and van den Hurk 1993). In a few genera, asexual reproduction may enhance aggregation (ten Hove 1979). This intense aggregation results in the formation of reefs in intertidal and shallow subtidal environments from the arctic to the tropics. The largest serpulid reefs are several meters in height and kilometers long and are typically found in sheltered bays (ten Hove and van den Hurk 1993).

Like the sabellid worms described earlier, serpulids utilize a funnel-shaped fan or crown-filament pump (Riisgård and Ivarsson 1990) to move water and filter particles. Particles between 2 and 12 µm are removed efficiently (Davies et al. 1989). Dales (1957), Klöckner (1976), Davies et al. (1989) and Riisgård and Ivarsson (1990) reported serpulid filtration rates of 4.7 to 10.2 l g^{-1} h^{-1} that are slightly higher than those values reported for sabellids (Table 1.1). These differences may be explained by allometry, as the sizes of the observed individual serpulids were smaller than those of the sabellids. Weight-specific rates typically decline with increasing size. It should also be noted that Davies et al. (1989) reported rates for a reef-building serpulid, *Ficopomatus enigmaticus* (=*Mercierella enigmatica*), and found no evidence of the food saturation effects that concerned Riisgård and Ivarsson (1990).

There are at least two published studies (Keene 1980; Davies et al. 1989) that examine the role of serpulid reefs in shallow coastal ecosystems. In both studies, the species of interest is *F. enigmaticus*. In the earlier study, Keene (1980) investigated the importance of the reefs of this species on the Lac de Tunis, a shallow hypereutrophic lagoon on the north coast of Tunisia. The lagoon has large inputs of raw and secondarily treated sewage from the metropolitan area of Tunis. The organic input coupled with poor flushing and circulation supports highly productive phytoplankton populations that in turn feed the dense serpulid reefs. Using static mesocosms, Keene (1980) determined that oxygen uptake and nutrient remineralization by the reefs were high with the latter enhancing system productivity. In the second study, serpulid reefs in a coastal marina near Cape Town, South Africa, were studied (Davies et al. 1989). Here *F. enigmaticus* was found to be capable of filtering the entire bay in about a day and removing approximately 130 kg of suspended particulate material per hour. Thus, these studies support the notion that, in poorly flushed, organically loaded, shallow coastal systems, the structural and functional aspects of serpulid reefs indicate their ability to transform the ecosystem.

1.3 Encrustations

In addition to monospecific reefs, benthic suspension feeders, such as ascidians (Petersen and Riisgård 1992; Petersen et al. 1995, 1997), barnacles (Crisp 1979; Crisp and Southward 1961; Hily 1991), sponges (Riisgård et al. 1993) and bryozoans (Buss 1979), are also found as living layers or encrustations on rocks or other firm substrates. Further, dense populations of suspension-feeding animals are sometimes found on unconsolidated substrates, i.e., solitary polychaetes (Riisgård 1991; Vedel and Riisgård 1993), clams (Jørgensen 1990), and echinoderms (Hily 1991).

One of the few studies of an encrusting suspension feeder that showed these animals to have an impact at the ecosystem level is that of the ascidian *Ciona intestinalis* (Table 1.1) living on the surfaces of seagrass blades in Kertinge Nor, Denmark (Petersen and Riisgård 1992; Riisgård et al. 1995). Observations on Kertinge Nor indicate that, depending on the season, there is sufficient filtration capacity by *C. intestinalis* to filter the water volume of the fjord in 1 to 10 days (Petersen and Riisgård 1992) potentially allowing the tunicates to control phytoplankton densities within the system.

1.4 Species Groups

When found in sufficient numbers, a group of species (Hily 1991; Lemmens et al. 1996) may also have a large enough effect to be considered ecosystem transformers. Water flow appears to be a major factor in determining the location, structure, and growth of suspension feeders in dense aggregations.

In barnacles that are feeding passively, increased flow positively influences growth and recruitment (Sanford et al. 1994) and the building of hummocks or mounds further accentuates water flow and growth of those animals at the top of the structure (Pullen and LaBarbera 1991; Bertness et al. 1998).

Hily (1991) published one of the first studies on a community of several suspension feeders influencing the water quality of the Bay of Brest, located on the northwest coast of France. The bay is a well-mixed estuarine environment that covers about 10,280 ha with an average depth of 10 m and has a water volume residence time of 15 to 30 days. The bottom of the bay is composed of mainly muddy and sandy substrates dominated by eleven suspension-feeding invertebrates. The main sources of suspended particles are phytoplankton and decomposing macrophytes. The barnacle *Balanus crenatus* was the dominant suspension feeder in muddy areas, the snail *Crepidula fornicata* was most important on slopes and channels, and the echinoderm *Ophiothrix fragilis* dominated the gravelly bottoms of the central bay. Suspension feeders were calculated to filter the volume of the bay in 4 to 6 days or in about 1/3 of the hydrodynamic residence time and equivalent to the doubling time of the phytoplankton (Fig. 1.3). At this rate of suspension feeding, these animals can control the density of the phytoplankton and influence primary production within the bay through the feedback of inorganic nutrients. These activities, as well as the removal of suspended detrital particles from the water column, act as a retention mechanism that further enriches the bay. Thus, the suspension feeders as a functional group are playing a major functional role in their ecosystem.

In a similar effort, Lemmens et al. (1996) determined the filtering capacities of the various habitats in Cockburn Sound, Western Australia. By filtration capacity, these authors mean how much water a given habitat filters per m^2 per day. This is a static comparison of the suspension feeders because they are assumed to interact with the same amount of water. However, Cockburn Sound is a coastal system subject to tides, waves, and currents that result in both vertical and horizontal water fluxes. Thus, without water volume residence time estimates, it is difficult to determine the net influences of the suspension feeders on suspended particles or the ecosystem. The authors argue that because of low suspension-feeder densities in unvegetated and *Heterozostera* habitats, filter feeder control of these systems is unlikely. But, in habitats dominated by the structure-building polychaete *Sabella spallanzanii*, filtration capacity may be sufficient to influence suspended particle concentrations. If this is the case, then *S. spallanzanii* may be a key player in its ecosystem.

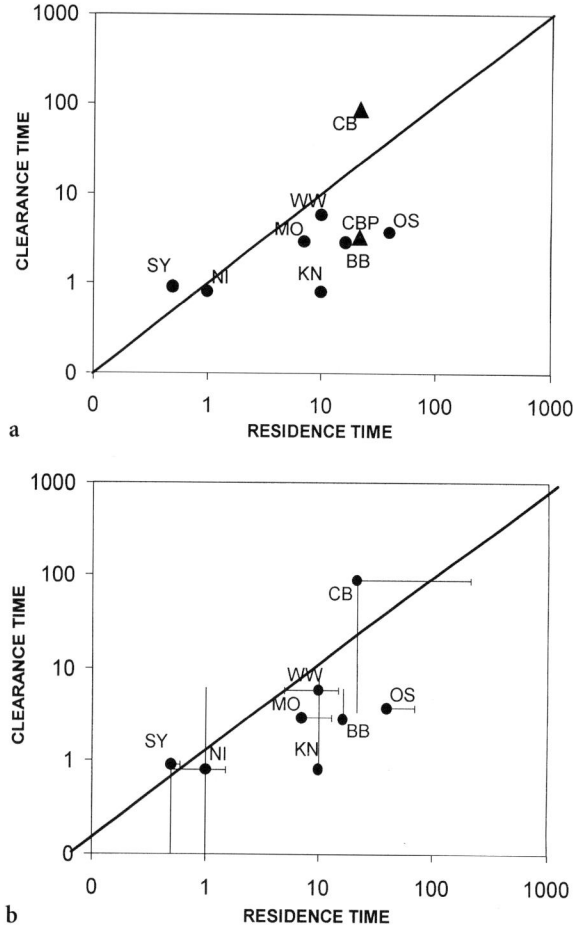

Fig. 1.3. a Water residence time plotted against system clearance time for a number of shallow water systems with populations of benthic suspension feeders. Symbol labels described in Table 1.3. **b** Similar to **a**, but only showing those systems for which range data were available

1.5 Discussion

Bivalves are common to many intertidal and coastal habitats, and are ecologically and economically the best-known and studied benthic suspension-feeding group (Dame 1996). These macrobenthic invertebrates are common to tidal waters with relatively high phytoplankton production or suspended organic material loading. There are numerous systems (Table 1.3) where oysters and mussels actually or potentially dominate ecosystem processes, but, in many cases, their dominance is related to intensive aquaculture. In natural systems, some species of oysters form reefs, while mussels typically aggregate in beds. These structures interact with the over-flowing waters to increase turbulent mixing, resuspend feces and pseudofeces, and increase the surface area of the reef or bed by occupying space in three dimensions.

Vermetid and sabellid aggregations are common to wave-swept, high-energy environments, that appear to dictate their geomorphology. Vermetids may have an advantage in turbulent environments because of their ability to change the construction and shape of their shell in response to water-flow dynamics. Sabellids are constantly building and repairing their tubes using resuspended sand particles. Because their environment is so energetic and often facing the open ocean, it is doubtful that these groups have much of an ecological impact on the transformation of suspended particles. However, the structures they build probably do influence local currents and wave patterns.

In contrast, serpulid reefs are found in low-energy, poorly flushed lagoons, bays and fjords. The density of serpulids in these environments implies that these worm reefs are translocating and transforming large quantities of suspended materials in order to support their structure. It also suggests that fjord habitats currently dominated by serpulid reefs may be adaptable to the raft culture of suspension-feeding bivalves similar to that now taking place at Carlingford Lough (Ball et al. 1997).

In all but a few cases, encrustations of ascidians, barnacles, bryozoans and sponges, and populations of solitary clams, echinoderms and polychaetes, only appear to be important at local scales. In ecosystems, epifaunal reefs or beds and/or combinations of suspension-feeding types may have the potential to transform shallow coastal environments (Crisp 1979; Buss and Jackson 1981; Hily 1991; Petersen and Riisgård 1992; Vedel and Riisgård 1993; Lemmens et al. 1996). In addition to encrusting rocks and other firm substrates including seagrass blades, many of these organisms are also found on the hard surfaces constructed by reef-building organisms or in the habitats generated by the reefs. In a sense, their functional similarity enhances the suspension-feeding mode of the reef or bed.

Heip et al. (1995) argued that benthic suspension-feeder biomass is determined by primary production at the scale of entire systems. However, the distribution of animals on smaller spatial scales is not determined by overall productivity but by numerous other factors, including predation, competition and physical environment (Dame 1996). Local distribution patterns, with suspension-feeder communities occurring at specific sites like slopes of tidal channels or across narrow inlets, illustrate the importance of localized hydrodynamic conditions in controlling seston flux (Verhagen 1985; Smaal et al. 1986; Wildish and Kristmanson 1997). Food depletion is a function of flow, and Herman et al. (1999) have argued that the maximum limit to the filtration capacity of the animals is a linear function of current velocity. It can be expected that animals with a high filtration capacity will be limited to a lower

Table 1.3. Comparison of system level clearance times

System	Suspension feeder biomass (g m⁻²)	Biomass range (g m⁻²)	Benthic surface area (10⁶ m²)	Suspension-feeder biomass (g m⁻³)	Water volume (10⁶ m³)	Clearance rate (l g⁻¹ day⁻¹)	Clearance rate (m³ m⁻² day⁻¹)	System clearance time (days)	Water volume residence time (days)
Königshafen, Sylt (SY)[a]	34.6	28–47	5.6	19.9	7.2	45.0	0.936	0.9–2.8	0.5–2.0
North Inlet (NI)[b]	38.5	5–60	8.8	15.4	22	81.6	3.142	0.8–6.1	0.5–1.5
Carlingford Lough (CF)[c]	0.3		39.5	0.07	196	163.2	0.057	87.5	66
Marennes-Oléron (M-O)[d]	21.0		135.7	4.2	675	81.6	1.714	2.9	5.0–10.0
South San Francisco Bay (SSF)[e]	15.0		490.0	2.5	2500	600.0	9.000	0.6	10–75
Narragansett Bay (NB)[f]	4.2		328.0	0.5	2724	62.4	0.259	32.1	26
Oosterschelde (OS)[g]	24.3	8–40	351.3	3.1	2740	87.0	2.114	3.7	10–150
Western Wadden Sea (WW)[h]	10.4		1386.2	3.7	4020	48.0	0.499	5.8	5–15
Ria de Arosa (RA)[i]	30.0		228.2	1.6	4335	51.0	1.530	12.4	23
Delaware Bay (DB)[j]	0.1		1942.0	0.009	19,420	163.2	0.015	680.8	97
Chesapeake Bay (CB)[k]	0.2		11,500.0	0.07	27,300	163.2	0.027	87.5	22
Marina da Gama (MD)[l]	10.1	5–85	0.32	4.500	0.025	206.1	2.062	1.1	Long (minimal exchange)
Kertinge Nor (KN)[m]	13.3	11–16	5.5	6.650	11.000		0.912	0.8–5	NA
Bay of Brest (BB)[n]	74.0		148.0	7.400	1480	48.0	3,552	2.8–6	16.7

[a-n] Source(s): [a] Asmus and Asmus 1991; Asmus et al. 1998; [b] Dame et al. 1980; [c] Ball et al. 1997; [d] Héral et al. 1988, Bacher 1989; [e] Cloern 1982; [f] Pilson 1985; [g] Smaad et al. 1986; [h] Dame et al. 1991; Van Stralen 1995; [i] Tenore et al. 1982; [j] Biggs and Howell 1984; [k] Newell 1988; [l] Davies et al. 1989; [m] Petersen and Riisgård 1992; [n] Hily 1991

maximum biomass than less active species (Herman et al. 1999). Food depletion in the water column is reduced by a high bottom roughness because this roughness generates turbulence that increases vertical mixing over potential suspension feeders. Bivalves can enhance turbulence by building rough and tall reefs and by producing outflow 'jets' as a by-product of their pumping (O'Riordan et al. 1995). In addition, the occurrence of animals in patches further increases the roughness (Fréchette et al. 1989; Herman et al. 1999). Bed roughness and the concurrently generated turbulence are enhanced by the morphology of the animal's shells and the topography generated by the dense packing of suspension feeders. In barnacles and mussels, the development of hummocks increases flow, benefits individual animal growth at the top of the hummocks and traps particles in the troughs between peaks (Seed and Suchanek 1992; Bertness et al. 1998; Thomason et al. 1998). Oysters and their reefs also develop structural attributes that increase elevation over the substrate and flow over the reef. These developments are also the product of individual shell structure, i.e., elongated growth forms in *C. virginica*, and increased reef height above the substrate (Bahr and Lanier 1981).

In a review of the evolution of all types of reefs, Wood (1999) argues that reefs made up of aclonal organisms, e.g., oysters, etc., are primitive because these systems are of low relief, relatively short duration and low community diversity. In contrast, clonal reefs, e.g., corals, are of high relief, long duration and high community diversity. She further contends that most aclonal systems are dominated by organisms that utilize the turbid shallow waters with their variable salinities and elevations as refuges from marine predators. In order to utilize these shallow environments, benthic suspension feeders usually exhibit synchronous spawning, gregarious settlement and rapid early growth in order to compete successfully against clonal forms for the limited hard surfaces in these soft-sediment environments.

Thus, suspension-feeding benthos from shallow water and intertidal environments need rapid and plastic growth, synchronized reproduction and gregarious recruitment because they live in a dynamic and unpredictable environment. These characteristics also enable them to control their major food source, the phytoplankton. The dominant benthic suspension feeder in each system in the lower portion of Fig. 1.3, oysters, mussels, clams, and ascidians, as well as the combination of species in the Bay of Brest, exhibit most of these characteristics. These suspension feeders have sufficient filtration capacity (density × filtration rate) to have a shorter system clearance time than water volume residence time which allows them to potentially control phytoplankton.

In the last century, there have been well-recorded and dramatic changes to the suspension-feeding benthos in Chesapeake Bay, Delaware Bay, San Francisco Bay, Marennes-Oléron and the Wadden Sea (Dame 1996 for a review). In the Chesapeake and Delaware Bays, the populations of *Crassostrea virginica* declined rapidly and the ecosystems changed to planktonic food web dominance (Newell 1988; Rothschild et al. 1994). In the San Francisco Bay, native species of benthic suspension feeders were replaced by several waves of trophically similar introduced species (Cohen and Carlton 1998). In Marennes-Oléron over-cultivation and disease caused a crash in oyster stocks that was eventually corrected through recultivation of introduced species (Héral 1993). Finally, in the Wadden Sea, there has been a change in dominance by a triad of oysters, mussels and worms, to mussels and cockles after oysters were over-harvested and subject to disease and the worm reefs were destroyed by fishing gear (Reise 1982; Reise et al. 1989). In those systems where the suspension-feeding benthos was lost or replaced by other species, at least two out of three possible external events occurred: large-scale climatic events, over-cultivation/over-harvesting, accompanied by disease, and nutrient loading.

Herman et al. (1999) have speculated that dense suspension-feeding benthic populations may induce strong positive feedback linkages that may invoke alternate stable states or phases. Utilizing Ulanowicz's (1997) terminology, the coupling of the benthic suspension-feeder component to the phytoplankton component via the positive effect of suspension-feeder grazing on the suspension feeders and suspension-feeder nutrient excretion on the phytoplankton is termed a direct relationship. If there are more than two components, the relationships between components are said to be indirect. When components are linked together by positive processes, they form a positive feedback loop that is said to be "autocatalytic" or "self-enhancing" (Ulanowicz 1997). In addition to exhibiting acceleration in some parameters, the positive feedback system can induce competition between different properties of components. In this case, there are at least two distinct groups of phytoplankton grazers present, the benthic suspension feeders and pelagic grazers. If something happens to the system that allows the pelagic grazers to provide greater enhancement to the system, then they either will grow to overshadow the benthic suspension feeders or will displace them altogether. Further, positive feedback systems also are typically fragile and exhibit threshold effects (DeAngelis et al. 1986). This shift in dominance and collapse of a strong positive feedback mechanism may explain why natural benthic suspension-feeding systems are so slow to return after fishing (Herman et al. 1999) has destroyed them. From the available evidence, it is clear that the system transforming benthic suspension feeders do not destroy their ecosystems because negative controls by internal and external factors also influence their systems. It appears that poor human management and strong climatic events are necessary for a system-level shift in the dominant component to occur.

References

Asmus H (1987) Secondary production of an intertidal mussel bed community related to its storage and turnover compartments. Mar Ecol Prog Ser 39:251–266

Asmus RM, Asmus H (1991) Mussel beds: limiting or promoting phytoplankton? J Exp Mar Biol Ecol 148:215–232

Asmus H, Asmus RM (1993) Phytoplankton-mussel bed interactions in intertidal ecosystems. In: Dame RF (ed) Bivalve filter feeders in estuarine and coastal ecosystem processes. Springer, Berlin Heidelberg New York, pp 57–84

Asmus RM, Asmus H (1998) Bedeutung der Organismengemeinschaften für den bentho-pelagischen Stoffaustausch. In: Gätje C, Reise K (eds) Öksystem Wattenmeer, Austausch Transport- und Stoffumwandlungsprozesse. Springer, Berlin Heidelberg New York, pp 257–302

Asmus H, Asmus RM, Reise K (1990) Exchange processes in an intertidal mussel bed: a Sylt-flume study in the Wadden Sea. Ber Biol Anst Helgoland 6:1–79

Asmus H, Asmus RM, Zubillaga FG (1995) Do mussel beds intensify the phosphorus exchange between sediment and tidal waters? Ophelia 41:37–55

Asmus RM, Jensen MH, Jensen KM, Kristensen E, Asmus H, Wille A (1998) The role of water movement and spatial scaling for measurement of dissolved inorganic nitrogen fluxes in intertidal sediments. Est Coast Shelf Sci 46:221–232

Bacher C (1989) Capacité trophique du bassin de Marennes-Oléron: couplage d'un modele de transport particulaire et d'un modele de croissance de l'huitre *Crassostrea gigas*. Aquat Living Res 48:199–214

Ball B, Raine R, Douglas D (1997) Phytoplankton and particulate matter in Carlingford Lough, Ireland: an assessment of food availability and the impact of bivalve culture. Estuaries 20:430–440

Bahr LM, Lanier WP (1981) The ecology of intertidal oyster reefs of the South Atlantic coast: a community profile. US Fish and Wildlife Service, Washington, DC. FWS/OBS-81/15. 105 pp

Bertness MD (1984) Ribbed mussels and *Spartina alterniflora* production in a New England salt marsh. Ecology 65:1794–1807

Bertness MD, Gaines SD, Yeh SM (1998) Making mountains out of barnacles: the dynamics of acorn barnacle hummocking. Ecology 79:1382–1394

Biggs RB, Howell BA (1984) The estuary as a sediment trap: alternate approaches to estimating filter efficiency. In: Kennedy VS (ed) The estuary as a filter. Academic Press, New York, pp 107–129

Bosence DWJ (1973) Recent serpulid reefs, Connemara, Eire. Nature 242:40–41

Buhr KJ (1976) Suspension-feeding and assimilation efficiency in *Lanice conochilega* (Polychaete). Mar Biol 38:373–383

Buss LW (1979) Bryozoan overgrowth interactions – the interdependence of competition for space and food. Nature 281:475–477

Buss LW, Jackson JBC (1981) Planktonic food availability and suspension-feeder abundance: evidence of in situ depletion. J Exp Mar Biol Ecol 49:151–161

Cadée GC, Hegeman J (1974) Primary production of phytoplankton in the Dutch Wadden Sea. Neth J Sea Res 8:240–259

Carlson DJ, Townsend DW, Hilyard A, Eaton JF (1984) Effect of an intertidal mudflat on plankton of the overlying water column. Can J Fish Aquat Sci 41:1523–1528

Carpenter SR, Kitchell JF (1988) Consumer control of lake productivity. BioScience 38:764–769

Cloern JE (1982) Does the benthos control phytoplankton biomass in South San Francisco Bay? Mar Ecol Prog Ser 9:191–202

Cohen AN, Carlton JT (1998) Accelerating invasion rate in a highly invaded estuary. Science 279:555–558

Coughlan J, Ansell AD (1964) A direct method for determining the pumping rate of siphonate bivalves. J Conseil 29:205–214

Craeymeersch JA, Herman, PMJ Meire PM (1986) Secondary production of an intertidal mussel (*Mytilus edulis* L.) population in the Eastern Scheldt (S.W. Netherlands). Hydrobiologia 133:107–115

Crisp DJ (1979) Dispersal and re-aggregation in sessile marine invertebrates, particularly barnacles. In: Larwood G, Rosen BR (eds) Biology and systematics of colonial organisms. Academic Press, New York, pp 319–327

Crisp DJ, Southward AJ (1961) Different types of cirral activity in barnacles. Philos Trans R Soc 243:273–307

Dales RP (1957) Some quantitative aspects of feeding in Sabellid and Serpulid worms. J Mar Biol Assoc UK 36:309–316

Dales RP (1961) Observations on the respiration of the polychaete *Schizobranchia insignis*. Biol Bull 121:82–91

Dame RF (1975) Energy flow in an intertidal oyster population. Est Coast Mar Sci 4:243–253

Dame RF (1979) The abundance, diversity and biomass of macrobenthos on North Inlet, South Carolina, intertidal oyster reefs. Proc Natl Shellfish Assoc 69:6–10

Dame RF (1996) Ecology of marine bivalves: an ecosystem approach. CRC Press, Boca Raton

Dame RF, Dankers N (1988) Uptake and release of materials by a Wadden Sea mussel bed. J Exp Mar Biol Ecol 118: 207–216

Dame RF, Dankers N, Prins T, Jongsma H, Smaal A (1991a) The influence of mussel beds on nutrients in the Western Wadden Sea and Eastern Scheldt estuaries. Estuaries 14:130–138

Dame RF, Libes S (1993) Oyster reefs and nutrient retention in tidal creeks. J Exp Mar Biol Ecol 171:251–258

Dame RF, Spurrier JD, Williams TM, Kjerfve B, Zingmark RG, Wolaver TG, Chrzanowski TH, McKellar HN, Vernberg FJ (1991b) Annual material processing by a salt marsh estuarine basin in South Carolina, USA. Mar Ecol Prog Ser 72:153–166

Dame RF, Spurrier JD, Wolaver TG (1989) Carbon, nitrogen and phosphorus processing by an oyster reef. Mar Ecol Prog Ser 54:249–256

Dame RF, Wolaver TG, Libes SM (1985) The summer uptake and release of nitrogen by an intertidal oyster reef. Neth J Sea Res 19:265–268

Dame RF, Zingmark RG, Haskin E (1984) Oyster reefs as processors of estuarine materials. J Exp Mar Biol Ecol 83:239–247

Dame RF, Zingmark RG, Stevenson H, Nelson D (1980) Filter feeding coupling between the water column and benthic systems. In: Kennedy VS (ed) Estuarine perspectives. Academic Press, New York, pp 521–526

Dankers N (1993) Integrated estuarine management – obtaining a sustainable yield of bivalve resources while maintaining environmental quality. In: Dame RF (ed) Bivalve filter feeders in estuarine and coastal ecosystem processes. NATO ASI series, series G, Ecological sciences, vol 33. Springer, Berlin Heidelberg New York, pp 479–512

Dankers N, Koelemaij K (1989) Variations in the mussel population of the Dutch Wadden Sea in relation to monitoring of other ecological parameters. Helgolander Wiss Meeresunters 43:529–535

Davies BR, Stuart V, de Villiers M (1989) The filtration activity of a serpulid polychaete population (*Ficopomatus enigmaticus* (Fauvel) and its effects on water quality in a coastal marina. Est Coast Shelf Sci 29:613–620

DeAngelis DL, Post WM, Travis CC (1986) Positive feedback in natural systems. Springer, Berlin Heidelberg New York

Doering PH, Kelly JR, Oviatt CA, Sowers T (1987) Effect of the hard clam *Mercenaria mercenaria* on benthic fluxes of inorganic nutrients and gases. Mar Biol 94:377–383

Doering PH, Oviatt CA (1986) Application of filtration rate models to field populations of bivalves: an assessment using experimental mesocosms. Mar Ecol Prog Ser 31:265–275

Doering PH, Oviatt CA, Kelly JR (1986) The effects of the filter-feeding clam *Mercenaria mercenaria* on carbon cycling in experimental marine mesocosms. J Mar Res 44: 839–861

Ferreira JG, Duarte P, Ball B (1997) Trophic capacity of Carlingford Lough for aquaculture – analysis by ecological modelling. Aquat Ecol 31:361–378

Fiala-Médoni A (1978) Filter-feeding ethology of benthic invertebrates (Ascidians). IV. Pumping rate. Filtration rate, filtration efficiency. Mar Biol 48:243–249

Fréchette M, Butman CA, Geyer WR (1989) The importance of boundary-layer flows in supplying phytoplankton to the benthic suspension feeder, *Mytilus edulis* L. Limnol Oceanogr 34:19–36

Gore RH, Scotto LE, Becker EJ (1978) Community composition, stability, and trophic partitioning in decapod crustaceans inhabiting some subtropical Sabellariid worm reefs. Bull Mar Sci 28:221–248

Granéli E, Olsson P, Carlsson P, Granéli W, Nylander C (1993) Weak 'top-down' control of dinoflagelate growth in the coastal Skagerrak. J Plankton Res 15:213–237

Gruet Y (1986) Spatio-temporal changes of sabellarian reefs built by the sedentary polychaete *Sabellaria alveolata* (Linne). Mar Ecol 7:303–319

Gutierrez J, Iribarne O (1999) Role of holocene beds of the stout razor clam *Tagelus plebeius* in structuring present benthic communities. Mar Ecol Prog Ser 185: 213–228

Hadfield MG, Kay EA, Gillette MU, Lloyd MC (1972) The Vermetidae of the Hawaiian Islands. Mar Biol 12:81–98

Haven DS, Morales-Alamo R (1966) Aspects of biodeposition by oysters and other invertebrate filter feeders. Limnol Oceanogr 11:487–498

Haven DS, Morales-Alamo R (1972) Biodeposition as a factor in sedimentation of fine suspended solids in estuaries. Geol Soc Am Mem 133:121–130

Heip CHR, Goosen NK, Herman PMJ, Kromkamp J, Middelburg J, Soetaert K (1995) Production and consumption of biological particles in temperate tidal estuaries. Annu Rev Ocean Mar Biol 33:1–149

Héral M (1993) Why carrying capacity models are useful tools for management of bivalve culture. In: Dame RF (ed) Bivalve filter feeders in estuarine and coastal ecosystem processes. Springer, Berlin Heidelberg New York, pp 455–477

Héral M, Deslous-Paoli J-M, Prou J (1988) Approche de la capacité trophique d'un écosystème conchylicole. J Cons Int Explor Mer Cm 1988/K, 22 pp

Herdman WA (1920) The marine biological station at Port Erin (Isle of Man). 34th Ann Rept, 32 pp

Herman PMJ (1993) A set of models to investigate the role of benthic suspension feeders in estuarine ecosystems. In: Dame RF (ed) Bivalve filter feeders in estuarine and coastal ecosystem processes. Springer, Berlin Heidelberg New York, pp 421–454

Herman PMJ, Middelburg JJ, Van de Koppel J, Heip CHR (1999) Ecology of estuarine macrobenthos. Adv Ecol Res 29:195–240

Herman PMJ, Scholten H (1990) Can suspension-feeders stabilize estuarine ecosystems? In: Barnes M, Gibson R (eds) Trophic relationships in the marine environment. Proc 24th EMBS Ed 35. Aberdeen Univ Press, Aberdeen, pp 104–116

Hily C (1991) Is the activity of benthic suspension feeders a factor controlling water quality in the Bay of rest? Mar Ecol Prog Ser 69:179–188

Horsted SJ, Nielsen TG, Riemann B, Pock-Steen J, Bjørnsen PK (1988) Regulation of zooplankton by suspension-feeding bivalves and fish in estuarine enclosures. Mar Ecol Prog Ser 48:217–224

Hughes RN (1979) Coloniality in Vermetidae (Gastropoda). In: Larwood G, Rosen BR (eds) Biology and systematics of colonial organisms. Academic Press, New York, pp 243–253

Jones CG, Lawton JH, Shachak M (1994) Organisms as ecosystem engineers. Oikos 69:373–386

Jørgensen CB (1966) Biology of suspension feeding. Pergamon Press, New York

Jørgensen CB (1990) Bivalve filter feeding: hydrodynamics, bioenergetics, physiology and ecology. Olsen and Olsen, Fredensborg, Denmark

Jørgensen CB, Kiørboe T, Møhlenberg F, Riisgård HU (1984) Ciliary and mucus-net filter feeding, with special reference to fluid mechanical characteristics. Mar Ecol Prog Ser 15:283–292

Keene WC (1980) The importance of a reef-forming polychaete, *Mercierella enigmatica* Fauvel, in the oxygen and nutrient dynamics of a hypereutrophic subtropical lagoon. Est Coast Mar Sci 11:167–178

Kimmerer WJ, Gartside E, Orsi JJ (1994) Predation by an introduced clam as the likely cause of substantial declines in zooplankton of San Francisco Bay. Mar Ecol Prog Ser 113:81–93

Kirby-Smith WW (1972) Growth of the bay scallop: the influence of experimental water currents. J Exp Mar Biol Ecol 8:7–18

Kirkwood JM, Burton HR (1988) Macrobenthic species assemblages in Ellis Fjord, Vestfold Hills, Antarctica. Mar Biol 97:445–457

Kirtley DW, Tanner WF (1968) Sabellariid worms: builders of a major reef type. J Sed Petrol 38:73–78

Klöckner K (1976) Ökologie von *Pomatoceros triqueter* (Serpulidae, Polychaeta). I. Reproduktionsablauf, Substratwahl, Wachstum und Mortalität. Helgolander Wiss Meeresunters 28:352–400

Koseff JR, Holen JK, Monismith SG, Cloern JE (1993) Coupled effects of vertical mixing and benthic grazing on phytoplankton populations in shallow, turbid estuaries. J Mar Res 51:843–868

Le Gall S, Bel Hassen M, Le Gall P (1997) Ingestion of a bacterivorous ciliate by the oyster *Crassostrea gigas*: protozoa as a trophic link between picoplankton and benthic suspension-feeders. Mar Ecol Prog Ser 152:301–306

Lemmens JWTJ, Clapin G, Lavery P, Cary J (1996) Filtering capacity of seagrass meadows and other habitats of Cockburn Sound, Western Australia. Mar Ecol Prog Ser 143:187–200

Lucas LV, Cloern JE, Koseff JR, Monismith SG, Thompson JK (1998) Does the Sverdrup critical depth model explain bloom dynamics in estuaries? J Mar Res 56:375–415

Møhlenberg F, Riisgård HU (1979) Filtration rate using a new indirect technique, in thirteen species of suspension feeding bivalves. Mar Biol 54:143–147

Moore CG, Saunders GR, Harries DB (1998) The status and ecology of reefs of *Serpula vermicularis* L. (Polychaeta: Serpulidae) in Scotland. Aquatic Conserv Mar Freshw Ecosyst 8:645–656

Multer HG, Milliman JD (1967) Geologic aspects of sabellarian reefs, southeastern Florida. Bull Mar Sci 17:257–267

Newell RIE (1988) Ecological changes in Chesapeake Bay: are they the result of overharvesting the American oyster, *Crassostrea virginica*? In: Lynch MP, Krome EC (eds) Understanding the estuary: advances in Chesapeake Bay research. Chesapeake Research Consortium, Solomon's, Maryland, pp 536–546

Nixon SW, Oviatt CA, Rogers C, Taylor K (1971) Mass and metabolism of a mussel bed. Oecologia 8:21–30

Officer CB, Smayda TJ, Mann R (1982) Benthic filter feeding: a natural eutrophication control. Mar Ecol Prog Ser 9:203–210

Olsson P., Granéli E., Carlsson P, Abreu P (1992) Structuring of a postspring phytoplankton community by manipulation of trophic interactions. J Exp Mar Biol Ecol 158:249–266

O'Riordan CA, Monismith SG, Koseff JR (1995) The effect of bivalve excurrent jet dynamics on mass transfer in a benthic boundary layer. Limnol Oceanogr 40:330–344

Petersen JK, Riisgård HU (1992) Filtration capacity of the ascidian *Ciona intestinalis* and its grazing impact in a shallow fjord. Mar Ecol Prog Ser 88:9–17

Petersen JK, Schou O, Thohr P (1995) Growth and energetics of the ascidian *Ciona intestinalis* (L.). Mar Ecol Prog Ser 120:175–184

Petersen JK, Schou O, Thohr P (1997) In situ growth of the ascidian *Ciona intestinalis* (L.) and the blue mussel *Mytilus edulis* in an eelgrass meadow. J Exp Mar Biol Ecol 218:1–11

Pilson MEQ (1985) On the residence time of water in Narragansett Bay. Estuaries 8:2–14

Porras R, Bataller JV, Murgui E, Torregrosa MT (1996) Trophic structure and community composition of polychaetes inhabiting some *Sabellaria alveolata* (L.) reefs along the Valencia Gulf coast, Western Mediterranean. Mar Ecol 17:583–602

Prins TC, Escaravage V, Smaal AC, Peeters JCH (1995a) Functional and structural changes in the pelagic system induced by bivalve grazing in marine mesocosms. Water Sci Technol 32:183–185

Prins TC, Escaravage V, Smaal AC, Peeters JCH (1995b) Nutrient cycling and phytoplankton dynamics in relation to mussel grazing in a mesocosm experiment. Ophelia 41:289–315

Prins TC, Escaravage V, Wetsteyn LPMJ, Peeters JCH (2000) Limitation of plankton development by mussel grazing in experimental marine ecosystems with a well-mixed water column (in prep)

Prins TC, Smaal AC (1990) Benthic-pelagic coupling: the release of inorganic nutrients by an intertidal bed of *Mytilus edulis*. In: Barnes M, Gibson RN (eds) Trophic relationships in the marine environment. Aberdeen Univ Press, Aberdeen, pp 89–103

Prins TC, Smaal AC (1994) The role of the blue mussel *Mytilus edulis* in the cycling of nutrients in the Oosterschelde estuary (The Netherlands). Hydrobiol 282/283:413–429

Prins TC, Smaal AC, Dame RF (1997) A review of the feedbacks between bivalve grazing and ecosystem processes. Aquat Ecol 31:349–359

Prins TC, Smaal AC, Dankers N, Pouwer AJ (1996) Filtration and resuspension of particulate matter and phytoplankton on an intertidal mussel bed in the Oosterschelde estuary (SW Netherlands). Mar Ecol Prog Ser 142:121–134

Pullen J, LaBarbera M (1991) Modes of feeding in aggregations of barnacles and the shape of aggregations. Biol Bull 181:442–452

Randløv A, Riisgård HU (1979) Efficiency of particle retention and filtration rate in four species of ascidians. Mar Ecol Prog Ser 1:55–59

Reise K (1982) Long-term changes in the macrobenthic invertebrate fauna of the Wadden Sea: are polychaetes about to take over? Neth J Sea Res 16:29–36

Reise K, Herre E, Sturm M (1989) Historical changes in the benthos of the Wadden Sea around the island of Sylt in the North Sea. Helgoländer Meeresunters 43:417–433

Riemann B, Nielsen TG, Horsted SJ, Bjørnsen PK, Pock-Steen J (1988) Regulation of phytoplankton biomass in estuarine enclosures. Mar Ecol Prog Ser 48:205-215

Riisgård HU (1988) The ascidian pump: properties and energy cost. Mar Ecol Prog Ser 47:129-134

Riisgård HU (1991) Suspension feeding in the polychaete *Nereis diversicolor*. Mar Ecol Prog Ser 70:29-37

Riisgård HU, Christensen PB, Olesen NJ, Petersen JK, Møller MM, Andersen P (1995) Biological structure in a shallow cover (Kertinge Nor, Denmark) - control by benthic nutrient fluxes and suspension-feeding ascidians and jellyfish. Ophelia 41:329-344

Riisgård HU, Ivarsson NM (1990) The crown-filament pump of the suspension-feeding polychaete *Sabella penicillus*: filtration, effects of temperature and energy cost. Mar Ecol Prog Ser 62:249-257

Riisgård HU, Manriquez P (1997) Filter-feeding in fifteen marine ectoprocts (Bryozoa): particle capture and water pumping. Mar Ecol Prog Ser 154:223-239

Riisgård HU, Thomassen S, Jakobsen H, Weeks JM, Larsen PS (1993) Suspension feeding in marine sponges *Halichondria panicea* and *Haliclona urceolus*: effects of temperature on filtration rate and energy cost of pumping. Mar Ecol Prog Ser 96:177-188

Rothschild BJ, Ault JS, Goulletquer P, Héral M (1994) Decline of the Chesapeake Bay oyster populations: a century of habitat destruction and overfishing. Mar Ecol Prog Ser 111:29-39

Safriel UN (1974) Vermetid gastropods and intertidal reefs in Israel. Science 186:1113-1115

Sanford E, Bermudez D, Bertness MD, Gaines SD (1994) Flow, food supply and acorn barnacle population dynamics. Mar Ecol Prog Ser 104:49-62

Schiparelli S, Cattaneo-Vietti R (1999) Functional morphology of vermetid feeding tubes. Lethaia 32:41-46

Seed R, Suchanek TH (1992) Population and community ecology of *Mytilus*. In: Gosling E (ed) The mussel *Mytilus*. Ecology, physiology, genetics, and culture. Elsevier, Amsterdam, pp 87-168

Shumway SE, Bogdanowicz E, Dean D (1988) Oxygen consumption and feeding rates of the sabellid polychaete, *Myxicola infundibulum* (Renier). Comp Biochem Physiol 90A:425-428

Smaal AC, Prins TC (1993) The uptake of organic matter and the release of inorganic nutrients by bivalve suspension feeder bed. In: Dame RF (ed) Bivalve filter feeders in estuarine and coastal ecosystem processes. Springer, Berlin Heidelberg New York, pp 273-298

Smaal AC, Verhagen JHG, Coosen J, Haas HA (1986) Interactions between seston quantity and quality and benthic suspension feeders in the Oosterschelde, The Netherlands. Ophelia 26:385-399

Smaal AC, Vonck APMA, Bakker M (1997) Seasonal variation in physiological energetics of *Mytilus edulis* and *Cerastoderma edule* of different size classes. J Mar Biol Assoc UK 77:817-838

Ten Hove HA (1979) Different causes of mass occurrence in serpulids. In: Larwood G, Rosen BR (eds) Biology and systematics of colonial organisms. Academic Press, New York, pp 281-298

Ten Hove HA, Van den Hurk P (1993) A review of recent and fossil serpulid "reefs"; actuopalaeontology and the "Upper Malm" serpulid limestones in NW Germany. Geol Mijnb 72:23-67

Tenore KR, Boyer LF, Cal RM, Corral J, Garcia-Fernandez C, Gonzalez N, Gonzalez-Gurrianan E, Hanson RB, Iglesias J, Krom M, Lopez-Jamar E, McClain J, Pamatmat

MM, Perez A, Rhoads DC, de Santiago G, Tiejen J, Westrich J, Windom HL (1982) Coastal upwelling in the Rias Bajas, NW Spain: contrasting the benthic regimes of the Rias de Arosa and de Muros. J Mar Res 40:701–772

Thomas LH, Stevens J-A (1991) Communities of constructional lips and cup reef rims in Bermuda. Coral Reefs 9:225–230

Thomason JC, Hills JM, Clare AE, Neville E, Richardson M (1998) Hydrodynamic consequences of barnacle colonization. Hydrobiologia 375/376:191–201

Ulanowicz RE (1997) Ecology, the ascendant perspective. Columbia Univ Press, New York

Vahl O (1973) Porosity of the gill, oxygen consumption and pumping rate in Cardium edule (L.). Ophelia 10:109–118

Van Stralen MR (1995) Growth and landings of cultivated mussels from 1952 onwards, and developments in the cockle stock in relation to food supply, eutrophication and other environmental factors in the Wadden Sea (in Dutch). RIVO report 95.016. Netherlands Institute of Fisheries Research, Yerseke

Vedel A, Riisgård HU (1993) Filter-feeding in the polychaete *Nereis diversicolor*: growth and bioenergetics. Mar Ecol Prog Ser 100:145–152

Verhagen JHG (1985) Tidal motion, and the seston supply to the benthic macrofauna in the Eastern Scheldt. DHL report R1310–14. Delft Hydraulics, Delft

Verwey J (1952) On the ecology of distribution of cockle and mussel in the Dutch Waddensea, their role in sedimentation and the source of their food supply. Arch Neerl Zool 10:171–239

Walne PR (1972) The influence of current speed, body size and water temperature on the filtration rate of five species of bivalves. J Mar Biol Assoc UK 52:345–374

Wells HW (1961) The fauna of oyster beds, with special reference to the salinity factor. Ecol Monogr 31:241–266

Wildish D, Kristmanson (1997) Benthic suspension feeders and flow. Cambridge Univ Press, Cambridge

Wilson DP (1971) Sabellaria colonies at Duckpool, North Cornwall, 1961–1970. J Mar Biol Assoc UK 51:509–580

Winter JE (1969) Über den Einfluss der Nahrungskonzentration und andere Faktoren auf Filtrierleistung und Nahrungsausnutzung der Muscheln *Arctica islandica* und *Modiolus modiolus* Mar Biol 4:87–135

Wood R (1999) Reef evolution. Oxford Univ Press, Oxford

Wright RT, Coffin RB, Ersing CP, Pearson D (1982) Field and laboratory measurements of bivalve filtration of natural marine bacterioplankton. Limnol Oceanogr 27:91–98

Zurburg W, Smaal AC, Héral M, Dankers N (1994a) In situ estimations of uptake and release of material by bivalve filter feeders in the bay of Marennes-Oléron (France). In: Dyer KR, Orth RJ (eds) Changes in fluxes in estuaries. Olsen and Olsen, Fredensborg, Denmark, pp 239–242

Zurburg W, Smaal AC, Héral M, Dankers N (1994b) Seston dynamics and bivalve feeding in the Bay of Marennes-Oléron (France). Neth J Aquat Ecol 28:459–466

2 Dynamics of Spatial and Temporal Complexity in European and North American Soft-Bottom Mussel Beds

J.A. COMMITO and N.M.J.A. DANKERS

2.1 Introduction

Mussel beds are conspicuous features of temperate and boreal coastlines. Beds of the edible or blue mussel, *Mytilus edulis*, can extend for kilometers along the shore. Soft-bottom beds have density, biomass, and respiratory flow values among the highest of any community known (Nixon et al. 1971; Seed 1976; Asmus 1987). In Europe, both rocky shore and soft-bottom mussel beds have been well studied for many years because of their ecological importance and economic value as a wild and cultured fishery. In North America it is probably fair to say that rocky shore mussel beds have received the lion's share of attention because of the role they have historically played as a model system, one that is especially amenable to experimental manipulations in the field. Investigations such as those by Paine (1966) and Dayton (1971) along the Pacific shore, generally involving *Mytilus californianus*, have become classics in the field of ecology and were among the first to demonstrate the power of field experiments. This focus on rocky shore mussels by North American ecologists has resulted in less knowledge about soft-bottom *Mytilus edulis* beds on western Atlantic shores than along the coastlines of European countries.

In this chapter, we examine some of the key characteristics of soft-bottom mussel beds, with an emphasis on the spatial and temporal dynamics of *Mytilus edulis*. In particular, we discuss the structure of mussel beds, both in aerial view (spatial pattern in the horizontal plane) and in vertical profile (surface topography). Mussel-bed structure is considered in light of some important recent advances in benthic ecology, including adult–larval interactions, costs and benefits of group living in sessile organisms, and flow-mediated regulation of communities by top-down and bottom-up processes.

2.2 The Dynamic Nature of Mussel-Bed Structure

In 1877, Karl Möbius used the term "biocoenosis" to describe the beds of oysters around the Wadden Sea island of Sylt. He described the positive and negative interactions among the biotic and abiotic components of these important suspension-feeding bivalve populations. In doing so, he formalized the new field of community ecology, a discipline that has been strongly influenced by benthic ecologists ever since. Like oysters, mussels are commercially valuable suspension-feeding bivalves found in shallow subtidal and intertidal sites, often in dense, well-defined populations. Reise and Schubert (1987) discussed the long-term changes in benthic community structure at subtidal Wadden Sea sites. They concluded that mussel beds in the 1920s had been restricted to the shallows, with oyster beds and tubicolous polychaete reefs in deeper water, whereas by the 1980s mussels occupied both shallow and deep depths. The reasons for these changes are uncertain, but Reise and Schubert argue that they result from a combination of coastal eutrophication, disruption of the bottom by dredging and trawling for oysters and mussels, and the intentional transplanting of small, shallow water mussels to deeper sites where they reach commercial size more quickly.

As is the case with subtidal mussel beds, there are intertidal sites in Europe with long-term data sets that simply do not exist for coastal locations in other parts of the world. These data demonstrate that some locations have persistent mussel beds, generally where the bottom is sheltered from the effects of storms and winter ice scouring. Examples include the protected embayment of Königshafen on the island of Sylt in the Wadden Sea (Nehls et al. 1997) and the River Exe estuary in Devon, England (McGrorty et al. 1990). True persistence stability occurs when a system stays the same in the face of external forces of change. If these sites are sheltered from storms and ice, then we cannot say for certain if they are exhibiting persistence stability. On the other hand, exposed sites in the Wadden Sea subjected to the removal of mussels by storms and ice and burial of mussels by sediment often show dramatic changes in the spatial distribution and abundance of mussel beds (Landahl 1988; Dankers and Koelemaij 1989; Dankers 1993; Nehls and Thiel 1993; Beukema and Cadée 1996). Fisheries activities have similar effects (Dankers and Koelemaij 1989; Dankers 1993). Mussel beds can become reestablished in the same location after the occurrence of a disturbance, especially where layers of old shell fragments remain (Dankers and Koelemaij 1989). Thus, beds may not be persistent, but they do demonstrate adjustment stability, or the ability to return to a former state after being perturbed. Figure 2.1 summarizes the major processes that occur on mussel beds, determine their abundance and distribution, and affect the soft-bottom species living in association with them.

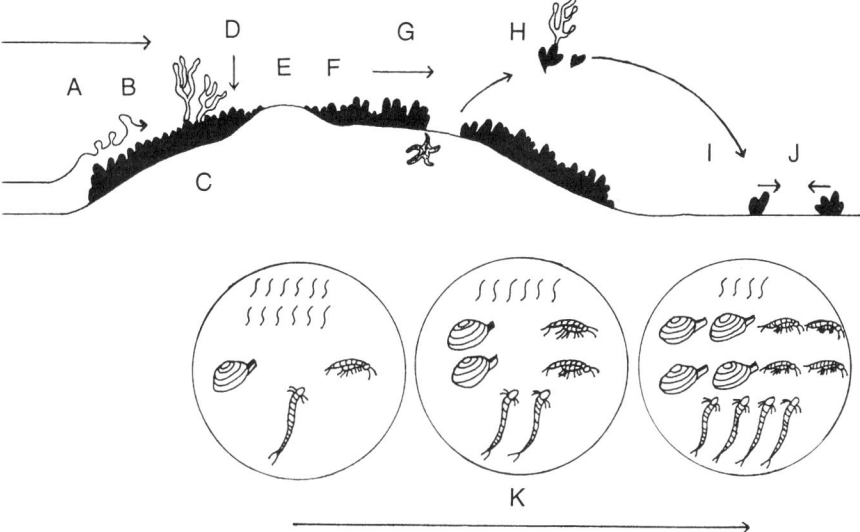

Fig. 2.1. Schematic diagram of important mussel-bed processes and characteristics. Numbers in parentheses refer to the relevant chapter sections. *A* Altered flow regime, from laminar to turbulent (2.5, 2.6); *B* increased delivery of food particles, sediment, oxygen, and larvae to bottom (2.2, 2.6); *C* raised bed profile due to trapping of sediment, feces, and pseudofeces (2.5); *D* increased sedimentation due to presence of attached algae (2.2); *E* burial due to storm deposition of sediment (2.2); *F* irregular surface topography, leading to lower rates of predation but higher rates of dislodgement (2.2, 2.3, 2.5, 2.6); *G* gradient of edge effects, with increased larval recruitment, growth, predation, and dislodgement rates at edge (2.2, 2.3, 2.4, 2.6); *H* dislodgement due to storms, ice scour, and commercial dredging (2.2); *I* postlarval dispersal and establishment of mussel clumps (2.2); *J* mussel movement towards each other to form larger clumps (2.2); *K* gradient of infaunal species, with lower densities of oligochaetes and higher densities of other species away from the center of the bed (2.4, 2.5)

Petraitis and Latham (1999) discuss experimental results showing how two strongly scale-dependent processes operate to create (ice scour) and maintain (predation) alternative community states in mussel-bed structure on the rocky shore of Maine, USA. Similar experimental manipulations in soft-bottom systems have rarely been performed specifically to examine the roles of disturbance, predation, and other factors that control the structure of mussel beds. At a soft-bottom site in the Kiel Fjord in the western Baltic Sea, Reusch and Chapman (1997) used a combination of observational, experimental, and modeling approaches to explain the persistence of shallow soft-bottom *Mytilus edulis* beds with a dynamic, patchy spatial structure. Transplant and removal experiments showed that mussel patches protected from predators expanded dramatically in size. Under control conditions, however, recruitment and growth rates of mussels were high enough to prevent the abundant predators, the seastar *Asterias rubens*, from eliminating mussel patches. Mussels and seastars were in dynamic equilibrium. Moreover, drift fences demonstrated that *Mytilus edulis* clumps were lifted from the bottom and carried by water currents to locations where they formed the nuclei of new beds.

Other workers have shown that the dynamic nature of soft-bottom mussel beds has a strong spatial scale component. Population performance depends on location within an estuary (*Mytilus edulis* – Devon, England: McGrorty et al. 1990; McGrorty and Goss-Custard 1991, 1993, 1995), proximity to the edge of a saltmarsh (ribbed mussels, *Geukensia demissa* – Rhode Island, USA: Bertness and Grosholz 1985), and position along a tidal height gradient in saltmarshes (Bertness and Grosholz 1985; North Carolina, USA: Stiven and Gardner 1992) and on soft-bottoms (McGrorty et al. 1990; McGrorty and Goss-Custard 1991, 1993, 1995; Kiel Fjord: Reusch and Chapman 1997).

On a smaller spatial scale, researchers have argued that independent, non-clonal organisms like mussels live in dense aggregations because there are benefits to living in close proximity to one's neighbors. Crab predation can be lower on soft-bottom mussels in large, dense patches and the centers of small patches than on solitary mussels and those living on the edges of patches (*Geukensia demissa* – Rhode Island, USA: Bertness and Grosholz 1985; North Carolina, USA: Lin 1991; Stiven and Gardner 1992; *Mytilus edulis* – northern California, USA: Okamura 1986; Devon, England: McGrorty et al. 1990). Presumably, crabs find it easier to attack soft-bottom mussels that are exposed rather than protected by an impenetrable matrix of living mussels, shell fragments, and byssal threads. One outcome is that differential predation on the edges slows down patch growth and coalescence, as demonstrated by the predator exclusion experiments of Reusch and Chapman (1997). Predation on mussel beds by bird predators such as oystercatchers, eiders, and gulls has not been the object of much attention in North America (Guillemette and Himmelman 1996), but it has been intensively studied in Europe (Meire and Ervynck 1986; Goss-Custard and Durell 1988; Bustnes and Erikstad 1990; Beukema 1993; Hilgerloh et al. 1997; McGrorty 1997; Nehls et al. 1997). In contrast to invertebrate predators, birds may not feed preferentially on the edges of patches because their mobility and size give them access to a greater spatial array of mussels, and their agonistic interactions force them to spread out over large areas.

Mytilus edulis can defend against crab, seastar, and lobster predators (and even respond to predator effluent) by growing thicker shells, producing more byssal threads, and actively moving towards each other into larger clumps with lower perimeter/area ratios, reducing their per mussel predation risk (e.g., west coast of Sweden: Reimer and Tedengren 1997; Limfjorden, Denmark: Dolmer 1998; North Norfolk coast, United Kingdom: Côté and Jelnikar 1999; Maine coast, USA: Leonard et al. 1999). Movement into clumps could speed up the rate at which mussel beds coalesce. Some studies have shown that predators may consume *Mytilus edulis* more effectively on smooth beds than on rough, irregular beds (Reusch and Chapman 1997; Dolmer 1998), so surface topography may be as important as clump size and shape.

Horizontal and vertical bed structure also affect removal and death of mussels by physical processes. For example, the removal of spatially aggregated *Geukensia demissa* by winter ice was inversely related to mussel density (Bertness and Grosholz 1985). *Mytilus edulis* individuals at the perimeter of mussel clumps on hard substrate were more susceptible to storm damage than were those in the center (Asamushi, Japan: Tsuchiya and Nishihira 1986). Studies on several species of mussels in Europe and on both coasts of North America have demonstrated that the risk of dislodgement was higher for individuals projecting above the bed surface or for mussels to which other organisms, especially algae, were attached (Witman and Suchanek 1984; Witman 1987; Dolmer and Svane 1994). Juvenile *Geukensia demissa* suffered lower rates of suffocation by shifting sediment when in the presence of adults because the larger individuals provide a safe location above the sediment surface to which juveniles can migrate (Bertness and Grosholz 1985; Lin 1991). Albrecht and Reise (1994) and Albrecht (1998) demonstrated that the presence of algae growing on Königshafen mussel beds contributed to increased sedimentation, which reduced mussel density and altered the bed surface profile. Algal abundance was under the control of *Littorina littorea* grazing, which increased with larger mussel patch size. As these studies all show, the removal or death of mussels depends upon patch size and shape in aerial view and surface roughness in profile.

Although mussel beds are fragmented by storms and ice scour, this change in patch size and shape sets the stage for beds to "heal" by recruitment of new mussels and growth of existing mussels. Flume studies with *Mytilus edulis* have shown that boundary-layer flow changes dramatically at the margins of mussel patches where they project above the sediment (Butman et al. 1994). Larval recruitment was greater to small patches and to the edges of large patches than to the centers of large patches of *Mytilus edulis* on sand bars in two Danish fjords (Svane and Ompi 1993). Flow-mediated delivery of phytoplankton to *Mytilus edulis* was increased as a result of bed roughness caused by the surface topography of the mussels themselves in an intertidal bed in the St. Lawrence River estuary, Québec, Canada (Fréchette et al. 1989). Mussels at the edges of patches typically grow faster and reach larger sizes than those in the center (*Geukensia demissa*: Bertness and Grosholz 1985; *Mytilus edulis* from California on hard substrate: Okamura 1986; *Mytilus edulis* on soft substrate: Svane and Ompi 1993). These patterns demonstrate that the benefits of group living (lower mortality due to protection from predation, ice scour, and sediment burial) may be counterbalanced by the costs of slower growth in the centers of dense mussel patches. They also show that higher edge-dependent rates of larval recruitment and growth might lead to rapid "in-fill" and the coalescence of individual patches into a more uniform whole.

In summary, it can be seen that stability of soft-bottom mussel beds is scale dependent with respect to space and time. So far, investigations have been conducted at large scales of entire seas over decades (Reise and Schubert 1987; Nehls and Thiel 1993; Dankers et al., in press), and in descending order, individual estuaries and embayments over 5 or 10 years (McGrorty et al. 1990; Nehls et al. 1997), kilometer-scale saltmarshes and stretches of shoreline over several years (Bertness and Grosholz 1985; Stiven and Gardner 1992; Reusch and Chapman 1997), meter-scale mussel patches from months to a few years (Bertness and Grosholz 1985; Stiven and Gardner 1992; Reusch and Chapman 1997), and centimeter-scale within-patch position effects from minutes to a few years (Bertness and Grosholz 1985; Okamura 1986; Fréchette et al. 1989; Fréchette and Lefaivre 1990; Svane and Ompi 1993; Fréchette and Despland 1999). Studies at small spatial and temporal scales have been carried out at field sites and in laboratories all over the world, whereas those conducted at the largest scales in time and space tend to be from northern Europe. These latter investigations provide the background that is needed to assess the impacts of long-term commercial harvesting, pollution, coastal development, and global warming over entire geographic regions.

2.3 Quantifying Mussel-Bed Structure Using Fractal Geometry

The persistence of mussel beds depends on the dynamic balance between removal of mussels and the addition of new mussel biomass by recruitment and growth. These processes are regulated at least in part by physical structure – the sizes, shapes, and surface topography of the bed components. How can this complexity be characterized quantitatively? Attention has increasingly been focused on the use of fractal geometry to characterize the shapes and distributions of plants and animals across spatial scales (e.g., Sugihara and May 1990; Hastings and Sugihara 1993). The application of fractals has recently found its way into marine benthic ecology investigations in Europe, North America, and Australia (Le Tourneux and Bourget 1988; Kaandorp 1991, 1994, 1999; Gee and Warwick 1994a,b; Davenport et al. 1996, 1999; Schwinghamer et al. 1996; Kostylev et al. 1997; Beck 1998; Snover and Commito 1998; Commito and Rusignuolo, 2000). A fractal outline is a jagged, non-differentiable curve, often associated with self-similar structure that is repeated at different scales. Fractal geometry allows a quantifiable dimension (the fractal dimension, D) to be given to natural objects like mussel beds that have complicated shapes. Indeed, Mandelbrot's (1982) computer-generated fractal shapes bear such a strong resemblance to aerial maps of mussel beds (such as the maps in Nehls and Thiel 1993 and Herlyn and Michaelis 1995) that they can easily be confused for the real thing.

Snover and Commito (1998) analyzed the fractal geometry of a soft-bottom *Mytilus edulis* bed in Maine, USA. They discovered that the complex, patchy, and seemingly disordered distribution of mussels within the bed was spatially ordered. They predicted that the fractal dimension of patch outlines would be low at low mussel density and percent cover values, high at intermediate values, and drop again as density and percent cover increased so that mussels essentially filled the horizontal plane. They found that D values varied as predicted, with concave-downward parabolic second-order regression curves (density: $r^2=0.94$; percent cover: $r^2=0.92$). They were also able to relate fractal dimension to Morisita's index, a parameter commonly used by ecologists to quantify spatial pattern (negative-slope regression line, $r^2=0.82$).

The application of the fractal dimension to mussel-bed surface topography has also proven to have predictive value. Commito and Rusignuolo (2000) made plaster casts of the same mussel bed and calculated D values for the vertical profiles of cross-sections through the casts. They predicted that D values for surface topography would be lower than D values in aerial view, simply because the bed profile cannot have irregularities as large and complex as the horizontal gaps and projections in mussel patches extending great distances. They also predicted that the fractal dimension would be similar to that of the familiar Koch curve, which (in their view!) has a shape somewhat like that of a mussel-bed surface. In fact, D was lower for every surface profile than for every aerial view. Moreover, D for the plaster casts ranged up to mean (1.24) and median (1.25) values quite close to the predicted Koch curve value of 1.26. At sites in Maine with high rates of mussel recruitment and growth, hummocking has been observed, leading to more complex bed surface topography, which Commito and Rusignuolo predict would cause higher D values. If so, then the fractal dimension could serve as a useful indicator of rates of recruitment and growth. Similarly, the fractal dimension of the bed surface could be valuable in analyzing many important shape-dependent parameters, including flow regimes over the bottom, the provision of habitat space within the three-dimensional matrix of mussel beds, and the ability of predators and herbivores to negotiate the bed surface.

Although the two Maine studies were carried out over a two order of magnitude range in spatial scale, they were still at the very local scale of less than 25 cm. At larger scales, the fractal dimension might serve as a useful monitoring tool for the progress of mussel-bed alteration and recovery. Schwinghamer et al. (1996) used fractals to measure the impact of dredging on seafloor habitat structure, so the practical application of fractal dimension in assessing large areas of the bottom has already begun. Sugihara and May (1990) have argued that the fractal dimension can be used to measure the stability of space-occupying ecological systems. They suggest that a stable system should exhibit large, uniform patches with low D values. As the patches break up, they become more complex, resulting in higher D values. According to this model, the fractal dimension varies inversely with persistence. Does this model apply to mussel beds? The jury is still out on this question. Nobody has explicitly tested the Sugihara and May model. The existing mussel-bed data are equivocal at best. No mussel-bed studies have found the lowest D values at the highest levels of density and percent cover, as predicted by the model. Rather, D increased as density and cover rose to intermediate or high levels (Snover and Commito 1998; Commito and Rusignuolo, 2000; see also Kostylev et al. 1997 for a rocky shore example with *Mytilus galloprovincialis* in Spain). This general result suggests that, with some modification, the Sugihara and May model could be of benefit in the analysis of the spatial and temporal dynamics of mussel beds.

2.4 Effects of Mussel Beds on Soft-Bottom Community Structure

One reason to study the spatial and temporal dynamics of soft-bottom mussel beds is because they cover large areas and may have an impact on benthic community structure. Here we mean "structure" to be the species composition and relative abundances of the soft-bottom community members, not to be confused with the physical architecture of mussel beds. Workers have shown that mussel beds are an important regulating force on rocky shores because they out-compete barnacles and other species for primary attachment space. Beds also provide habitat for species that are normally rare because the complex matrix of byssal threads and trapped objects creates space for animals to live. Patches of mussels serve as habitat islands in soft-bottom (Dittmann 1990; Dankers 1993) and hard-bottom systems, with positive exponential and nonlinear power relationships between patch size and species richness (*Mytilus edulis* – Asamuchi, northern Japan: Tsuchiya and Nishihira 1986; *Brachidontes rostratus* – Victoria, southeastern Australia: Peake and Quinn 1993).

Mussel beds also affect infaunal community structure. In her important paper describing the types of adult–larval interactions that can occur in dense infaunal assemblages, Woodin (1976) summarized the research of many European and North American benthic ecologists. She created a series of testable hypotheses about the impacts that deposit feeders, tube builders, and suspension feeders have on soft-bottom community structure. The response was an immediate flurry of research activity as soft-bottom ecologists, perhaps suffering from "rocky shore envy", sought to test her hypotheses in the field and laboratory. Numerous experiments were carried out on deposit feeders and tubicolous species, especially in North America, where the tradition of rocky shore field manipulations was strong. But, surprisingly, little work was done on the third group – the suspension feeders. One reason might be that epibenthic predation by fish and crabs has devastating effects on the densities of soft-bottom suspension feeders in the southeastern United States, where much of the soft-bottom ecology research in America was being conducted at the time. Woodin had predicted that "no infaunal forms should consistently attain their highest densities among densely packed suspension-feeding bivalves" and that "suspension-feeding forms … reduce the probability of successful larval settlement by any larvae including their own" so that "their assemblages should persist but be strongly age-class dominated". If ecologists thought that densely packed suspension-feeding bivalves were uncommon due to high predation rates, the hypothesis was not likely to be tested. However, in the northeastern United States, the suspension-feeding bivalve *Mytilus edulis* forms large beds that can certainly be considered "densely packed".

Mussel beds have proven to be an excellent model system for testing Woodin's hypotheses in the field. She argued that bivalves ingest larvae from the water column and bury recently settled larvae in feces and pseudofeces, as described earlier by workers such as Thorson (1966) and Mileikovsky (1974). Mussels are known to filter enormous quantities of water, and they produce copious amounts of feces and pseudofeces (Bayne et al. 1976; Seed 1976), so they seem to meet the criteria for adult–larval interactions. As described above, beds are often long-lived, so they can persist as Woodin suggested, and their cumulative effects have time to be expressed. Commito (1987) sampled inside and outside an intertidal mussel bed in Maine with the specific intent of testing Woodin's hypotheses. He found a polymodal distribution of *Mytilus edulis* size classes, indicating that established adults were not able to prevent subsequent larval settlement at the sampling scale of 0.02 m^2. Such polymodal size- and age-class distributions are frequently reported in the literature (e.g., *Mytilus edulis* – Exe estuary, Devon, England: McGrorty and Goss-Custard 1991; *Mytilus galloprovincialis* – northern Adriatic, Italy: Ceccherelli and Rossi 1984). Furthermore, total infaunal density was threefold higher inside the bed than outside, mostly due to the enhanced abundance of the oligochaete *Tubificoides benedeni*. This oligochaete had higher absolute and relative abundances inside the mussel bed than outside, accounting for 97.8 % of bed infauna. It is often found in high densities wherever the silt-clay fraction of the sediment is high and oxygen levels are low (Hunter and Arthur 1978), the conditions that exist in mussel beds. In addition, oligochaetes produce cocoons out of which juveniles crawl. There are no free-living larvae in the water column where they would be exposed to ingestion by mussels. Thus, both of the hypotheses were falsified. On the other hand, non-oligochaete species as a group were less abundant inside the bed than outside, indicating that mussel beds are not the optimal environment for infaunal species with free-swimming larvae.

Based on these results, Commito (1987) suggested a modification of the original Woodin hypotheses. Densely packed suspension-feeding bivalves may indeed lower the density of species with free-swimming larvae or an inability to withstand low oxygen conditions in sediment to which fine-grained feces and pseudofeces have been added. However, they may harbor or even enhance the abundance of infaunal species whose dispersal stages can avoid ingestion by suspension feeders or suffocation in their fecal and pseudofecal material. These species include oligochaetes that produce cocoons as well as taxonomic groups with asexual fragmentation (some oligochaetes and polychaetes), brooding (some polychaetes, bivalves, crustaceans), and post-larval dispersal of relatively large juveniles (many types of macrofauna, e.g., Commito et al. 1995).

Dittmann (1990) conducted an intensive comparison of the intertidal macrofauna and meiofauna in samples taken from the edge and center of a *Mytilus edulis* bed and the adjacent sandflat at Königshafen on the Wadden Sea island of Sylt. She demonstrated that the sandflat macrofauna had substantial numbers of oligochaetes (*Tubificoides benedeni* and two similar species), but was dominated by polychaetes with free-swimming larvae. The center of the bed had a shift in species composition. Polychaetes were less abundant and oligochaetes were more abundant, leading to the same absolute and relative dominance by oligochaetes as in the *Mytilus edulis* bed studied by Commito (1987). Of the polychaetes that did live in the mussel bed, there was a dramatic shift towards *Capitella capitata* and *Heteromastus filiformis*, deposit-feeding species very similar to the oligochaetes in size, morphology, and feeding mode. Subsequent work on *Mytilus edulis* beds from two sites on the west coast of Denmark (Svane and Setyobudiandi 1996) and the East Frisian Islands in the German Wadden Sea has shown similar patterns for macrofauna (Günther 1996; Kröncke 1996). As she found for the macrofauna, Dittmann observed that the mussel-bed meiofauna had a different relative species composition and higher density compared to the bare sandflat, results similar to those found by Radziejewska (1986) for a *Mytilus edulis* bed in Pomeranian Bay, southwestern Baltic Sea.

It can be argued that a simple "inside-outside" sampling study is a weak test of the adult–larval interaction hypothesis. The observed differences might be due to correlations with some third factor, rather than a cause-and-effect relationship caused by the activities of the mussels themselves. Field manipulations can be extremely useful in situations like these. Commito and Boncavage (1989) performed a mussel density manipulation at the same site in Maine studied by Commito (1987). They hypothesized that mussels enhance the abundance of oligochaetes. All mussels were removed from 1-m² plots and put back to create $0\times$, $1\times$, and $2\times$ ambient mussel density treatments and unmanipulated controls. After 3 months, there were significant positive Spearman rank correlations between oligochaete and *Mytilus edulis* abundances. Moreover, the $0\times$ treatment plots had half as many oligochaetes per core as did the other three treatments, which did not differ among themselves. The results indicate that mussels enhance oligochaete abundance, and that this relationship is nonlinear. Above a certain density of mussels, further increases do not result in greater oligochaete abundance.

Ragnarsson and Raffaelli (1999) recently performed a manipulation of *Mytilus edulis* density at an intertidal site in the Ythan estuary, Aberdeenshire, Scotland. They removed mussels from 1-m² plots and also created 0.6-m² patches of mussels transplanted to bare sediment. Overall density and number of taxa declined in the mussel removal patches, particularly the oligochaetes and amphipods. These two groups also increased in the mussel transplant (addition) patches. These results were consistent with the observations and predictions from the Maine studies (Commito 1987; Commito and Boncavage 1989). The silt content in the transplanted patches skyrocketed within 2 weeks, leading to an increase in deposit-feeding polychaetes and a decline in suspension-feeding bivalves, as observed by Dittmann (1990) in the Wadden Sea.

The studies described above were all at intertidal sites and examined the effects of mussels that form dense beds or mats. What about mussels that live in other habitat types or, more importantly, do not form densely packed aggregations? Jaramillo et al. (1992) investigated subtidal, soft-bottom areas inhabited by *Choromytilus chorus* and *Mytilus chilensis* in the Quele River estuary in southern Chile. They showed that macrofaunal density (primarily polychaetes), species richness, and the Shannon index of diversity were lower inside mussel areas than outside. Cummings et al. (1998) sampled patches of the pinnid horse mussel, *Atrina zelandica*, at a subtidal location in northern New Zealand and found no significant differences in macrofauna between samples taken inside and outside the patches at a muddy site. However, at their sandy site, they found effects similar to those from Chile. Density (primarily amphipods and polychaetes), species richness, and the Shannon index of diversity were lower in pinnid patches than in bare sediment.

The two Chilean mytilids and the pinnid were subtidal and had lower mussel densities than are found in intertidal *Mytilus edulis* beds, yet they still produced an impact on soft-bottom community structure. In both cases, they generally lowered the abundance of associated infauna. This result is expected for polychaetes, but amphipods do not produce free-swimming larvae and ought to be able to avoid ingestion by mussels. However, unlike cocoon-producing oligochaetes that thrive in organically enriched sediment, amphipods probably do not do well in fine sediments with low oxygen levels. Interestingly, the New Zealand site where amphipods were less abundant inside pinnid patches was the site that also showed dramatically higher silt-clay fractions and organic content and smaller median grain size inside pinnid patches. Sediment characteristics were virtually identical inside and outside pinnid patches where the infaunal densities were not significantly different. Thus, the results from Chile and New Zealand seem to be consistent with expectations from the intertidal investigations carried out on dense beds of *Mytilus edulis*. Clearly, however, we need more studies, including experimental manipulations, to reveal the differences and similarities in mussel-bed impacts from different parts of the world.

2.5 Mechanisms of Mussel-Bed Impacts on Soft-Bottom Community Structure

The mechanisms that cause mussel beds to alter the relative densities and species composition in soft-bottom systems have not been extensively studied. The explanation offered above is that some species of infauna can avoid ingestion and suffocation by mussels because they do not have free-swimming larvae, can withstand the low oxygen and high sulfide levels of mussel bed sediment enriched with mussel feces and pseudofeces, and are able to exploit that material as a food source. These species have enhanced densities in mussel beds, while other species experience density reductions. However, this explanation has not been explicitly tested and fails to include other possibilities described by Commito (1987) and Commito and Boncavage (1989).

Recent studies of invasive mussels provide some insights into the mechanisms causing mussel-bed effects on infauna. They have been directed specifically at the impacts of invaders on existing community structure. They emphasized experimental manipulations, including experiments designed to separate active effects caused by the biological activities of the mussels from the passive effects due to the presence of physical structure. For example, rubber mussel mimics and empty mussel valves had many of the same effects on community structure as live zebra mussels, *Dreissena polymorpha*, on hard substrates (Slepnev et al. 1994; Ricciardi et al. 1997).

Mytilus edulis impacts might be similar to those observed by Crooks (1998) and Crooks and Khim (1999) for an introduced mussel, *Musculista senhousia*, in California. This marine species forms dense byssal mats in relatively short-lived patches in the soft-bottom intertidal zone. The presence of live or mimic mussels and natural or mimic mats caused an increase in the silt-clay fraction and percent organic material of bottom sediment. These results suggest that active biodeposition by mussels is not the only, or even the primary, reason why sediment composition is usually finer and more organically enriched in mussel mats and beds than adjacent bare sediment. They are also consistent with results demonstrating that physical structure, in this case algae attached to mussels, causes sediment to accumulate rapidly on *Mytilus edulis* beds in the Wadden Sea (Albrecht and Reise 1994; Albrecht 1998).

In addition, the experiments showed that structure of any type had effects on species abundances. Mats generally had stronger effects than did the mussels themselves. Compared to bare areas, mats caused a higher abundance and proportion of infauna with direct development, including oligochaetes and tanaid crustaceans, and a lower abundance and proportion of planktonic developers. These results are consistent with results from the studies described earlier for *Mytilus edulis*. Compared to mussel mimics, the live mussels had only a slight enhancement effect on the direct developers. This result suggests that active in situ biodeposition may not be as important as passive sediment trapping in creating conditions favorable to direct developers. Compared to mussel mimics, the live mussels had only a slight reduction effect on the planktonic developers. This result suggests that direct ingestion of larvae by mussels may not be as important as some surface roughness factor that inhibits successful larval settlement. It also suggests that feces and pseudofeces may not be as important as transported bedload sediment in smothering recently settled larvae.

These experimental results suggest that mussels of any species probably have important passive effects that alter environmental conditions for residents of soft-bottom communities. This result should not be surprising, especially in light of recent advances in our understanding of near-bottom flow. It is possible that flow is the master parameter that affects all the possible mechanisms listed above, as well as current-induced storm effects that rip beds apart and transport mussel clumps to new sites. If so, then the structure of mussel beds in the horizontal and vertical planes deserves more attention than it has thus far received. The calculations used to estimate flow regimes are based on the height of roughness elements above the bottom (Eckman 1990; Ke et al. 1994; Abelson and Denny 1997). Mussel beds have complex surface topographies at many spatial scales, and we know little about water flow over mussel beds. A field study of the pinnid *Atrina zelandica* in New Zealand by Green et al. (1998) showed that seabed drag coefficients were higher over three horse mussel-bed sites than over a site where the bottom had no live mussels but consisted of shells, seaweeds, and crab burrows. The four sites had seabed drag coefficient values greater than the value generally applied to an abiotic, flat, cohesionless bed. Despite precise site maps with each mussel's position, orientation, shell height, and shell width, the investigators were not able to establish the reasons why the three mussel-bed sites had seabed drag coefficients that differed from each other by as much as an order of magnitude. The authors concluded that site differences in spatial distribution and density must have caused the wide variation in flow regime.

Butman et al. (1994) used live *Mytilus edulis* to simulate a mussel bed in a laboratory flume and determined the contours of turbulent stress in front of and over the bed at different flow velocities. They used young mussels all of about the same length (2.7±0.2 cm), a simplifying first step, and it would be useful to expand the range of mussel sizes and bed surface configurations to determine how flow changes over different types of beds. Several field and laboratory studies by Fréchette and co-workers on intertidal *Mytilus edulis* beds in the St. Lawrence River estuary, Québec, Canada, have demonstrated that boundary layer flow regimes are strongly affected by bed surface topography, such that the delivery, depletion, and resuspension of particulate organic matter (chlorophyll *a* and phaeopigments) are enhanced because of roughness provided by the mussels themselves (Fréchette and Bourget 1985a,b; Fréchette et al. 1989; Fréchette and Lefaivre 1990; Fréchette and Grant 1991; Fréchette et al. 1992; Fréchette and Despland 1999). Similarly, Widdows et al. (1998) showed that *Mytilus edulis* bed structure regulated biodeposition and erosion rates at an intertidal site in the outer Humber estuary in England. Studies of *Mytilus edulis* beds in two Dutch estuaries (Dame et al. 1991) and the German Wadden Sea (Asmus and Asmus 1991; Asmus et al. 1992) have shown that differences in benthic-pelagic flux rates are strongly flow-dependent and occur over even small spatial scales, suggesting that differences in bed structure might be important.

2.6 Top-Down vs Bottom-Up Control of Soft-Bottom Mussel-Bed Community Structure

We have presented some information indicating that the physical structure of mussel beds plays a role in determining the delivery and depletion of larvae, food particles, and oxygen, as well as the rates of biodeposition and erosion. These are resource-based, "bottom-up", control processes. On the other hand, mussel-bed structure also affects the abundances of herbivores and carnivores and their rates of consumption. These are consumer-based, "top-down", control processes. We need to understand the interactions between mussel-bed structure and water flow if we want to learn how top-down and bottom-up factors combine to regulate the population performance of mussels and soft-bottom community structure where mussels are present. To our knowledge, a comprehensive investigation of this type has not been conducted for soft-bottom mussel assemblages.

However, recent studies on the rocky shore, Maine, USA, where *Mytilus edulis* dominates the lower intertidal and shallow subtidal zones, are instructive (Leonard et al. 1998, 1999). They demonstrate clearly that water flow is a dominant control agent in both these suspension-feeder communities. High-flow *Mytilus edulis* sites had greater recruitment of all species with planktonic larvae than did low flow sites, leading to more mussel cover and higher densities of snail grazers (*Littorina littorea*), snail predators (*Nucella lapillus*), and crab predators (*Carcinus maenas*). At high-flow sites, barnacle (*Semibalanus balanoides*) and *Nucella lapillus* growth rates were faster, but mussel growth was slower, possibly because of reduced ability to remove phytoplankton from the water column at high flow speeds. Despite the greater abundance of predators, there were lower predation intensities and per capita predation rates, possibly because predators had difficulty following chemical cues under conditions of high flow. Even the *Mytilus edulis* size-class structure was related to flow, with a unimodal size distribution at high-flow sites, but a bimodal size distribution at low-flow sites due to selective predation on mid-sized mussels. These results suggest that high-flow sites were regulated by resource-based factors, while consumer-based factors played a more important role at low-flow sites.

Similar patterns were observed on natural, experimentally constructed, and harvested *Crassostrea virginiana* (oyster) reefs at shallow subtidal sites in North Carolina, USA (Lenihan and Peterson 1998; Lenihan 1999). The flow environment explained 81% of the variability in oyster growth and mortality on experimentally constructed reefs. Location and physical structure – both vertical profile and horizontal patchiness – influenced flow speed and determined the balance between bottom-up and top-down control of oyster performance.

To the extent that soft-bottom mussel beds share physical and biological characteristics with oyster reefs and rocky shore mussel beds, they are controlled by many of the same ecological processes. The structure of soft-bottom mussel beds is variable in time and space. It results from the interplay among natural processes that kill and remove mussels (including storm damage, ice scour, sedimentation, predation), fill in bare space (larval settlement, juvenile and adult crawling, growth), and form the nuclei of new patches elsewhere (larval settlement, clump dispersal). This mix of physical disturbances and bottom-up and top-down processes is under the control of the water flow regime, which, in turn, depends on the location and structure of the mussel patches that make up the bed.

2.7 Conclusions

Many features of soft-bottom mussel beds are common across species and geographic locations. Mussel beds are highly dynamic, and fractal geometry can reveal predictable patterns in their complex spatial distribution and surface topography. Field and laboratory investigations, including density manipulation experiments, show that individual mussel performance, predation rates, resident infauna, and fluxes of sediment, larvae, and post-larvae are controlled by mussel abundance and bed structure.

Most mussel beds are affected by human activities in the form of anthropogenic nutrient enrichment to coastal waters (certainly a resource-based, bottom-up factor) and dredging of seed and adult mussels (which is similar to natural physical disturbances as well as top-down predation). How these processes interact to change the ecology of mussel beds and the fauna associated with them is not known. Continued dredging of natural beds and the susceptibility of cultured beds to predators and storms calls for enlightened management (Dankers and Zuidema 1995). Management informed by the best possible science means that increasing attention will be paid by scientists and managers alike to the spatial and temporal complexity of soft-bottom mussel beds.

References

Abelson A, Denny M (1997) Settlement of organisms in flow. Annu Rev Ecol Syst 28:317–339

Albrecht AS (1998) Soft bottom versus hard rock: community ecology of macroalgae on intertidal mussel beds in the Wadden Sea. J Exp Mar Biol Ecol 229:85–109

Albrecht A, Reise K (1994) Effects of *Fucus vesiculosus* covering intertidal mussel beds in the Wadden Sea. Helgoländer Meeresunters 48:243–256

Asmus H (1987) Secondary production of an intertidal mussel bed community related to its storage and turnover compartments. Mar Ecol Prog Ser 39:251–266

Asmus RM, Asmus H (1991) Mussel beds: limiting or promoting phytoplankton? J Exp Mar Biol Ecol 148:215–232

Asmus H, Asmus RH, Prins TC, Dankers N, Francé G, Maaß B, Reise K (1992) Benthic-pelagic flux rates on mussel beds: tunnel and tidal flume methodology compared. Helgoländer Meeresunters 46:341–361

Bayne BL, Thompson RJ, Widdows J (1976) Physiology. I. In: Bayne BL (ed) Marine mussels: their ecology and physiology. Cambridge Univ Press, Cambridge, pp 121–206

Beck MW (1998) Comparison of the measurement and effects of habitat structure on gastropods in rocky intertidal and mangrove habitats. Mar Ecol Prog Ser 169:165–178

Bertness MD, Grosholz E (1985) Population dynamics of the ribbed mussel, *Geukensia demissa*: the costs and benefits of an aggregated distribution. Oecologia 67:192–204

Beukema JJ (1993) Increased mortality in alternative bivalve prey during a period when the tidal flats of the Dutch Wadden Sea were devoid of mussels. Neth J Sea Res 31: 395–406

Beukema JJ, Cadée GC (1996) Consequences of the sudden removal of nearly all mussels and cockles from the Dutch Wadden Sea. P.S.Z.N.: Mar Ecol 17:279–289

Bustnes JO, Erikstad KE (1990) Size-selection of common mussels, *Mytilus edulis*, by common eiders, *Somateria mollissima*,: energy maximization or shell weight minimization? Can J Zool 68:2280–2283

Butman CA, Fréchette M, Geyer WR, Starczak VR (1994) Flume experiments on food supply to the blue mussel *Mytilus edulis* L. as a function of boundary-layer flow. Limnol Oceanogr 39:1755–1768

Ceccherelli VU, Rossi R (1984) Settlement, growth and production of the mussel *Mytilus galloprovincialis*. Mar Ecol Prog Ser 16:173–184

Commito JA (1987) Adult-larval interactions: predictions, mussels and cocoons. Estuar Coast Shelf Sci 25:599–606

Commito JA, Boncavage EM (1989) Suspension-feeders and coexisting infauna: an enhancement counter example. J Exp Mar Biol Ecol 125:33–42

Commito JA, Rusignuolo BR (2000) Structural complexity in mussel beds: the fractal geometry of surface topography. J Exp Mar Biol Ecol 255:133–152

Commito JA, Thrush SA, Pridmore RD, Hewitt JE, Cummings VJ (1995) Dispersal dynamics in a wind-driven benthic system. Limnol Oceanogr 40:1513–1518

Côté IM, Jelnikar E (1999) Predator-induced clumping behavior in mussels (*Mytilus edulis* Linnaeus). J Exp Mar Biol Ecol 235:201–211

Crooks JA (1998) Habitat alteration and community-level effects of an exotic mussel, *Musculista senhousia*. Mar Ecol Prog Ser 162:137–152

Crooks JA, Khim HS (1999) Architectural vs. biological effects of a habitat-altering, exotic mussel, *Musculista senhousia*. J Exp Mar Biol Ecol 240:53–75

Cummings VJ, Thrush SF, Hewitt JE, Turner SJ (1998) The influence of the pinnid bivalve *Atrina zelandica* (Gray) on benthic macroinvertebrate communities in soft-sediment habitats. J Exp Mar Biol Ecol 228:227–240

Dame R, Dankers N, Prins T, Jongsma H, Smaal A (1991) The influence of mussel beds on nutrients in the western Wadden Sea and eastern Scheldt estuaries. Estuaries 14: 130–138

Dankers N (1993) Integrated estuarine management – obtaining a sustainable yield of bivalve resources while maintaining environmental quality. In: Dame RF (ed) Bivalve filter feeders in estuarine and coastal ecosystem processes. NATO ASI series G, vol 33. Springer-Verlag, Berlin Heidelberg New York

Dankers N, Koelemaij K (1989) Variations in the mussel population of the Dutch Wadden Sea in relation to monitoring. Helgoländer Meeresunters 43:529–535

Dankers N, Zuidema DR (1995) The role of the mussel (*Mytilus edulis*) and mussel culture in the Dutch Wadden Sea. Estuaries 18:71–80

Dankers N, Herlyn M, Sand Kristensen P, Michaelis H, Millat G, Nehls G, Ruth M (1999) Blue mussels and blue mussel beds in the littoral. In: De Jong (ed) Wadden sea quality status report. Wadden Sea Ecosystem no. 9. CWSS, Wilhelmshaven 141–145

Davenport J, Pugh PJA, McKechnie J (1996) Mixed fractals and anisotropy in subantarctic marine macroalgae from South Georgia: implications for epifaunal biomass and abundance. Mar Ecol Prog Ser 136:245–255

Davenport J, Butler A, Cheshire A (1999) Epifaunal composition and fractal dimensions of marine plants in relation to emersion. J Mar Biol Assoc UK 79:351–355

Dayton PK (1971) Competition, disturbance and community organization: the provision of and subsequent utilization of space in a rocky intertidal community. Ecol Monogr 41:351–389

Dittmann S (1990) Mussel beds – amensalism or amelioration for intertidal fauna? Helgoländer Meeresunters 44:335–352

Dolmer P (1998) The interactions between bed structure of *Mytilus edulis* L. and the predator *Asterias rubens* L. J Exp Mar Biol Ecol 228:137–150

Dolmer P, Svane I (1994) Attachment and orientation of *Mytilus edulis* L. in flowing water. Ophelia 40:63–74

Eckman JE (1990) A model of passive settlement by planktonic larvae onto bottoms of differing roughness. Limnol Oceanogr 35:887–901

Fréchette M, Bourget E (1985a) Energy flow between the pelagic and benthic zones: factors controlling particulate organic matter available to an intertidal mussel bed. Can J Fish Aquat Sci 42:1158–1165

Fréchette M, Bourget E (1985b) Food-limited growth of *Mytilus edulis* L. in relation to the benthic boundary layer. Can J Fish Aquat Sci 42:1166–1170

Fréchette M, Despland E (1999) Impaired shell gaping and food depletion as mechanisms of asymmetric competition in mussels. Écoscience 6:1–11

Fréchette M, Lefaivre D (1990) Discriminating between food and space limitation in benthic suspension feeders using self-thinning relationships. Mar Ecol Prog Ser 65: 15–23

Fréchette M, Grant J (1991) An in situ estimation of the effect of wind-driven resuspension on the growth of the mussel *Mytilus edulis* L. J Exp Mar Biol Ecol 148:201–213

Fréchette M, Butman CA, Geyer WR (1989) The importance of boundary-layer flows in sampling phytoplankton to the benthic suspension feeder, *Mytilus edulis* L. Limnol Oceanogr 34:19–36

Fréchette M, Aitken AE, Pagé L (1992) Interdependence of food and space limitation of a benthic suspension feeder: consequences for self-thinning relationships. Mar Ecol Prog Ser 83:55–62

Gee JM, Warwick RM (1994a) Metazoan community structure in relation to fractal dimensions of marine macroalgae. Mar Ecol Prog Ser 103:141–150

Gee JM, Warwick RM (1994b) Body-size distribution in a marine metazoan community and the fractal dimensions of macroalgae. J Exp Mar Biol Ecol 178:247–259

Goss-Custard JD, Durell SEA, Le V (1988) The effect of dominance and feeding method on the intake rates of oystercatchers, *Haematopus ostralegus*, feeding on mussels. J Anim Ecol 57:827–844

Green MO, Hewitt JE, Thrush SF (1998) Seabed drag coefficient over natural beds of horse mussels (*Atrina zelandica*). J Mar Res 56:613–637

Guillemette M, Himmelman JH (1996) Distribution of wintering common eiders over mussel beds: does the ideal free distribution apply? Oikos 76:435–442

Günther CP (1996) Development of small *Mytilus* beds and its effects on resident intertidal macrofauna. P.S.Z.N. I. Mar Ecol 17:117–130

Hastings HM, Sugihara G (1993) Fractals: a user's guide for the natural sciences. Oxford Univ Press, Oxford

Herlyn M, Michaelis H (1995) Bestandaufnahme und Populationsbiologie von *Mytilus edulis* L. Methoden der quantitativen Erfassung von Miesmuschelvorkommen. Forschungstelle Küste Norderney 03F0023-A

Hilgerloh G, Herlyn M, Michaelis H (1997) The influence of predation by herring gulls *Larus argentatus* and oystercatchers *Haematopus ostralegus* on a newly established mussel *Mytilus edulis* bed in autumn and winter. Helgoländer Meeresunters 51:173–189

Hunter J, Arthur DR (1978) Some aspects of the ecology of *Peloscolex benedeni* Udekem (Oligochaeta: Tubificidae) in the Thames estuary. Estuar Coastal Mar Sci 6:197–208

Jaramillo E, Bertrán C, Bravo A (1992) Community structure of the subtidal macroinfauna in an estuarine mussel bed in southern Chile. P.S.Z.N. I. Mar Ecol 13:317–331

Kaandorp JA (1991) Modelling growth forms of the sponge *Haliclona oculata* (Porifera: Demospongiae) using fractal techniques. Mar Biol 110:203–215

Kaandorp JA (1994) Fractal modelling: growth and form in biology. Springer-Verlag, Berlin Heidelberg New York

Kaandorp JA (1999) Morphological analysis of growth forms of branching marine sessile organisms along environmental gradients. Mar Biol 134:295–306

Ke X, Collins MB, Poulos SE (1994) Velocity structure and sea bed roughness associated with intertidal (sand and mud) flats and saltmarshes of the Wash, UK. J Coastal Res 10:702–715

Kostylev V, Erlandsson J, Johanneson K (1997) Microdistribution of the polymorphic snail *Littorina saxatilis* (Olivi) in a patchy rocky shore habitat. Ophelia 47:1–12

Kröncke I (1996) Impact of biodeposition on macrofaunal communities in intertidal sandflats. P.S.Z.N. I. Mar Ecol 17:159–174

Landahl J (1988) Sediment-level fluctuation in a mussel bed on a "protected" sand-gravel beach. Estuar Coast Shelf Sci 26:255–267

Lenihan HS (1999) Physical-biological coupling on oyster reefs: how habitat structure influences individual performance. Ecol Monog 69:251–276

Lenihan HS, Peterson CH (1998) How habitat degradation through fishery disturbance enhances impacts of hypoxia on oyster reefs. Ecol Appl 8:128–140

Leonard GH, Levine JM, Schmidt PR, Bertness MD (1998) Flow-driven variation in intertidal community structure in a Maine estuary. Ecology 79:1395–1411

Leonard GH, Bertness MD, Yund PO (1999) Crab predation, waterborne cues, and inducible defenses in the blue mussel, *Mytilus edulis*. Ecology 80:1–14

Le Tourneux F, Bourget E (1988) Importance of physical and biological settlement cues used at different spatial scales by the larvae of *Semibalanus balanoides*. Mar Biol 97:57–66

Lin J (1991) Predator-prey interactions between blue crabs and ribbed mussels living in clumps. Estuar Coast Shelf Sci 32:61–69

Mandelbrot BB (1982) The fractal geometry of nature. Freeman, San Francisco

McGrorty S (1997) Winter growth of mussels *Mytilus edulis* as a possible counter to food depletion by oystercatchers *Haematopus ostralegus*. Mar Ecol Prog Ser 153:153–165

McGrorty S, Goss-Custard JD (1991) Population dynamics of the mussel *Mytilus edulis*: spatial variations in age-class densities of an intertidal estuarine population along environmental gradients. Mar Ecol Prog Ser 73:191–202

McGrorty S, Goss-Custard JD (1993) Population dynamics of the mussel *Mytilus edulis* along environmental gradients: spatial variations in density-dependent mortalities. J Anim Ecol 62:415–427

McGrorty S, Goss-Custard JD (1995) Population dynamics of *Mytilus edulis* along environmental gradients: density-dependent changes in adult mussel numbers. Mar Ecol Prog Ser 129:197–213

McGrorty S, Clarke RT, Reading CJ, Goss-Custard JD (1990) Population dynamics of the mussel *Mytilus edulis*: density changes and the regulation of the population in the Exe estuary, Devon. Mar Ecol Prog Ser 67:157–169

Meire PM, Ervynck A (1986) Are oystercatchers (*Haematopus ostralegus*) selecting the most profitable mussels (*Mytilus edulis*)? Anim Behav 34:1427–1435

Mileikovsky SA (1974) On predation of pelagic larvae and early juveniles of marine bottom invertebrates by adult benthic invertebrates and their passing alive through their predators. Mar Biol 26:303–311

Möbius K (1877) Die Auster und die Austernwirtschaft. Wiegund, Hempel/Parey, Berlin

Nehls G, Thiel M (1993) Large-scale distribution patterns of the mussel *Mytilus edulis* in the Wadden Sea of Schleswig-Holstein: do storms structure the ecosystem? Neth J Sea Res 31:181–187

Nehls G, Hertzler I, Scheiffarth (1997) Stable mussel *Mytilus edulis* beds in the Wadden Sea – they're just for the birds. Helgoländer Meeresunters 51:361–372

Nixon SW, Oviatt CA, Rogers C, Taylor K (1971) Mass and metabolism of a mussel bed. Oecologia 8:21–31

Okamura B (1986) Group living and the effects of spatial position in aggregations of *Mytilus edulis*. Oecologia 69:341–347

Paine RT (1966) Food web complexity and species diversity. Am Natur 100:65–75

Peake AJ, Quinn GP (1993) Temporal variation in species-area curves for invertebrates in clumps of an intertidal mussel. Ecography 16:269–277

Petraitis PS, Latham RE (1999) The importance of scale in testing the origins of alternative community states. Ecology 80:429–442

Radziejewska T (1986) On the role of *Mytilus edulis* aggregations in enhancing meiofauna communities off the southern Baltic coast. Ophelia [Suppl] 4:211–218

Ragnarsson SA, Raffaelli D (1999) Effects of *Mytilus edulis* L. on the invertebrate fauna of sediments. J Exp Mar Biol Ecol 241:31–43

Reimer O, Tedengren M (1997) Predator-induced changes in byssal attachment, aggregation and migration in the blue mussel, *Mytilus edulis*. Mar Fresh Behav Physiol 30:251–266

Reise K, Schubert A (1987) Macrobenthic turnover in the subtidal Wadden Sea: the Norderaue revisited after 60 years. Helgoländer Meeresunters 41:69–82

Reusch TBH, Chapman ARO (1997) Persistence and space occupancy by subtidal blue mussel patches. Ecol Monogr 67:65–87

Ricciardi A, Whoriskey FG, Rasmussen JB (1997) The role of the zebra mussel (*Dreissena polymorpha*) in structuring macroinvertebrate communities on hard substrata. Can J Fish Aquat Sci 54:2596–2608

Schwinghamer JY, Guigné JY, Siu WC (1996) Quantifying the impact of trawling on benthic habitat structure using high resolution acoustics and chaos theory. Can J Fish Aquat Sci 53:288–296

Seed R (1976) Ecology. In: Bayne BL (ed) Marine mussels: their ecology and physiology. Cambridge Univ Press, Cambridge, pp 13–66

Slepnev AY, Protasov AA, Videnina YL (1994) Development of a *Dreissena polymorpha* population under experimental conditions. Hydrobiol J 30:26–33

Snover ML, Commito JA (1998) The fractal geometry of *Mytilus edulis* spatial distribution in a soft-bottom system. J Exp Mar Biol Ecol 223:53–64

Stiven AE, Gardner SA (1992) Population processes in the ribbed mussel *Geukensia demissa* (Dillwyn) in a North Carolina salt marsh tidal gradient: spatial pattern, predation, growth and mortality. J Exp Mar Biol Ecol 160:81–102

Sugihara G, May RM (1990) Applications of fractals in ecology. Trends Ecol Evol 5:79–86

Svane I, Ompi M (1993) Patch dynamics in beds of the blue mussel *Mytilus edulis* L.: effects of site, patch size, and position within a patch. Ophelia 37:187–192

Svane I, Setyobudiandi I (1996) Diversity of associated fauna in beds of the blue mussel *Mytilus edulis* L.: effects of location, patch size, and position within a patch. Ophelia 45:39–53

Thorson G (1966) Some factors influencing the recruitment and establishment of marine benthic communities. Neth J Sea Res 3:267–293

Tsuchiya M, Nishihira M (1986) Islands of *Mytilus edulis* as a habitat for small intertidal animals: effect of *Mytilus* age structure on the species composition of the associated fauna and community organization. Mar Ecol Prog Ser 31:171–178

Widdows J, Brinsley MD, Salkeld PN, Elliott M (1998) Use of annular flumes to determine the influence of current velocity and bivalves on material flux at the sediment-water interface. Estuaries 21:552–559

Witman JD (1987) Subtidal coexistence: storms, grazing, mutualism, and the zonation of kelps and mussels. Ecol Monogr 57:167–187

Witman JD, Suchanek TH (1984) Mussels in flow: drag and dislodgement by epizoans. Mar Ecol Prog Ser 16:259–268

Woodin SA (1976) Adult-larval interactions in dense infaunal assemblages: patterns of abundance. J Mar Res 34:25–41

3 Suspension Feeders on Sandy Beaches

E. JARAMILLO and M. LASTRA

3.1 Introduction

Sandy beaches are the most common coastal habitat on temperate and tropical coasts; comprising about 75% of the world's coastal zone (Bascom 1980). Exposed sandy beaches support a diverse and abundant macroinfauna (Brown and McLachlan 1990). In terms of species richness, sandy beach macroinfauna is dominated by crustaceans, bivalves and polychaetes. In terms of abundance, peracarid crustaceans are usually the dominant organisms. However, large suspension feeders, such as anomuran decapods (*Emerita* spp.) and bivalves (*Mesodesma* and *Donax* spp.), are usually the top contributors to biomass (e.g. Hutchings et al. 1983; Donn 1990; Dugan et al. 1994; Jaramillo et al. 2001).

Basically, sandy beaches can be described in terms of wave and sediment characteristics, also called beach morphodynamics. A dimensionless index, Dean's parameter (also called parameter Ω), reflects the interaction between wave height, wave period and sediment fall velocity of sand particles from the sediments at the breaker zone (Short and Wright 1983; Short 1996); i.e. $\Omega = Hb/T \times$ sand fall velocity, where Hb is the height in centimeters of waves at the breaker zone, T is the wave period in seconds, while sand fall velocity (cm s^{-1}) is derived from the mean grain size of sands from the breaker zone and empirical data given by Gibbs et al. (1971). Using this parameter, three major types of beaches can be described: reflective, intermediate and dissipative.

Reflective beaches are characterized by a virtual absence of surf zone, coarse sand, small waves (usually lower than 1 m) and steep profiles. At the other end of the spectrum, dissipative beaches have a wide surf zone (i.e., waves dissipate most of their energy before reaching the beach face), fine sands, large waves (usually higher than 2 m) and flat profiles. Intermediate beaches lie between both extremes, with bar–trough systems, rip currents and variable seasonal conditions (Short and Wright 1983; Short 1996). The swash climate of exposed sandy beaches varies with the morphodynamic states. For example, reflective beaches have swashes with short periods, dissipative ones have longer swash periods, and intermediate beaches show swash periods intermediate between that of reflective and dissipative beaches (McArdle and McLachlan 1991). The following Dean's values have been mentioned (e.g. Short and Wright 1983) for the beach types: ≤ 1.0=reflective beaches, 1.0–6.0= intermediate beaches, and >6.0=dissipative beaches.

Dean's parameter is a useful tool to categorize sandy beaches on microtidal coasts. However, when tide ranges are greater than 2–3 m, the role of tides in beach morphodynamics increases (Masselink and Short 1993). To account for this, McLachlan et al. (1993) created the beach state index (BSI) where BSI=log [(Hb × MTR/T × sand fall velocity × ET)+1]. MTR is the maximum tide range and ET is the maximum theoretical equilibrium tide for which the earth covered in water is 0.8 m (McLachlan et al. 1993). Based upon a comparative study of about 70 beaches, McLachlan et al. (1993) suggested the following scale for BSI: <0.5=reflective beaches, 0.5–1.0=low to medium energy intermediate beaches, 1.0–1.5=high energy intermediate–dissipative beaches, 1.5–2.0=fully dissipative beaches, and >2.0=ultra-dissipative macrotidal beaches.

Beach morphodynamics have been considered a key factor in the community structure of the sandy beach macroinfauna around different zoogeographic regions of the world. A comparative study of sandy beach macroinfauna communities from all around the coast of South Africa, Australia and the coast of Oregon showed a linear increase in species richness and an exponential increase in abundance and biomass of the macroinfauna from reflective to dissipative conditions (Fig. 3.1, McLachlan 1990). Since the BSI allows for comparisons of sandy beaches located at different zoogeographical regions which have very large differences in tidal range, McLachlan et al. (1993, 1996, 1998) subsequently studied the sandy beach macroinfauna in South Africa, Oregon, Australia, south central Chile, and Oman showing that the patterns already found also hold globally. However, recent seasonal studies carried out along the coast line of Chile (from about 19 to 42° S, circa 2500 km) show that these sorts of relationships are indeed quite variable. Thus, the higher abundance and biomass values are found at intermediate beaches with Dean's values close to 3 (Fig. 3.2a). These values of abundance and biomass are usually higher than the values predicted by the worldwide model given by McLachlan et al. (1996) in which the BSI is used (Fig. 3.2b).

In the first part of this study we compare the community structure of the sandy beach macroinfauna vs. beach morphodynamic types on different coasts. We will explore the role of the most distinctive suspension feeder along Chilean sandy beaches – the anomuran crab *Emerita analoga* – in terms of the observed trend for this coast by comparing unpublished information of sandy beach surveys carried out along the Chilean coast and Spain (Galicia)

Suspension Feeders on Sandy Beaches

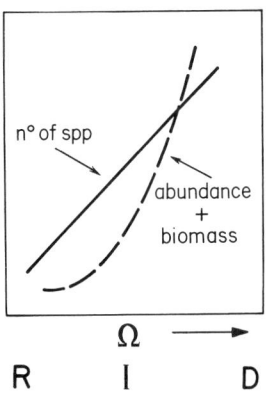

Fig. 3.1. General trends of species richness, abundance and biomass of the sandy beach macroinfauna vs. morphodynamic beach type. R, I, D reflective, intermediate and dissipative beaches (see text)

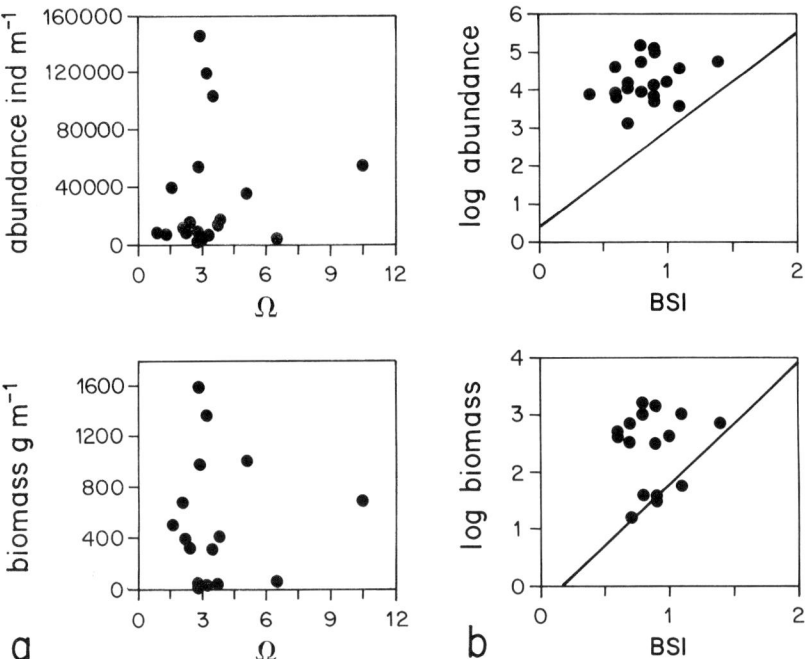

Fig. 3.2. Sandy beach macroinfauna from the Chilean coast. Abundance and biomass of the total macroinfauna vs. beach types defined by Dean's parameter (Ω) (**a**) and beach state index (BSI) (**b**). The regression lines are from McLachlan et al. (1996); $y = 0.39 + 2.55x$ for abundance and $y = -0.34 + 2.12x$ for biomass

with published data from Australia (McLachlan et al. 1996) and Oman (McLachlan et al. 1998). In the second part we also examine unpublished and published information to examine whether morphodynamic beach types affect the population biology and behavior of large suspension feeders such as *Emerita* and *Donax*. Finally, we analyze across- and along-shore zonation of these suspension feeders to examine the possible role of biological interactions in shaping spatial distributions on exposed sandy beaches.

3.2 Beach Morphodynamic Types vs. Community Structure of the Macroinfauna

E. analoga is the most common suspension feeder inhabiting sandy beaches along the Chilean coast. It primarily occupies the lower shore levels (swash–resurgence zone), although stranded animals are sometimes also found at the mid-shore levels or retention zone. Along this coast, the abundance of this species is usually higher than 50 %, while biomass values may well represent more than 80 % of the whole macroinfauna (Fig. 3.3). The dominance of this species results in significant correlations for total abundance and biomass vs. abundance and biomass of *E. analoga* on these beaches (log total abundance= $-3.27+1.61 \times$ log abundance of *E. analoga*, $r=0.85$, $n=19$, $p<0.01$ and log total biomass=$-1.04+1.31 \times$ log biomass of *E. analoga*, $r=0.95$, $n=16$, $p<0.01$). Dugan et al. (2000) have reported *E. analoga* as the most abundant species on 22 out of 36 beaches surveyed along the coast of California (5 to 98 % of the total macrofaunal abundance). Similarly, biomass of this crab accounted for 22–99 % of the total biomass on those beaches (Dugan et al. 2000).

The total abundance of the macroinfauna on Chilean beaches is higher than the worldwide model presented by McLachlan et al. (1996) (Fig. 3.4); in other words, similar abundances were found on Chilean beaches with lower BSI values than those studied at other latitudes such as Australia, Spain and Oman. Biomass vs. BSI shows a similar pattern to that of the abundance, and, in some cases, macroinfaunal biomasses on Chilean beaches were higher than those found on Omani beaches, being both similar in BSI (Fig. 3.4). We conclude from this that for the sandy beaches of Chile and also those in California (see Dugan and Hubbard 1996), factors other than beach morphodynamics influence patterns of macroinfaunal community structure in exposed sandy beaches. Those factors would apply primarily to large suspension feeders such as *E. analoga*; e.g. upwelling waters along the coast of Chile and California. Interestingly, sandy beaches located on coasts with upwelling waters (e.g., sandy beaches of Galicia, Spain; de la Huz 1999) but with low abundances and biomass of large suspension feeders (or even absent)

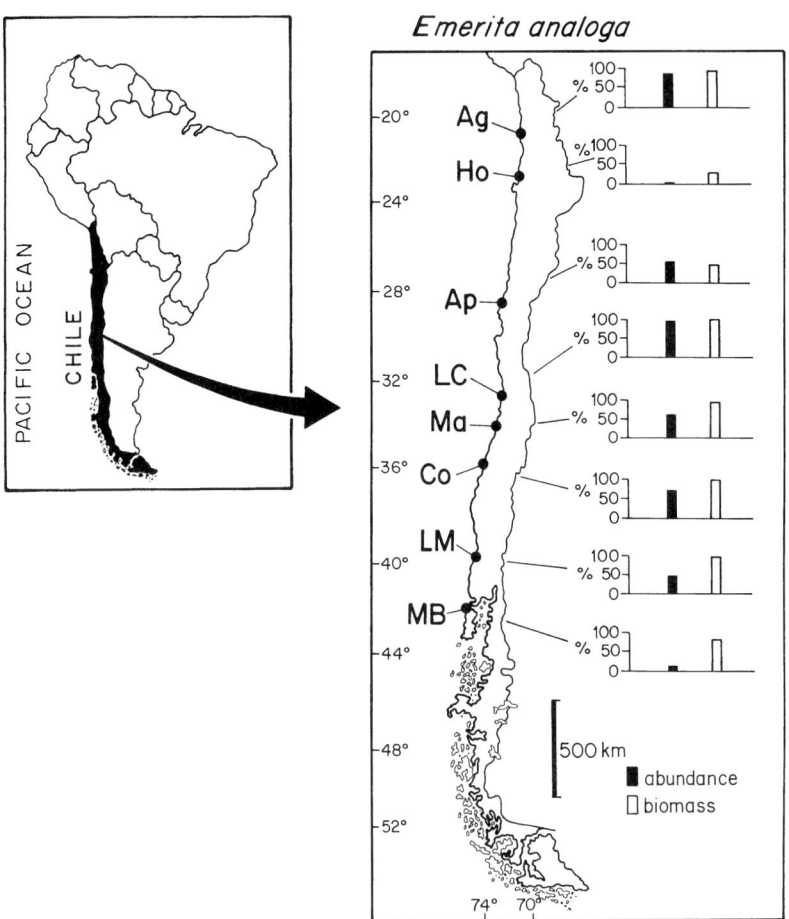

Fig. 3.3. Relative values for *Emerita analoga* (abundance and biomass) at exposed sandy beaches located along the Chilean coast. The names of the study beaches are: *Ag* Aguila; *Ho* Hornitos; *Ap* Apolillado; *LC* Las Cruces; *Ma* Matanzas; *Co* Cobquecura; *LM* La Misión; *MB* Mar Brava

agree quite well with the McLachlan model of increasing abundance and biomass from low to high BSI values. At this point it is worthwhile mentioning that earlier studies on the Chilean coast (Jaramillo and McLachlan 1993) were similar to the results of McLachlan et al. (1996), in the sense that abundance and biomass of the macroinfauna increased from reflective to dissipative beaches. However, those studies included three reflective beaches located close to an estuarine outlet where *E. analoga* was absent or in very low abundance. As shown experimentally by Jaramillo (1987), water salinities lower than 20 ppt result in significant mortality of this species. This would

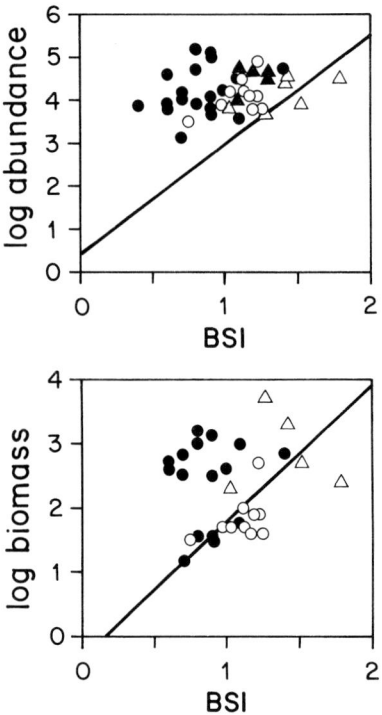

Fig. 3.4. Values of abundance and biomass of the total macroinfauna at Chilean, Australian, Omani and Spanish beaches. The Chilean data are taken from Jaramillo (2001), that of Australia from McLachlan et al. (1996), and those of Oman from McLachlan et al. (1998). The data from Spain are unpublished data of M. Lastra. The regression lines originate from the equations given by McLachlan et al. (1996)

- beaches from Chile
- beaches from Australia
- beaches from Oman
- beaches from Spain

explain the low macroinfaunal abundances found by Jaramillo and McLachlan (1993) and the linear increase in total macroinfauna along a range of morphodynamic beach types in that area of the Chilean coast. Indeed, it would be interesting to evaluate if the abundance and biomass of the total macroinfauna inhabiting exposed sandy beaches of the North and South American coasts that have species of *Emerita* agree or not with the worldwide model presented by McLachlan et al. (1996). Apart from *E. analoga*, five species of *Emerita* have been collected along this coast: *E. rathbunae* along the tropical coast of the Pacific Ocean, *E. talpoida* from the Atlantic coast of the USA and the Gulf of Mexico, *E. benedicti* primarily found inside that gulf, *E. portoricensis* from the West Indies in the Caribbean, and *E. brasiliensis* from the Atlantic coast of South America (Tam et al. 1996).

3.3 Beach Morphodynamic Types and Abundances and Population Biology of *E. analoga*

Seasonal variability in population abundances of *E. analoga* at two regions of the Chilean coast is shown in Fig. 3.5. In northern Chile, the abundances were higher at an intermediate beach (average Dean's parameter=2.9) as compared with a site with quite reflective characteristics (average Dean's parameter=1.4). On the other hand, two beaches examined in south central Chile, a dissipative site (average Dean's parameter=6.6) and an intermediate beach (average Dean's parameter=4.7), supported similar population abundances of *E. analoga*. This suggests that beach types affect population abundances of this suspension feeder just when beach types are close to or at the reflective condition. This assertion is supported by the fact that, at those sites sampled in south central Chile, growth parameters of *E. analoga* are quite similar (E. Jaramillo, unpubl. data).

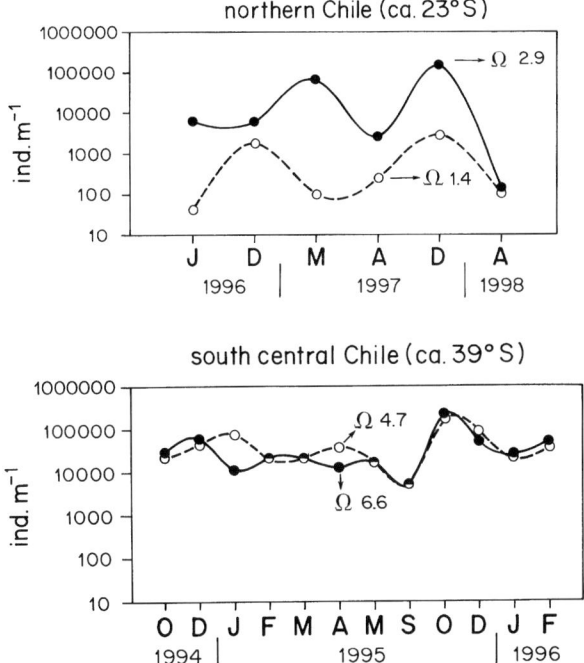

Fig. 3.5. Seasonal variability in population abundances of *Emerita analoga* in two areas of the Chilean coast. The values of Dean's parameter (Ω) are means of the whole study period at each beach. The sampling months at the beaches of northern Chile were July (J) and December (D) 1996, March (M), August (A) and December (D) 1997, and August (A) 1998

3.4 Tidal Movements and Burrowing Behavior

It has long been known that large suspension feeders of sandy beaches (e.g. *Emerita, Donax*) move up and down the shore with tidal variation (Branch and Branch 1993). However, not much is known about the effects of different beach types upon that sort of movement. Jaramillo et al. (2000) compared tidal variability of *E. analoga* between a dissipative and a reflective beach of southern Chile (ca. 42° S) and found that the movement of this species was greater at the dissipative site. Along a tidal cycle that included two lows and one high tide (ca. 12 h) the population mode of *Emerita* moved up and down following the limits of the swash zone as it moved up and down the beach. In contrast, in the reflective site, most of the animals remained below the lowest swash level through the tidal cycle suggesting that the width of the swash zone may affect the extent of tidal movement in this suspension feeder. In other words, this supports the swash exclusion hypothesis (McLachlan et al. 1993) in the sense that crabs remained below the effluent line on that reflective beach.

Exposed sandy beaches are coastal habitats dominated by sediment instability and strong hydrodynamic forces. One of the main adaptative mechanisms to cope with these factors is burrowing behavior; i.e., in exposed beaches organisms must burrow fast enough to avoid drifting in strong waves and currents. Burrowing rates of sandy beach organisms are usually affected by body size of organisms (see, e.g., McLachlan et al. 1995). Jaramillo et al. (2000) measured burrowing rates of *E. analoga* on both a dissipative and a reflective beach in southern Chile and found that crabs burrowed at similar rates in sediments from both beaches, which suggests that this crab is a sediment generalist (cf. Alexander et al. 1993). Laboratory experiments using *Donax trunculus* from Spain (M. Lastra, unpubl. data) showed ontogenic changes in clams burrowing in different sediment sizes. Those results show that individuals measuring 5–25 mm in shell length burrow faster in medium and fine sands, while bigger animals burrow faster in fine and very fine sands. These results suggest that small (juvenile) clams are able to cope with more reflective morphodynamic conditions (e.g., coarser sands) than adults. To determine whether this trend represents a plastic response of the individuals that inhabit changing sedimentary environments, further studies are needed.

3.5 Across- and Along-Shore Zonation

The patchy distribution of sandy beach organisms has been mainly studied in bivalves (e.g. Sastre 1985; Defeo et al. 1986; Jaramillo et al. 1994; Lastra and

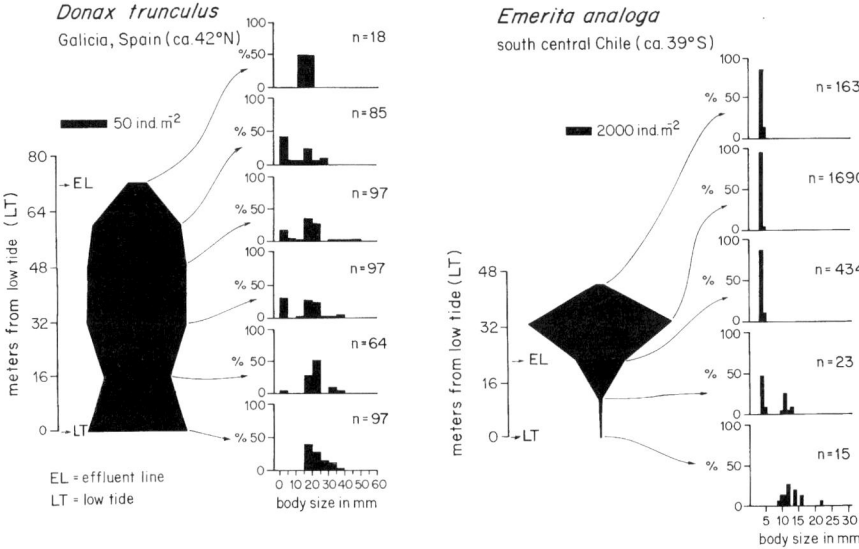

Fig. 3.6. Across-shore distribution of *Donax trunculus* and *Emerita analoga* at an exposed beach in Galicia, Spain, and south central Chile

McLachlan 1996) and anomuran crustaceans such as *Emerita* (Efford 1965; Cubit 1969; Perry 1980); the across-shore distribution of the bivalve *Donax trunculus* and that of *Emerita analoga* in Spain and Chile, respectively. Both of them show space partitioning across the intertidal. This is most obvious with *E. analoga*; i.e., smaller (younger) crabs mostly inhabit the upper levels of the distribution (i.e., at the top of the swash zone). During the times when juvenile *E. analoga* are more abundant on the sandy beaches of south central Chile – spring to early autumn – their highest abundances are usually found in areas where the abundance of larger crabs is lower (Hinrichsen and Rivera 1994). Thus, results presented in Fig. 3.6 suggest that the spatial segregation of juveniles and adults of *E. analoga* on sandy beaches of south central Chile may be due to some sort of biological interactions. Dugan and Hubbard (1996) reached similar conclusions for Californian populations of this species; i.e., measures of crab size increased while population abundance decreased among 12 populations on 50 km of coast.

Physical and biological factors have been proposed as causes of spatial partitioning in the across- and along-shore distribution of large sandy beach suspension feeders, among them sediment size (Maze and Laborda 1988) and intraspecific competition (Guillou and Le Moal 1978; Ansell and Lagardere 1980). Although there is not much experimental evidence for the role of these factors, some published results suggest that biological interactions may well be involved. Field experiments carried out by McLachlan (1998) on a sandy beach in South Africa showed that the burrowing rates of *Donax serra* and *D. sordidus* were affected at high densities of the former species, a result which suggests that in high density beds of these clams interactions would occur. Similarly, in a field experiment carried out by Defeo (1996) on a sandy beach in Uruguay, density-dependent processes were considered important for *Mesodesma mactroides* and *Donax hanleyanus*. More experiments are clearly needed to elucidate these aspects.

3.6 Conclusions

This review has explored the role of large suspension feeders such as *E. analoga* in the observed trends of sandy beach community structure in coastal areas like Chile and California, where these animals are the most abundant organisms. In these areas, the relationship between abundance and biomass of the total macroinfauna does not match the worldwide model in which morphodynamic beach type is considered to be the key factor in shaping macroinfaunal community structure. Although few studies have been carried out, it seems that beach morphodynamic type affects the behavioral aspects of these organisms, such as tidal migration and burrowing behavior. The patchy distribution and often size-structured aggregations of large suspension feeders, such as *E. analoga*, *Donax* and *Mesodesma*, also suggest that biological interactions (an almost unexplored field) may indeed be important in shaping these patterns on exposed sandy beaches.

Acknowledgements. We thank Karsten Reise for the invitation to present this study at the Workshop Ecological Comparisons of Sandy and Muddy Shores at Sylt, Germany. E.J. and M.L. acknowledge the assistance of Heraldo Contreras, Marcia González, Cristián Duarte (Instituto de Zoología, Universidad Austral de Chile), María del Rosario de la Huz and María Esther Pita Siebert (Departamento de Ecología y Biología Animal, Universidad de Vigo) while preparing this manuscript. E.J. also appreciates the financial support of CONICYT (Chile), Proyecto FONDAP Oceanografía & Biología Marina, for research funds used in Chile, Oman and Spain, and to the Alfred Wegener Institute for Polar and Marine Biology (AWI, Bremerhaven, Germany) and Universidad Austral de Chile (DID) for financial aid to attend the workshop. M.L. acknowledges financial support from Xunta de Galicia and Universidad de Vigo for research funds used on the coast of Spain, and to AWI which provided the funds to attend the workshop. We also thank Anton McLachlan and Jenifer Dugan for reviewing a former manuscript.

References

Alexander RR, Stanton RJ, Dodd JR (1993) Influence of sediment grain size on the burrowing of bivalves – correlation with distribution and stratigraphic persistence of selected neogene clams. Palaios 8:289–303

Ansell AD, Lagardere F (1980) Observations on the biology of *Donax trunculus* and *D. vittatus* at Ile dÓleron (French Atlantic coast). Marine Biology 57:287–300

Bascom W (1980) Waves and beaches. Anchor Press/Darbleday, New York, 366 pp

Branch M, Branch G (1993) The living shores of Southern Africa. Strik Publishers, Cape Town, South Africa, 272 pp

Brown AC, McLachlan A (1990) Ecology of sandy shores. Elsevier Science Publishers BV, Amsterdam, The Netherlands, 328 pp

Cubit J (1969) Behaviour and physical factors causing migration and aggregation of the sand crab *Emerita analoga* (Stimpson). Ecology 50:118–123

Defeo O (1996) Experimental management of an exploited sandy beach bivalve population. Rev Chil Hist Nat 69:605–614

Defeo O, Layerle C, Masello A (1986) Spatial and temporal structure of the yellow clam *Mesodesma mactroides* (Deshayes 1854) in Uruguay. Medio Ambiente (Chile) 8:48–57

De la Huz MR (1999) Estudio de playas expuestas de Galicia: morfología y macrofauna. Tesis Licenciatura, Universidad de Vigo, Vigo, España, 82 pp

Donn TE (1987) Longshore distribution of *Donax serra* in two log-spiral bays in the eastern Cape, South Africa. Mar Ecol Prog Ser 35:217–222

Donn TE (1990) Morphometrics of *Donax serra* Roding (Bivalvia: Donacidae) populations with contrasting zonation patterns. J Coastal Res 6893–901

Dugan JE, Hubbard DM (1996) Local variation in populations of the sand crab *Emerita analoga* on sandy beaches in southern California. Rev Chil Hist Nat 69:579–588

Dugan JE, Hubbard DM, Wenner AM (1994) Geographic variation in life history of the sand crab, *Emerita analoga* (Stimpson) on the California coast: relationships to environmental variables. J Exp Mar Biol Ecol 181:255–278

Dugan JE, Hubbard DM, Martin DL, Engle JM, Richards DM, Davis GE, Lafferty KD, Ambrose RF (2000) Macrofauna communities of exposed sandy beaches on the southern California mainland and Channel Islands. Proceedings of the 5th California Islands Symposium. OCS Study, MMS 99-0038:339–346

Efford IE (1965) Aggregation in the sand crab *Emerita analoga* (Stimpson). J Anim Ecol 34:63–75

Gibbs RJ, Mathews MD, Link DA (1971) The relationship between sphere size and settling velocity. J Sediment Petrol 41:7–18

Guillou J, Le Moal Y (1978) Variabilité spatio-temporal des populations de *Donax* en la bahie de Douarnemez. Haliotis 9:77–88

Hinrichsen CA, Rivera VJ (1994) Variabilidad espacio – temporal de la macroinfauna intermareal en una playa arenosa del centro sur de Chile y su relación con la variabilidad abiótica del sustrato. Tesis Biología Marina, Universidad Austral de Chile, Valdivia, Chile, 32 pp

Hutchings L, Nelson G, Horstman DA, Tarr R (1983) Interaction between coastal plankton and sand mussels along the Cape coast, South Africa. In: McLachlan A, Erasmus T (eds) Sandy beaches as ecosystems. Junk Publishers, The Hague, pp 481–500

Jaramillo E (1987) Community ecology of Chilean sandy beaches. PhD dissertation, University of New Hampshire, Durham, NH, USA, 216 pp

Jaramillo E (2000) The sand beach ecosystem of Chile. In: Seeliger U, Kjerfve B (eds) Coastal marine ecosystems of Latin America. Ecological Studies 144. Springer-Verlag, Berlin Heidelberg New York, pp 219–227

Jaramillo E, McLachlan A (1993) Community and population responses of the macrofauna to physical factors over a range of exposed sandy beaches in south central Chile. Estuar Cost Shelf Sci 37:615–624

Jaramillo E, Pino M, Filún L, González M (1994) Longshore distribution of *Mesodesma donacium* (Bivalvia: Mesodesmatidae) on a sandy beach of the south of Chile. Veliger 37:192–200

Jaramillo E, Dugan E, Contreras H (2000) Abundance, population structure, tidal movement and burrowing rate of *Emerita analoga* (Stimpson, 1857) (Anomura, Hippidae) at a dissipative and a reflective sandy beach in south central Chile. Mar Ecol, PSZNI 21:113–127

Lastra M, McLachlan A (1996) Spatial and temporal variations in recruitment of *Donax serra* Roding (Bivalvia: Donacidae) on an exposed sandy beach of South Africa. Rev Chil Hist Nat 69:631–639

Masselink G, Short AD (1993) The effect of tide range on beach morphodynamics and morphology: a conceptual beach model. J Coastal Res 9:785–800

Maze R, Laborda AJ (1988) Aspectos de la dinámica de población de *Donax trunculus* (Linnaeus, 1758) (Bivalvia: Donacidae) en la Rìa del Barquero (Lugo, España). Invest Pesqueras 52:299–312

McArdle S, McLachlan A (1991) Dynamics of the swash zone and effluent line on sandy beaches. Mar Ecol Prog Ser 76:91–99

McLachlan A (1990) Dissipative beaches and macrofauna communities on exposed intertidal sands. J Coastal Res 6:57–71

McLachlan A (1998) Interactions between two species of *Donax* on a high energy beach: an experimental approach. J Molluscan Stud 64:492–495

McLachlan A, Jaramillo E, Donn TE, Wessels F (1993) Sandy beach macrofauna communities and their control by the physical environment: a geographical comparison. J Coastal Res (Special Issue) 15:27–38

McLachlan A, Jaramillo E, Defeo O, Dugan J, de Ruyck A, Coetzee P (1995) Adaptations of bivalves to different beach types. J Exp Mar Biol Ecol 187:147–160

McLachlan A, De Ruyck A, Hacking N (1996) Community structure on sandy beaches: patterns of richness and zonation in relation to tide range and latitude. Rev Chil Hist Nat 69:451–467

McLachlan A, Fisher M, Al-Habsi HN, Al-Shukairi SS, Al-Habsi A (1998) Ecology of sandy beaches in Oman. J Coastal Conserv 4:181–190

Perry DM (1980) Factors influencing aggregation patterns in the sand crab *Emerita analoga* (Crustacea: Hippidae). Oecologia 45:379–384

Sastre MP (1985) Aggregated patterns of dispersion in *Donax denticulatus*. Bull Mar Sci 36:220–224

Short A (1996) The role of wave height, period, slope, tide range and embaymentisation in beach classifications: a review. Rev Chil Hist Nat 69:589–604

Short AD, Wright L (1983) Physical variability of sandy beaches. In: McLachlan A, Erasmus T (eds) Sandy beaches as ecosystems. Junk Publishers, The Hague, pp 133–144

Tam YK, Kornfield I, Ojeda FP (1996) Divergence and zoogeography of mole crabs, *Emerita* spp. (Decapoda: Hippidae) in the Americas. Mar Biol 125:489–497

4 Switching Between Deposit and Suspension Feeding in Coastal Zoobenthos

H.U. RIISGÅRD and P. KAMERMANS

This chapter, consisting of an overview followed by two subchapters each dealing with a specific example, attempts to provide current insight into possible switching between deposit and suspension feeding in coastal zoobenthos. The examples are two common inhabitants of shallow waters, the polychaete *Nereis diversicolor* and the bivalve *Macoma balthica*. The examples describe different aspects, such as adaptations needed for true suspension feeding, conditions that may favour switching between suspension and deposit feeding, and relative time spent on suspension feeding.

4.1 Overview

H. U. RIISGÅRD

Feeding mode is a term used to describe the mechanism of food transport from the surroundings into the organism. One may distinguish between four feeding modes. *Suspension feeders* or *filter feeders* extract their food from suspended particles in the sea and rely principally upon phytoplankton for nourishment. A suspension-feeding benthic animal strains food particles either passively or it creates a feeding current directed towards a filter. *Deposit* or *detritus feeders* utilise mainly the living components of the ingested sediment, i.e. bacteria, fungi, microalgae and the microfauna, whereas the predominant amount of dead organic matter (detritus) is not digested but primarily utilised by microbial decomposers (Fenchel et al. 1975; Fenchel and Finlay 1995). *Surface deposit feeders* take their food from the sediment surface, and *subsurface deposit feeders* ('burrowers') seek nourishment below the sediment surface.

Classification of feeding types into deposit feeders and suspension feeders is frequently used, although it has become clear that some organisms are capable of using more than one of these feeding methods. It can be a problem to discriminate between suspension feeders and deposit feeders because there is a gradual transition from the food present in the top-layer of the sediment, on which surface deposit feeders feed, to the suspended matter in the water just above the sediment surface on which benthic suspension feeders feed. This makes discrimination based on stomach content analysis alone doubtful.

In dynamically variable benthic environments organisms capable of switching their feeding behaviour may be common. For example, a variety of polychaetes representing some spionids, nereids, fabricine sabellids, and oweniids possess such an ability to shift their feeding mode (Fauchald and Jumars 1979). Buhr and Winter (1977) found that the polychaete *Lanice conchilega* is capable of replacing surface deposit feeding by suspension feeding because the worm may retain suspended particles directly from the water column by means of its tentacles. Switching feeding behaviour has been reported in tellinids among the bivalves (Brafield and Newell 1961) and amphipods among the crustaceans (Mills 1967; Fenchel et al. 1975; Taghon et al. 1980). Thus, some benthic organisms appear versatile – 'opportunistic' – in their feeding modes (Cadée 1984), but the relative significance of their different feeding methods remains unclear. Furthermore, the ratio of suspension feeding to deposit feeding may vary considerably among different populations of the same species. Switching from deposit to suspension feeding may be an adaptation particularly useful in shallow waters where the concentration of suspended food particles may vary widely. Dauer et al. (1981) used the term 'interface' feeders to refer to species that can switch between deposit and suspension feeding.

Zoobenthos live and feed on the sea floor in close contact with the overlying water currents and resulting fluxes of particulate organic material. Animals feeding upon this particulate material, as deposited or in near-bottom suspension, may be expected to respond to near-bottom currents and sediment transport in their feeding behaviour (Miller et al. 1992). The development of additional, alternative suspension-feeding mechanisms among various deposit-feeding zoobenthos is of primary importance when the amount of available food is limited and when coexisting species compete for food. The apparently widespread ability of deposit feeders to utilise alternative feeding mechanisms may lead to resource partitioning, and several niche dimensions related to feeding may allow a certain diversity of coexisting species (Fenchel et al. 1975). The ability to use alternative feeding mechanisms thus seems to be an attribute of many deposit-feeding animals. This paper attempts to give an outline of present knowledge about switching to suspension feeding by deposit-feeding coastal zoobenthos.

4.1.1 Polychaetes

Polychaetes are among the most frequent and abundant marine invertebrates. Fauchald and Jumars (1979) summarised the current information about food and feeding habits of polychaetes, including predictions on the most likely feeding habits in the form of hypotheses.

Arenicolids. Suspension feeding in arenicolids has for some time been a controversial subject. Thus, the principal food source of the lugworm *Arenicola marina* is still a point of contention. Generally, the lugworm is regarded as a subsurface deposit feeder (see also Chap. 6) that is nourished by swallowing relatively large amounts of sediments with low nutritional value. *A. marina* lives in 20–40 cm deep J-shaped burrows in the sediment. With its head down, the worm ingests sediment. As a result the sand above sinks downwards, forming a funnel. The pumping activity of the lugworm causes a tail-to-head-directed ventilation current through its tube, resulting in an upward flow of oxygenated water in the sediment in front of the head (Riisgård et al. 1996a; Riisgård and Banta 1998). It has been suggested that *A. marina* may live as a suspension feeder by using the sand immediately in front of the head as a particle-retaining filter for restraining suspended food particles in the water (Krüger 1959, 1962, 1964, 1971). Suspension-feeding macroinvertebrates are characterised by pumping large amounts of water per ml of oxygen consumed. Thus, true filter-feeding macro-invertebrates (except for sponges) pump 10 to 350 l of water per ml of oxygen consumed, which may be compared with 0.4 l of water per ml of oxygen consumed in *A. marina* (Table 4.1). This very low F/R-value means that the lugworm cannot survive as a true suspension feeder. However, it does not exclude the possibility that some fraction of its nutrition may come from suspended material drawn down into the sediment while irrigating. Resuspended organic-rich sediment may represent a potential food source, but subsurface deposit feeding must be considered the major mode of feeding in the arenicolids (Riisgård and Banta 1998).

Chaetopterids. The suspension-feeding mode of the Chaetopteridae family is well known. In all species a single or several mucus nets are used to capture suspended food particles and the feeding current is set up by notopodial cilia or muscular motion (Barnes 1964, 1965). However, at low particle concentrations, *Spiochaetopterus* sp. will search the sediment surface with the palps. Thus, surface deposit feeding seems to be important for species living in bathyal and abyssal regions, such as *Spiochaetopterus costarum* and *Phyllochaetopterus limnicolus* in the Pacific Ocean (Fauchald and Jumars 1979). In *P. prolifica* two feeding modes other than the mucus filter-bag feeding method

Table 4.1. Volumes of water filtered (F, l) per ml O_2 consumed (R) for various suspension-feeding invertebrates (adapted from Riisgård and Larsen 2000), and for comparison supplemented with two deposit feeders, the lugworm *Arenicola marina* and the clam *Macoma balthica*

Taxanomic groups and species	F/R (l ml^{-1} O_2)	Reference
Polychaetes		
Sabella penicillus	354	Riisgård and Ivarsson (1990)
Chaetopterus variopedatus	50	Riisgård (1989)
Nereis diversicolor	40	Riisgård (1991)
Arenicola marina	0.4	Riisgård et al. (1996a)
Bivalves		
Mytilus edulis	15–50	Riisgård et al. (1980)
Mytilus edulis	18	Clausen and Riisgård (1996)
Macoma balthica	0.4–1.6[a]	De Wilde (1975), Hummel (1985b), Kamermans (1994a)
Ascidians		
Ciona intestinalis	82	Petersen et al. (1995)
Ciona intestinalis	13	Jørgensen (1955)
Copepods		
Acartia tonsa	37	Kiørboe et al. (1985)
Sponges		
Halichondria panicea	2.7	Thomassen and Riisgård (1995)
Mycale sp.	19.6	Reiswig (1974)
Verongia gigantea	4.1	Reiswig (1974)
Verongia fistularis	9.7	Reiswig (1974)
Tethya crypta	22.8	Reiswig (1974)
Bryozoans		
Celleporella hyalina	68	Riisgård and Manríquez (1997)
Lancelets		
Brachiostoma lanceolatum	79	Riisgård and Svane (1999)

[a] Filtration rate (F) of *Macoma balthica* is 5 ml h^{-1} for an 8 mg body dry wt. individual (Kamermans 1994a and pers. comm.), or 6–23 ml h^{-1} for a 38 mg body dry wt. individual (Hummel 1985b). Respiration (R) is 0.014 ml O_2 h^{-1} for an approx. 25 mg dry wt *M. balthica* (De Wilde 1975). F/R=0.4–1.6 l water filtered per ml O_2 consumed

may take place (Fig. 4.1b). When currents are moderately strong, the palps are often erected to capture suspended particles (i.e. 'passive suspension feeding'), but in slowly moving waters surface deposit feeding with the aid of the palps takes place (Fauchald and Jumars 1979). Miller et al. (1992) reported that *S. oculatus* is a deposit feeder in still water and a facultative, palp-coiling suspension feeder in oscillatory flows, the percentage of time spent on suspension feeding increasing with flow.

Dinophilids. These tiny, interstitial polychaetes are also, as an alternative to deposit feeding, capable of catching suspended particles with a ciliary mucoid mechanism that enables them to feed on bacteria, protozoans, unicellular algae, diatoms, and organic debris (Fauchald and Jumars 1979).

Nereids. Nereids are common in shallow water and all species have jawed eversible pharynges. With few exceptions they are all omnivorous (Fauchald and Jumars 1979) (Fig. 4.1g), but only one species, *Nereis diversicolor,* seems capable of suspension feeding with the aid of a mucus net-bag suspended within its tube (Riisgård 1991; Riisgård et al. 1992, 1996b; Vedel and Riisgård 1993). The importance of suspension feeding in *N. diversicolor* may vary from one population to the next, and the habit may perhaps be entirely missing in some populations, see detailed account in 'Example I' (Sect. 4.2).

Oweniids. The only oweniid so far investigated seems to be *Owenia fusiformis* which has a shallow, lobed tentacular crown. This worm is capable both of suspension feeding and surface deposit feeding. During deposit feeding, the lips are used to pick up particles directly. Quantitative investigations comparing time devoted to each mode are still lacking (Fauchald and Jumars 1979).

Sabellids. A suspension-feeding habit is the most important one in this family (Riisgård and Ivarsson 1990). In addition, *Manayunkia aestuarina* is also capable of feeding by turning over and touching the substratum with its tentacular crown in a form of surface deposit feeding (Fig. 4.1h). The method may be of primary importance in freshwater and brackish water sabellids (Fauchald and Jumars 1979).

Spionids. Spionids are generally considered to be surface deposit feeders, using their ciliated palps to select food particles from the sediment surface (Dauer 1991, 1994; Dauer and Ewing 1991). However, certain species can switch from deposit to suspension feeding by catching plankton on the palps when the water current velocity and/or the concentration of suspended material becomes sufficiently high (Fig. 4.1 c) (Self and Jumars 1978; Dauer et al. 1981; Shimeta and Jumars 1991; Qian and Chia 1997; Shimeta and Koehl 1997; Williams and Mcdermott 1997); some examples are *Polydora cornuta* (Dauer et al. 1981), *Boccardia polybranchia* (Qian and Chia 1997), *Scolelepis squamata* (Dauer 1983), *Spiophanes bombyx* (Dauer et al. 1981), *Spio setosa* (Dauer et al. 1981), *Marenzelleria viridis* (Dauer 1997), and *Paraprionospio pinnata* (Dauer 1985).

Pygospio elegans is the most versatile of the spionids studied to date. It can filter by building a mucus net within its tube, it can catch plankton with the

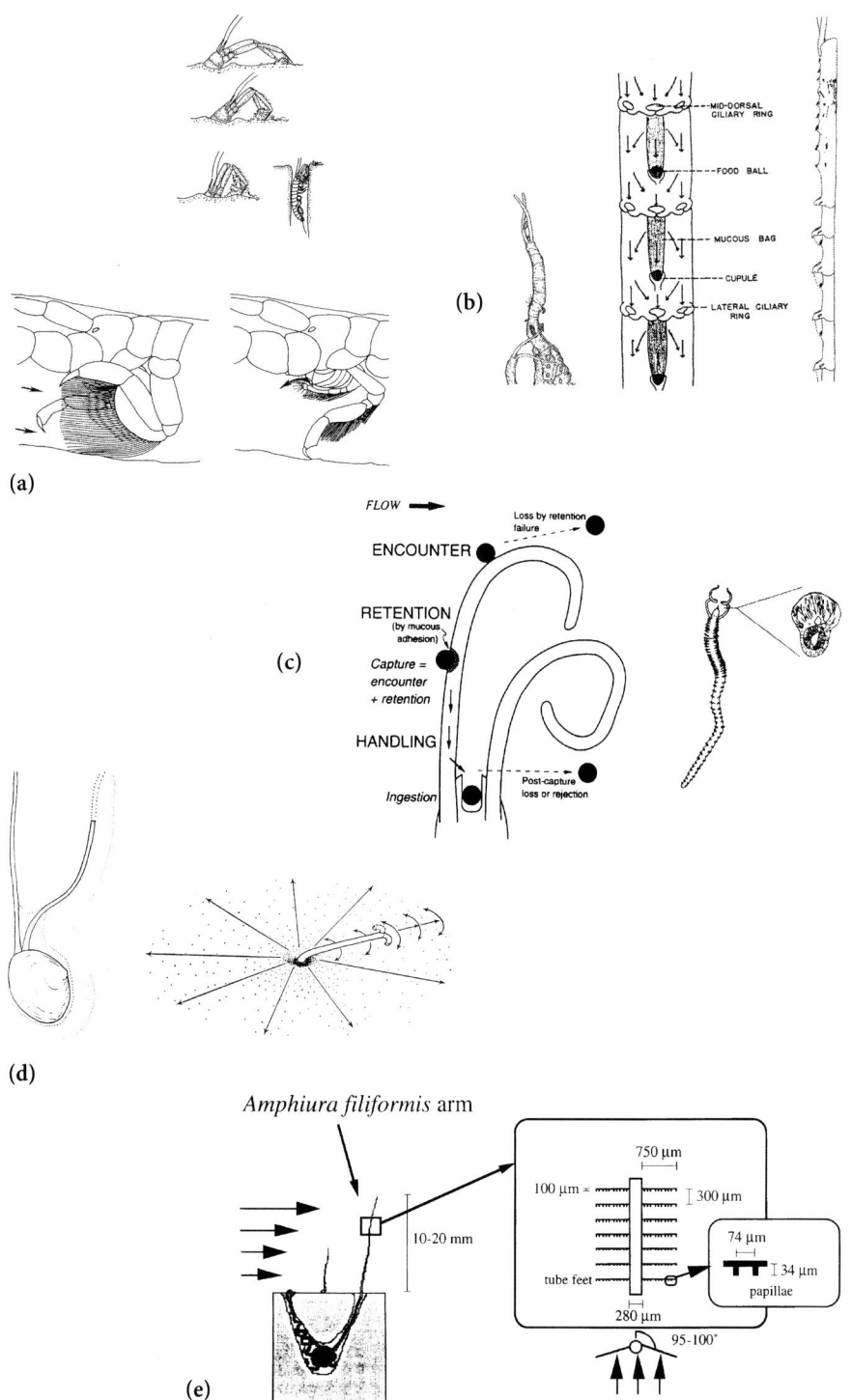

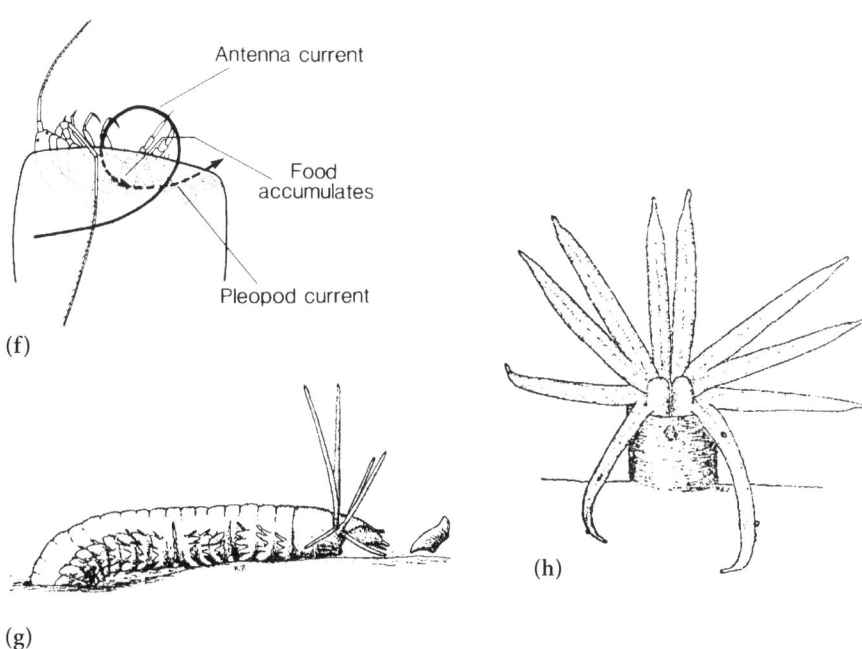

Fig. 4.1a–d. a *Upper figures* Deposit-feeding *Corophium volutator* collects particles from the sediment surface with a scraping motion of the enlarged second antennae (from Meadows and Reid 1966); *lower figure, left* suspension-feeding *Corophium* where particles are brought into the tube by the pleopodal current and caught on the filter of setae of the second gnathopods; *lower figure, right* surface deposit-feeding *Corophium* where particles are brought to the mouthparts from the antennal bolus by the first gnathopods and manipulated (from Miller 1984). b *Left figure* Phyllochaetopterus prolifica utilising two feeding modes other than the suspension-feeding mucus-bag method: when the currents are moderately strong, the palps are held erect until impacted by a food particle; the uppermost individual is about to engulf a particle caught in this fashion; the other individual is deposit feeding on the fouled tube of the first individual (from Fauchald and Jumars 1979); *middle figure* dorsal view of three segments showing the position of mucus-bags and the formation of food balls (from Barnes 1964); *right figure* lateral view of anterior and middle body region of *Spiochaetopterus oculatus* with mucus-bags indicated (from Barnes 1965). c *Left figure* component steps in suspension feeding by a spionid polychaete (from Shimeta and Koehl 1997); *right figure* dorsal surface of entire worm *Scolelepis squamata* and a section of palp (from Dauer 1983). d Position of *Scrobicularia plana* and inhalant siphonal activity while surface deposit feeding (from Hughes 1969). e Extended *Amphiura filiformis* arms during passive suspension feeding; *horizontal arrows* indicate current direction and velocity gradient, and *schematic drawings* show morphometrics of the tube feet and angle between the tube feet and the current (from Loo et al. 1996). f Feeding position and currents in the amphipod *Ampelisca* (from Mills 1967). g *Nereis* worm comes out from its burrow to approach a food object (from Fauchald and Jumars 1979). h The sabellid polychaete *Manayunkia aestuarina* with a pair of branchial filaments being used in deposit feeding (from Fauchald and Jumars 1979).

help of its palps, or it can feed on surface deposits (Hempel 1957). Supplementary suspension feeding may take place in most species, but it is difficult to distinguish from feeding on resuspended bottom materials. Taghon et al. (1980) reported on a novel suspension-feeding mechanism in three species of spionid polychaetes (*Pseudopolydora kempi japonica, Boccardia proboscidea, P. elegans*). It was found that the feeding behaviour of these spionids varied with the water velocity. At moderate flows the worms cease deposit feeding and form their feeding tentacles into helices that are lifted into the water and extended downstream to capture material in suspension. This behaviour is apparently a response to increased flux of suspended matter rather than to water current velocity alone. Brine shrimps introduced upstream were captured along the entire length of the tentacles and transported to the mouth by ciliary action. Direct interception by the tentacle is the primary means by which these annelids feed on larger particles in suspension (for detailed description of 'direct interception' capture mechanisms, see Shimeta and Jumars 1991). Miller et al. (1992) reported that the spionid *Spio setosa* is a deposit feeder in still water, but a facultative, palp-coiling suspension feeder in oscillatory flows. Taghon and Greene (1992) tested the hypothesis that switching from deposit feeding to suspension feeding in *P. kempi japonica* and *B. pugettensis* as the flux of suspended particles increases is energetically profitable because suspended particles have greater food value.

Terebellids. Lanice conchilega feeds on material in bed-load transport. Laboratory experiments carried out by Buhr (1976) revealed that, in addition to surface deposit feeding, suspension feeding may play a role in the nutrition of this polychaete. The anterior end of the worm is equipped with a series of extensible tentacles that are oriented at right angles to the current. This orientation makes them function as baffles in the current, allowing transported material to drop in the quiet areas behind the fans. The feeding behaviour of *L. conchilega* has been discussed by Buhr and Winter (1977). From the high population densities observed in the outer part of the Weser Estuary, Germany, they found it unlikely that *L. conchilega* engages solely in surface deposit feeding. The fringed ends of the tubes of populations in this habitat form a dense network, such that only a limited amount of detritus reaches the bottom. Consequently, food requirements have to be fulfilled by an alternative. Food uptake may thus be performed not only from the bottom surface, but also from the fringed ends of the tubes.

4.1.2 Echinoderms

The brittle-star *Amphiura filiformis* feeds on suspended material in flowing water, but shifts to deposit feeding in stagnant water (Loo et al. 1996). In still water it was observed that many *A. filiformis* extend their arms into the water, but at moderate flow velocities the proportion of individuals with at least one extended arm increases. From video recordings in flume experiments, *A. filiformis* was observed to capture suspended food items. Particles encountered and retained on the tube feet are transported between adjacent tube feet in the proximal direction along the arm. Small papillate protrusions on the tube feet act to improve retention efficiency by increasing adhesion to encountered particles (Fig. 4.1e). Reorientation of active arms at increasing flow velocities may be an attempt to adjust the height of arm extension to match an optimal flow velocity in the boundary layer with respect to encounter rate and retention efficiency. During transport, captured particles become entangled in mucus and are rolled into a bolus by the tube feet. Eventually, the bolus is transported by tube feet below the sediment surface to the mouth. Occasionally, individuals may also pick deposited particles from the sediment surface, and this behaviour is most common in still water (Loo et al. 1996). Buchanan (1964) also observed in directional bottom-current flows that *A. filiformis* holds the arms up into the current flow with a rheotactic response to current direction and feeds by trapping suspended particles. *A. chiajei* does not show this response and feeds exclusively on deposited matter on the sediment surface. Miller et al. (1992) found that the brittle-star *Amphipholis squamata* in still water holds the arms on the sediment surface, but in a moderate oscillatory current the arms are held in flow, waving to and fro. In high flows, however, suspension feeding ceases.

4.1.3 Bivalves

Only a small group of bivalves is dealt with in this section, viz. the tellinids. Yonge (1949) discerned a continuum in the Tellinaceae from real deposit feeders (e.g. *Macoma*) to real suspension feeders (e.g. *Donax*). Although primarily a deposit feeder (Fig. 4.1d), *Scrobicularia plana* may obtain some of its food by filtering suspended matter from the ambient water (Hughes 1969). Bradfield and Newell (1961) proposed that *Macoma balthica* is a deposit feeder at low water and a suspension feeder when covered by the tide. Thus, *M. balthica* and *S. plana* may not rely on deposit feeding only, because they are able to filter food from near-bottom resuspended material while the inhalant siphon is just at the sediment surface (Brafield and Newell 1961; Hughes 1969; de Wilde 1975; Earll 1975; Hummel 1985a).

Reid and Reid (1969) examined eight species of *Macoma* which were classified into three feeding categories: two deposit feeders, five suspension feeders, and one feeding on the surface films of bacteria on sand grains. Kamermans (1994a) found that high concentrations of suspended algal cells, between 180 to 1300 µg chl*a* l^{-1} (chl*a*: chlorophyll *a*), are required to sustain growth in *M. balthica* when limited to suspension feeding only. Exclusive suspension feeding for *M. balthica* is unlikely because the requisite phytoplankton concentrations are far above in situ values. Thus, a mean value of 5 µg chl*a* l^{-1} applies for coastal Danish waters (Riisgård 1998), and in, e.g., the Marsdiep tidal inlet (Dutch Wadden Sea), the annual average is about 8 µg chl*a* l^{-1} while the monthly averages vary from 1 µg chl*a* in winter to 30 µg chl*a* during the spring peak period (Cadée and Hegeman 1993, see also Beukema and Cadée 1997). Other studies by Specht and Lee (1989) suggest that a related species *M. nasuta* is not a very efficient suspension feeder because of a low weight-specific pumping rate compared with rates for obligate suspension-feeding bivalves. Further, in contrast to Reid and Reid (1969), Hylleberg and Gallucci (1975) found that *M. nasuta* is primarily a deposit feeder that indiscriminately sucks the top millimeter of the sediment surfaces rich in settled organic material, and selection of food material subsequently takes place on the gills and palps. Meyhöfer (1985) measured both filtration rate (F, l h^{-1}) and gill area (G, cm^2) as a function of size (W, g body soft weight) of *M. nasuta* and found: $F=0.08W^{0.86}$ and $G=2.85W^{0.63}$. This shows that the weight-specific filtration rate is about 100 × lower than found for the blue mussel *Mytilus edulis*, and other true suspension-feeding bivalves (e.g. Møhlenberg and Riisgård 1979; Riisgård 2000). Further, the approach velocity of water to the *M. nasuta* gill surface can be estimated as $F/G=0.1$ mm s^{-1} which is about 10 times lower than found for *M. edulis* (Riisgård and Larsen 2000). All together this indicates that the gills of *M. nasuta* are not adapted to suspension feeding in the traditional sense. For a detailed account on present knowledge on switching between deposit and suspension feeding in *M. baltica*, see 'Example II' (Sect. 4.3).

Wilson (1990) examined the mode of feeding by both *Tellina tenuis* and *T. fabula* and found that they could function as suspension feeders, although Trevallion (1971) suggested that suspension feeding alone would not suffice to meet the energy demands of *T. tenuis*.

4.1.4 Amphipods

The amphipod sibling species *Ampelisca vadorum* and *A. abdita* are widespread in the shallow marine waters of North America, and two methods of feeding are common in these amphipods (Mills 1967). Occasionally, the second antennae pick up sand grains from the sediment surface and cast them inwards to the midline of the body to be held by the gnathopods, and the microflora is then scraped off with the mouthparts and discarded (Fig. 4.1f). An animal may also tip forward and grasp sand grains with its mouthparts. The second method of feeding involves currents set up by the pleopods and second antennae that whirl rapidly besides the body. The beat of the pleopods causes a strong current to move posteriorly along the midline of the animal. The second gnathopods and first two pairs of peraeopods are heavily set with plumose setae, and suspended material is retained on this filter-screen. Every few seconds the first gnathopods close towards the midline, scraping food material off the posterior limbs and passing it forwards to the mouthparts. The feeding method suggests that detritus and the microflora associated with sand grains make up the main food source. Thus, suspension feeding does occur, but it is combined at times with swirling up of sand grains from the bottom, almost a form of deposit feeding.

Two feeding modes of the amphipod *Corophium volutator* have been described by Medows and Reid (1966). *C. volutator* feeds on surface sediment particles or on suspended matter by bringing these food items into its shallow U-shaped or semicircular tube in the sediment (Fig. 4.1a). In the surface deposit-feeding mode, *C. volutator* gather particles from the sediment surface with its enlarged second antennae. In the suspension-feeding mode, *C. volutator* extends its body along the tube and feeds by filtering particles from a pleopodal current with a basket formed by the finely plumose setae of the second gnathopods. Fenchel et al. (1975) found that *C. volutator* ingests particles between 4 and 60 µm and this particle size comprises nearly the total volume of the gut content. It was found highly probable that *C. volutator* can directly utilise suspended particles, e.g. small planktonic algae. The setae of the second gnathopod indicate that the fine bristles form a regular network which retains particles larger than about 4 µm. Miller (1984) studied suspension and deposit feeding by *C. insidiosum*, *C. salmonis*, and *C. spinicorne*. Collection of food particles and subsequent manipulation of the particles seem to be identical for all studied species of the genus *Corophium*.

4.1.5 Soft Corals

Soft corals are usually assumed to be passive suspension feeders that are dependent on suspended particles being carried to them by a current. However, Slattery et al. (1997) have observed that the colonial nephtheid soft coral *Gersemia antarctica* can switch to deposit feeding in the soft sediment communities of McMurdo Sound, Antarctica. Deposit feeding in *G. antarctica* involves bending by means of hydrostatic inflation of the entire colony against the sediment surface. Gut content analyses revealed a mixed diet including benthic diatoms, foraminiferans, and particulate organic matter. The deposit-feeding mode in *G. antarctica* has apparently evolved to supplement the capture of planktonic prey and seems to be of particular importance in the Antarctic.

4.1.6 Most Examples Among Passive Suspension Feeders

From the above examination of the literature dealing with possible switching between deposit and suspension feeding, it is striking that most examples are found among passive suspension feeders that strain food particles from the near-bottom current without metabolic energetic costs. This attracts attention to possible prerequisites and limitations of adaptation to active suspension feeding.

4.1.7 Adaptation to Suspension Feeding

To assess the adaptation of a suspension feeder to the biotope it is of interest to know the food energy uptake in relation to the total metabolic energy requirement of the animal. The latter may conveniently be expressed as the respiration (R) in amount of oxygen consumed. The former may be expressed simply as volume of water pumped through the filter device (F) times the food particle concentration, and will thus depend on the prevailing phytoplankton concentration. The ratio F/R expresses the litres of water pumped per ml O_2 consumed and may be used as a tool to characterise true suspension feeding on phytoplankton. A typical F/R-value of 10 l of water pumped per ml of oxygen consumed was reported by Jørgensen (1975) as the minimum to ensure the performance of suspension feeders inhabiting inshore waters. Using the above F/R-value as a guide, it is possible to discriminate between suspension feeders and deposit feeders (see Table 4.1). It can be seen that *Nereis diversicolor* is able to compete with obligate suspension feeders, like the blue mussel *Mytilus edulis*, but the same is obviously not the case for *Macoma balthica* or *Arenicola marina*.

The low concentrations of phytoplankton in the sea is the key to understanding the characteristics of active, obligate suspension feeders which seem to have evolved according to 'a principle of minimal scaling' according to which the dimensions of the filter-pumps are sufficient to enable continuous feeding at low rates, rather than discontinuous feeding at a correspondingly high rate (Jørgensen 1975). Studies during recent years have shown that active suspension feeders operate low-energy pumps that continuously process the surrounding water through filters appropriately dimensioned to cope with the phytoplankton concentrations of the biotope (Riisgård and Larsen 1995). The increasing evidence that suspension feeding is based on 'minimal scaling' may be the key to understanding why switching between deposit and suspension feeding in *Nereis diversicolor* is likely to be an exception: Referring to metabolic powers estimated for *Mytilus edulis* by Riisgård and Larsen (1995) it can be argued that other measures of efficiency than the overall filter-pump efficiency (i.e. useful pumping power/total metabolic rate) are relevant. Thus, the metabolic efficiency of gills/whole organism is about 20%, and, using this figure as a measure of energy, it is obvious that the energetic costs of suspension feeding by means of enlarged specialised gills is considerable. Clearly, the enlarged gill structures are expensive to maintain irrespective of whether the mussel is pumping water or not, and this means that energy for functions other than pump work can only be justified when the part of the organism responsible for the pumping action is dimensioned for continuous feeding. From these arguments it can be realised why deposit-feeding bivalves have small gills and a low specific 'filtration' rate and no ability to switch to suspension feeding. Further, it may be realised that the adaptation of *N. diversicolor* to an occasional suspension feeder (see Example I, Sect. 4.2) is exceptionally cheap in terms of no conspicuous anatomic alterations and accompanying metabolic investments.

4.2 Example I: Switching to Suspension Feeding in *Nereis diversicolor*

H. U. RIISGÅRD

The polychaete *Nereis diversicolor* is a common benthic species of shallow areas in north-western Europe, where it often penetrates far into estuaries (Theede et al. 1973; Wolff 1973; Chambers and Milne 1975). The ragworm is almost entirely restricted to the littoral zone where it lives in a U-shaped burrow in the sediment. *N. diversicolor* has been described as a carnivore, a scavenger, a suspension feeder and a surface-deposit feeder, feeding partly by ingesting detritus and microphytobenthos around the openings of the burrow (Wells and Dales 1951; Goerke 1966, 1971; Evans 1971; Smith 1977; Olafsson and Persson 1986; Rönn et al. 1988).

The occurrence of a suspension-feeding mechanism in *Nereis diversicolor* was first described by Harley (1950), and later confirmed by Goerke (1966), but quantitative measurements have only recently been made. *N. diversicolor* may meet its metabolic requirements on a pure diet of suspended phytoplankton at naturally low (i.e. usually <about 5 µg chl$a$ l^{-1}) concentrations, just as a typical obligate suspension feeder. Combined with its abundance, this worm may contribute to the control of phytoplankton production in many shallow brackish water areas – a role previously overlooked and thus undervalued (Riisgård 1991; Riisgård et al. 1992, 1996b; Vedel et al. 1994; Nielsen et al. 1995; Vedel 1998). The aim of the present account is to give an overview of the current knowledge concerning suspension feeding in *N. diversicolor* and its ecological implications.

4.2.1 Suspension-Feeding Behaviour

Observations by Riisgård (1991) of feeding behaviour in *Nereis diversicolor* were carried out in glass tubes immersed in seawater (Fig. 4.2). Within 5 to 20 min after adding a suspension of algal cells, the worm moved to one end of the glass tube where it fixed mucus threads to the glass wall forming the circular opening of a net-bag. The net was completed as the worm slowly retreated down the tube moving its anterior end from side to side in semicircles. During construction of the funnel-shaped net-bag, and for a period (5 to 15 min) after the bag was completed, the ragworm pumped water through the net by vigorously undulating its body. Particles suspended in the inhalant water were retained by the net and, after the period of pumping, the worm moved forward to swallow the net-bag and its entrapped food particles.

4.2.2 Mucus-Net and Particle-Retention Efficiency

The structure of the food-trapping net of *Nereis diversicolor* was studied by Riisgård et al. (1992). Electron micrographs of the filter-net structure showed

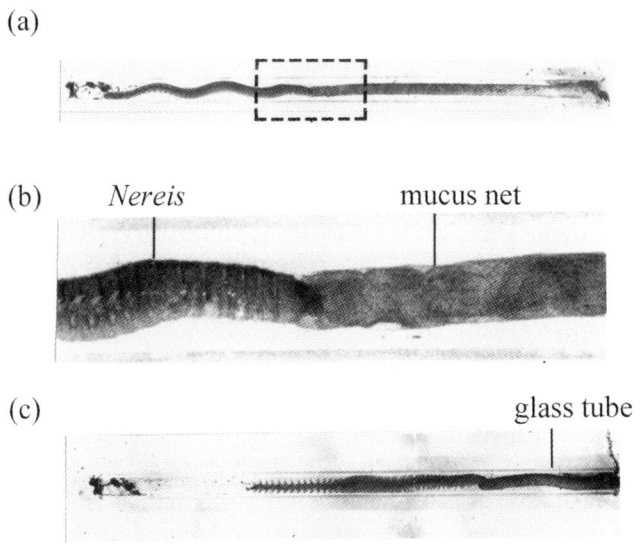

Fig. 4.2. *Nereis diversicolor* lodged in a glass tube. **a** The undulating body movements produce a water current that is filtered through a mucous net-bag made visible by means of suspended carmine powder. **b** Magnification of anterior end of the same worm. **c** The worm is swallowing the mucous net-bag 3 min after photograph **a** was taken. (Riisgård 1991)

that the net is composed of an irregular mesh made up of long, relatively thick filaments (up to 300 nm) inter-connected with a variety of shorter and thinner filaments with diameters ranging from 5 to about 25 nm. The average mesh size, measured directly on the EM pictures, was found to be between 0.5 and 1.0 µm, but due to shrinkage during preparation for EM these values represent only approximately 75% of the actual dimension of the intact net. Subsequent measurements performed by Riisgård et al. (1992) showed that particles as small as 2 to 3 µm can be cleared from the water with near 100% efficiency. This efficiency is comparable to that found in other suspension-feeding invertebrates (Møhlenberg and Riisgård 1978; Riisgård 1988; Jørgensen et al. 1984).

4.2.3 Filtration Rates

The filtration rate of *Nereis diversicolor* individuals was measured by Riisgård (1991) using three different methods. Results were given as pumping rate, measurement of clearance of suspended algal cells, and by means of the so-called 'suction' method. Further, the filtration rate of *N. diversicolor* was measured as a function of dry weight. Using this relationship and knowing the density and size distribution at the collecting site the population pumping rate was estimated to be about $10 \text{ m}^3 \text{ m}^{-2} \text{ d}^{-1}$, i.e. a volume corresponding to 10 times the above water column daily. Vedel et al. (1994), Riisgård et al. (1996b), and Vedel (1998) found similar population filtration capacities. In the shallow Odense Fjord, Denmark, the population filtration capacity was found to increase with rising water temperature from nearly $2 \text{ m}^3 \text{ m}^{-2} \text{ d}^{-1}$ in late March to $7 \text{ m}^3 \text{ m}^{-2} \text{ d}^{-1}$ in early August, or a water volume equivalent to 3 to 14 times the water column per day (Vedel 1998).

4.2.4 Energy Cost of Pumping

In *Nereis diversicolor* the pumping action is a result of the undulating movements of the body in the confining tube. Observations indicate *N. diversicolor* has three different phased posteriorly directed propagating waves, one of which creates an effective stroke responsible for the pumping action. During the effective wave, the worm's body makes firm contact with the tube wall on the dorsal and ventral sides, forming good seals. On the lateral sides, the parapodia make contact with the wall of the tube, to form less perfect seals that may leak with increasing backward pressure, resembling a positive displacement pump. Back-pressure characteristics of the *Nereis* pump were measured, and hence the pump was modelled as a leaking positive displacement unit by Riisgård et al. (1992) and Riisgård and Larsen (1995). The pumping rate was expressed as: $Q=Q_P-Q_L$, where Q_P=volume flow of a leak-free pump and Q_L=volume flow of leakage. In the expression, $Q_P=ALf$, A, L and f denote piston area, stroke length and stroke frequency of the pumping action, respectively.

The engineering principles of pump analysis and modelling of *Nereis diversicolor* and other suspension-feeding macroinvertebrates have been reviewed by Riisgård and Larsen (1995). Only the energy cost of pumping will briefly be mentioned here. The useful pumping power ('power output') of the *Nereis* pump was calculated as the product of pump pressure and pumping rate to be $P=2.10\,\mu W$. The total metabolic rate of the worm was measured at $R=12.6\,\mu l\,O_2\,h^{-1}$, equivalent to $70\,\mu W$, and the overall pump efficiency ('pump work') was estimated to be $P/R=3\%$ of the total metabolic energy expenditure. This low value is comparable with the obligate suspension-feeding polychaete *Chaetopterus variopedatus* (Riisgård 1989) and other true suspension-feeding macroinvertebrates (Riisgård and Larsen 1995).

4.2.5 Adaptation to Suspension Feeding

To obtain enough food to satiate the minimal energy requirements the performance of suspension feeders inhabiting inshore waters must generally exceed 10 l of water per ml of oxygen consumed (Jørgensen 1975). From this reference, and with a thorough knowledge of particle retention efficiency, the adaptation of suspension feeders to different biotopes with different typical phytoplankton levels may be evaluated and compared. By relating the pumping rates to respiration, Riisgård (1991) found that *N. diversicolor* pumps approximately 40 l of water per ml oxygen consumed. This shows that *N. diversicolor* fulfils the conditions for subsistance exclusively as a filter feeder, provided that the phytoplankton concentration is sufficiently high. But this is not always the case. It is therefore of interest to know how much time *N. diversicolor* spends on filter feeding versus surface deposit feeding, or other alternative foraging forms.

4.2.6 Time Spent on Suspension Feeding

To investigate to what extent suspension feeding of *Nereis diversicolor* in the field varies over the season, Vedel et al. (1993, 1994) and Vedel (1998) used an infrared phototransducer system to obtain continuous, long-term measurements of the characteristic undulating body movements of the filter-feeding activity of *N. diversicolor* in glass tubes immersed in the natural sediment. The phototransducer system consists of an infrared light-emitting diode and a phototransistor detector. The phototransistor detected variations in reflected infrared light intensity caused by the undulating body activity of the filter-feeding worm. Vedel et al. (1993, 1994) found that suspension feeding was 'triggered' by the presence of phytoplankton in the water, and that *N. diversicolor* was filter feeding 50–100 % of the time during the productive May–August period. In early spring and autumn, filter feeding occurred for approximately 5–20 % of the total time. However, no firm rules for time-share between different feeding modes can be made due to frequent depletion of phytoplankton below the 'trigger level' in the near-bottom water. This statement supports the view of 'opportunistic feeding' presented by Cadée (1984).

4.2.7 Phytoplankton Reduction in Near-Bottom Water

A thin layer of phytoplankton drained water overlying the sediment, especially on days when wind-induced mixing of the water column is low, indicates that the grazing impact can be restricted (Riisgård et al. 1996b; Vedel 1998). Near-bottom reduction of phytoplankton caused by a dense population of suspension-feeding *Nereis diversicolor* has been observed in the field by Riisgård et al. (1996b). Water samples were simultaneously collected at different heights above the bottom where *N. diversicolor* were present, and it appeared that a phytoplankton-reduced near-bottom water layer of 5–10 cm in thickness developed on calm days. That such depletion of phytoplankton in near-bottom waters plays a significant role for this worm was demonstrated in field-growth experiments performed with worms transferred to glass tubes placed at different heights above the bottom (Riisgård et al. 1996b). A nearly ten-fold increase in growth from elevating worms just 10 cm above the seafloor indicates that extremely meagre food conditions may be prevalent near the seafloor due to intraspecific competition for food.

A diffusion model was used by Larsen and Riisgård (1997) to describe the development of vertical phytoplankton profiles, over suspension-feeding *Nereis diversicolor* in stagnant water. The model was based on sinks located at inhalant openings, and Fick's law with an effective diffusivity that decreased with distance above the bottom due to biomixing generated by exhalant and inhalant feeding currents. For *N. diversicolor*, having inhalant and exhalant openings flush with the sediment surface and only a moderate exhalant jet velocity of about 1 cm s^{-1}, the thickness of a well-mixed boundary layer was limited by the low values of diffusivity prevailing at heights greater than about 5 cm above the bottom.

The development of 4–8 cm thick near-bottom phytoplankton-depleted layers and the influence of temperature and wind-mixing have recently been further investigated by Vedel (1998) who also tried to evaluate the ecological implications of the dense population of *Nereis diversicolor* in the Odense Fjord. The worm's filtration capacity of 3–14 times the water column per day was sufficiently high to control the phytoplankton level in the eutrophicated fjord, but wind-driven vertical mixing appeared to be a crucial factor for realisation of the potential grazing impact. Wind speeds of <8 m s^{-1} were not sufficient to mix the water column, and, therefore, phytoplankton depletion occurred near the bottom. Because the wind speed seems to be below the critical level during a considerable part of the year (about 70% of the time in March to August 1995), near-bottom phytoplankton depletion and switching from suspension feeding to an alternative feeding mode (predation, scavenging, surface deposit feeding) may be a frequent phenomenon in the Odense Fjord. From this it may be predicted that the time spent on suspension feeding by *N. diversocolor* and its actual grazing impact may vary considerably between different localities, and be dependent on currents, wind and wave action, and other local conditions. So far, no such knowledge exists for, e.g., the Wadden Sea where this worm is very common.

4.3 Example II: Switching to Suspension Feeding in *Macoma balthica*

P. KAMERMANS

The tellinid bivalve *Macoma balthica* is a common soft-bottom species along both sides of the North Atlantic (Beukema and Meehan 1985; Kamermans et al. 1999). Yonge (1949) classified all *Macoma* species as deposit feeders. Since then, variability in feeding behaviour has been observed within this group. *M. balthica* (L.) is now generally considered a facultative deposit/suspension feeder (Bradfield and Newell 1961; Hughes 1969; Bubnova 1972; Rasmussen 1973; Hummel 1985a; Olafsson 1986; Thompson and Nichols 1988; Kamermans 1994b; Lin and Hines 1994; Peterson and Skilleter 1994; Skilleter and Peterson 1994).

Deposit-feeding behaviour entails movements of the extended inhalant siphon on the sediment surface and the burrow entrance while sucking in material (e.g. Yonge 1949; Gilbert 1977). Defining suspension-feeding behaviour is more difficult. Some authors have suggested that *M. balthica* may be suspension feeding when the inhalant siphon is short and immobile (Bradfield and Newell 1961; de Wilde 1975). On the other hand, Levinton (1991) and Kamermans and Huitema (1994) suggested that this behaviour may be resting. Other authors considered a rotating movement of the inhalant siphon in the water column as suspension feeding (Rasmussen 1973; Hummel 1985a; Olafsson 1986; Rosenberg 1993). However, Hulscher (1973) described the rotating movement as a way to whirl up the top layer of the sediment in order to facilitate intake by the siphon. In addition, Kamermans and Huitema (1994) observed that the rotating behaviour was always carried out just prior to ejecting pseudofaeces and considered it to be part of the deposit-feeding behaviour.

4.3.1 Switching to Suspension Feeding

Different factors that control the choice between feeding on food particles suspended in the water column or on food particles deposited on the sediment surface have been proposed for *Macoma*. These are:

1. High current velocities can cause drag on the siphons that prevents the characteristic grazing movements of the siphon. This may lead to a switch to suspension feeding (Levinton 1991; Peterson and Skilleter 1994).
2. High food availability in the water column compared to the surface of the sediment may induce a switch to suspension feeding (Cadée 1984; Hummel 1985a; Olafsson 1986; Lin and Hines 1994).
3. High feeding on siphon tips by carnivorous bottom feeders such as fishes, shrimps and crabs may cause a switch to suspension feeding in which the siphons are not exposed above the sediment surface (Levinton 1971; de Vlas 1985; Kamermans and Huitema 1994; Skilleter and Peterson 1994).
4. High rates of lethal predation by birds and crabs may favour suspension feeding. Burying deeply is important as protection against predation by birds and crabs (Reading and McGrorty 1978; Blundon and Kennedy 1982). The length of the siphon determines the maximum depth at which the animals can be buried in the sediment (Zwarts 1986). Zwarts and Wanink (1989) suggested that a feeding mode in which a short inhalant siphon protrudes above the sediment may allow the bivalve to use the extra length of the siphon to increase its burying depth.

The next sections discuss the data presented by the above-mentioned authors to detect a possible switch to suspension feeding in *Macoma*. Before doing so it is necessary to demonstrate that *Macoma* is capable of suspension feeding. Kamermans (1994a) placed *Macoma balthica* on a grid and exposed them to high concentrations of suspended algal cells only. The concentration of algal cells was reduced compared to the control. The 4-week experiment yielded growth rates that were higher than field rates. These results indicate that *Macoma* is capable of suspension feeding. In addition, when food is not limited, a suspension-feeding *M. balthica* can grow faster than *M. balthica* in the field. However, the data presented in Table 4.1 show that this species can not filter as efficiently as obligate suspension feeders. This observation suggests that suspension feeding is only utilised under suboptimal conditions such as high current flow, low food availability on the sediment surface and high predation pressure on siphons or whole individuals.

4.3.2 Current Velocity

Levinton (1991) studied the effect of high current velocity on the feeding behaviour of three Pacific *Macoma* species. In a flume experiment, a decrease in deposit-feeding behaviour was observed with an increase in water velocity. The deposit-feeding radius was used as a measure of deposit-feeding behaviour. This was the distance from the siphon hole to the tip of the deposit-feeding siphon. He suggested that the drag on the siphon was preventing deposit feeding at high velocities. At velocities between 10.5 and 12 cm s^{-1} protruding siphons quivered and then flopped over. At higher velocities, the bivalves did not show their siphons above the sediment surface which suggests suspension feeding, or non-feeding. However, evidence of an actual switch to suspension feeding was not provided. Peterson and Skilleter (1994) studied feeding behaviour in *Macoma balthica* in relation to current velocities in the field. Two identical sets of *M. balthica* were introduced into two adjacent field sites with different flow regimes in a tributary of the Neuse River estuary, North Carolina. Flow velocity was measured with a current meter. The amount of sediment present in the mantle cavity of the bivalves was used as a measure of deposit-feeding activity. The site with a mean current velocity of 20 cm s^{-1} showed lower levels of deposit feeding than the site with a mean current velocity of 5 cm s^{-1}. Again, the data indicate a reduction in deposit feeding, but do not provide evidence of a switch to suspension feeding.

4.3.3 Food Availability

In the Wadden Sea, Hummel (1985a) compared the algal species composition of the stomach contents of *Macoma balthica* with the algal composition of simultaneously collected plankton and surface sediments. The algal species composition from the stomach was then used as an indicator of feeding behaviour. Results indicated that the bivalves on the tidal flats took twice as much food originating from the overlying water than from the bottom. This behaviour was related to the high chlorophyll concentrations in the water. Lin and Hines (1994) demonstrated in a laboratory experiment that an increase in the amount of food in the water to a chlorophyll *a* concentration of 75 µg l^{-1} resulted in an increase in the proportion of *M. balthica* individuals that showed suspension-feeding behaviour. In this case, suspension-feeding behaviour was defined as holding the inhalant siphon straight up and relatively still in the water column. Olafsson (1986) observed differences in feeding mode between *M. balthica* populations from two contrasting habitats in the Baltic Sea. Individuals from both populations were placed in aquariums under conditions that were similar to those at the site of collection. Individuals from a muddy site with stagnant water showed predominantly deposit-feeding behaviour, while suspension-feeding behaviour was observed in individuals from a sandy site with moderate currents. In this study, suspension-feeding behaviour was defined as holding the siphon relatively still and straight up in the water as well as swirling it around in a circular fashion. Olafsson (1986) suggested that suspension feeding was a response to increased food availability in the water column in the sandy habitat with moderate currents. A cage experiment in the Baltic Sea showed density-dependent growth in the population in the muddy habitat, but not in the sandy habitat (Olafsson 1986). These results support deposit feeding in the muddy habitat and suspension feeding in the sandy habitat.

On a local scale, food limitation is expected to occur more readily in deposit feeders than in suspension feeders, because for suspension feeders food is continuously supplied by the passing water (Levinton 1972). Growth reduction of suspension feeders occurs only at very high densities of suspension feeders such as cockle and mussel banks (Jensen 1992; Kamermans 1993). One can ask why the deposit-feeding *Macoma balthica* of the muddy site in the Baltic, that were experiencing growth limitation, did not switch to suspension feeding. Experiments in outdoor flow-through basins carried out by Kamermans et al. (1992) showed density-dependent growth in *M. balthica*, but not in the suspension feeder *Cerastoderma edule*. In this experiment, the sediment was sandy and the fresh seawater was constantly supplied, but density-dependent growth was found which suggests deposit-feeding behaviour. Following Olafsson's suggestion, suspension-feeding behaviour, and thus no density-dependent growth, was expected in *M. balthica* in the basins. The continuation of deposit feeding may be explained by the low food availability in the water column in the basins. During the basin experiment chl*a* levels in the water ranged between 2 and 20 µg l^{-1}, which is much lower than the switching level of 75 µg l^{-1} observed by Lin and Hines (1994). Olafsson (1986) does not provide data on food availability in the water column at the two sites where he performed the cage experiments. Collectively, these observations indicate that food supply in the water column was probably much lower at the muddy site than at the sandy site, which may explain the deposit-feeding behaviour of *M. balthica* at the muddy site.

4.3.4 Feeding on Siphon Tips

Feeding on siphon tips by carnivorous bottom feeders such as fish, shrimps and crabs is another factor suggested to influence the feeding behaviour of *Macoma balthica* (Levinton 1971; de Vlas 1985). Deposit feeding requires exposure of the inhalant siphon above the sediment in order to be able to graze on the surrounding sediment. For suspension feeding the siphon is directed towards the water column. It is not known at which height the opening of the inhalant siphon should be above the sediment for the most favorable suspension feeding. Keeping the siphon level with the sediment surface to allow contact with the overlying water column could be sufficient. A switch to a suspension-feeding mode involving only a short siphon exposed above the sediment has therefore been interpreted as a defence against feeding on siphon tips (Levinton 1971; de Vlas 1985). Kamermans and Huitema (1994) carried out a series of experiments which included both removing part of the inhalant siphon of *M. balthica* and exposing the bivalves to siphon-nipping shrimps. They demonstrated a reduction in the sediment surface area that could be grazed by *M. balthica* following siphon nipping, and a decrease in siphon activity in *M. balthica* under nipping conditions. These results demonstrate a decrease in deposit feeding when exposed to siphon nipping and suggest that it is favourable to switch to suspension feeding under such conditions. A switch from deposit feeding under conditions without nippers to suspension feeding when subjected to nipping was, however, not observed (Kamermans and Huitema 1994). Exclosure and fish-inclosure experiments also showed that siphon nipping can induce a reduction in deposit-feeding activity (Peterson and Skilleter 1994; Skilleter and Peterson 1994). The authors used the amount of sediment in the mantle cavity as a measure of deposit-feeding activity, but did not demonstrate a switch to suspension feeding.

4.3.5 Protection Against Lethal Predation

Another advantage of the fact that the siphon does not need to be stretched over the sediment when suspension feeding is the possibility for the bivalve to increase its burying depth. In this way the clam can increase its protection against lethal predation by birds (Zwarts and Wanink 1989) and crabs (Blundon and Kennedy 1982). To date, studies that determine how the risk of lethal predation influences the feeding behaviour of *Macoma balthica* are absent in the literature.

4.4 Conclusions

We conclude that the concept of switching feeding mode, and assumed occurrence of this ability among coastal zoobenthos, should be re-assessed. Most examples of switching between deposit and suspension feeding are found among passive suspension feeders that strain food particles from the near-bottom current. So far, only *Nereis diversicolor* has proven to be able to make a genuine switch from a surface deposit feeder to an active and efficient suspension feeder that operates an energy-consuming filter-pump that competes well with filter-pumps operated by obligate suspension-feeding invertebrates. For *Macoma balthica* it is concluded that suspension feeding is not an important feeding mode because the water-processing capacity of this species is insignificant compared with obligate suspension feeders, and no convincing evidence for a switch to suspension feeding was provided for any of the four factors proposed in the literature to induce a change in feeding behaviour.

Acknowledgements. Thanks are due to Drs. P. S. Larsen, J.-M. Gili, G. Cadée and an anonymous referee for constructive comments. Mrs. Katrine Worsaae provided supplementary pieces of information on spionids.

References

Barnes RH (1964) Tube-building and feeding in the chaetopterid polychaete, *Spiochaetopterus oculatus*. Biol Bull 127:397–412
Barnes RD (1965) Tube-building and feeding in chaetopterid polychaetes. Biol Bull 129: 217–233
Beukema JJ, Cadée GC (1997) Local differences in macrobenthic response to enhanced food supply caused by mild eutrophication in a Wadden Sea area: food is only locally a limiting factor. Limnol Oceanogr 42:1424–1435
Beukema JJ, Meehan BW (1985) Latitudinal variation in linear growth and other shell characteristics of *Macoma balthica*. Mar Biol 90:27–33
Blake JA (1996) Family Spionidae Grube, 1850. In: Blake JA, Hilbig B, Scott PH (eds) Taxonomic atlas of the benthic fauna of the Santa Maria Basin and the western Santa Barbara Channel, vol 6 (the annelida part 3 Polychaeta: Orbiniidae to Cossuridae), Santa Barbara Museum of Natural History, Santa Barbara, CA, pp 81–223
Blundon JA, Kennedy VS (1982) Refuges for infaunal bivalves from blue crab, *Callinectes sapidus* (Ratbun), predation in Chesapeake Bay. J Exp Mar Biol Ecol 65:67–81
Bradfield AE, Newell GE (1961) The behaviour of *Macoma balthica* (L.). J Mar Biol Assoc UK 41:81–87
Bubnova NP (1972) The nutrition of the detritus-feeding mollusks *Macoma balthica* (L.) and *Portlandia arctica* (Gray) and their influence on bottom sediments. Okeanologiya 12:899–905

Buchanan JB (1964) A comparative study of some features of biology of *Amphiura filiformis* and *Amphiura chiajei* (Ophiuroidea) considered in relation to their distribution. J Mar Biol Assoc UK 44:565-576

Buhr KJ (1976) Suspension-feeding and assimilation efficiency in *Lanice conchilega* (Polychaeta). Mar Biol 38:373-383

Buhr KJ, Winter JE (1977) Distribution and maintenance of a *Lanice conchilega* association in the Weser estuary (FRG), with special reference to the suspension-feeding behaviour of *Lanice conchilega*. In: Keegan BF, Ceidigh PO, Boaden PJS (eds) Biology of benthic organisms. Pergamon Press, New York, pp 101-113

Cadée GC (1984) 'Opportunistic feeding', a serious pitfall in trophic structure analysis of (paleo)faunas. Lethaia 17:289-292

Cadée GC, Heegeman J (1993) Persisting high levels of primary production and declining phosphate concentrations in the Dutch coastal area (Marsdiep). Neth J Sea Res 31:147-152

Chambers MR, Milne H (1975) Life cycle and production of *Nereis diversicolor* O. F. Müller in the Ythan Estuary, Scotland. Est Coast Mar Sci 3:133-144

Clausen I, Riisgård HU (1996) Growth, filtration and respiration in the mussel *Mytilus edulis*: no regulation of the filter-pump to nutritional needs. Mar Ecol Prog Ser 141:37-45

Dauer DM (1983) Functional morphology and feeding behavior of *Scolelepis squamata* (Polychaeta: Spionidae). Mar Biol 77:279-285

Dauer DM (1985) Functional morphology and feeding behavior of *Paraprionospio pinnata* (Polychaeta: Spionidae). Mar Biol 85:143-151

Dauer DM (1991) Functional morphology and feeding behavior of *Polydora commensalis* (Polychaeta: Spionidae). Ophelia [Suppl] 5:607-614

Dauer DM (1994) Functional ciliary groups of the feeding palps of Spionid polychaetes. In: Dauvin J-C, Laubier L, Reish DJ (eds) Actes de la 4ème Conférence internationale des Polychètes. Mémoires Museum natn. Hist Nat 162:81-84

Dauer DM (1997) Functional morphology and feeding behavior of *Marenzelleria viridis* (Polychaeta: Spionidae). Bull Mar Sci 60:51-516

Dauer DM, Ewing RM (1991) Functional morphology and feeding behavior of *Malacoceros indicus* (Polychaeta: Spionidae). Bull Mar Sci 48:395-400

Dauer DM, Maybury CA, Ewing RM (1981) Feeding behavior and general ecology of several spionid polychaetes from the Chesapeake Bay. J Exp Mar Biol Ecol 54:21-38

de Vlas J (1985) Secondary production by siphon regeneration in a tidal flat population of *Macoma balthica*. Neth J Sea Res 19:147-164

de Wilde PAWJ (1975) Influence of temperature on behaviour, energy metabolism and growth of *Macoma balthica* (L.). In: Barnes H (ed) 9th European marine biology symposium. Aberdeen Univ Press, Aberdeen, pp 239-256

Earll R (1975) Temporal variation in the heart activity of *Scrobicularia plana* (da Costa) in constant and tidal conditions. J Exp Mar Biol Ecol 19:257-274

Evans SM (1971) Behaviour in polychaetes. Q Rev Biol 46:379-405

Fauchald K, Jumars PA (1979) The diet of worms: a study of polychaete feeding guilds. Oceanogr. Mar Biol Annu Rev 17:193-284

Fenchel T, Finlay BJ (1995) Ecology and evolution in anoxic worlds. Oxford Univ Press, Oxford

Fenchel T, Kofoed LH, Lappalainen A (1975) Particle size-selection of two deposit feeders: the amphipod *Corophium volutator* and the prosobranch *Hydrobia ulvae*. Mar Biol 30:119-128

Gilbert MA (1977) The behaviour and functional morphology of deposit feeding in *Macoma balthica* (Linne, 1758), in New England. J Moll Stud 43:18-27

Goerke H (1966) Nahrungsfiltration von *Nereis diversicolor* O. F. Müller (Nereidae, Polychaeta). Veröff Inst Meeresforsch Bremerh 10:49-58

Goerke H (1971) Die Ernährungsweise der *Nereis* Arten (Polychaeta, Nereidae) der Deutschen Küsten. Veröff Inst Meeresforsch Bremerh 13:1-50

Harley MB (1950) Occurrence of a filter-feeding mechanism in the polychaete *Nereis diversicolor*. Nature (London) 165:734-735

Hempel C (1957) Über den Röhrenbau und die Nahrungsaufnahme einiger Spioniden (Polychaeta sedentaria) der deutschen Küsten. Helgol Wiss Meeresunters 6:100-135

Hughes RN (1969) A study of feeding in *Scrobicularia plana*. J Mar Biol Assoc UK 49: 805-823

Hulscher JB (1973) Burying-depth and trematode infection in *Macoma balthica*. Neth J Sea Res 6:141-156

Hummel H (1985a) Food intake of *Macoma balthica* (Mollusca) in relation to seasonal changes in its potential food on a tidal flat in the Dutch Wadden Sea. Neth J Sea Res 19:52-76

Hummel H (1985b) Food intake and growth in *Macoma balthica* (Mollusca) in the laboratory. Neth J Sea Res 19:77-83

Hylleberg J, Gallucci VF (1975) Selectivity in feeding by the deposit-feeding bivalve *Macoma nasuta*. Mar Biol 32:167-178

Jensen KT (1992) Dynamics and growth of the cockle, *Cerastoderma edule*, on an intertidal mud-flat in the Danish Wadden Sea: effects of submersion time and density. Neth J Sea Res 28:335-345

Jørgensen CB (1955) Quantitative aspects of filter feeding in invertebrates. Biol Rev 30:391-454

Jørgensen CB (1975) Comparative physiology of suspension feeding. Annu Rev Physiol 30:391-454

Jørgensen CB, Kiørboe T, Møhlenberg F, Riisgård HU (1984) Ciliary and mucus-net filter feeding, with special reference to fluid mechanical characteristics. Mar Ecol Prog Ser 15:283-292

Kamermans P (1993) Food limitation in cockles (*Cerastoderma edule* (L.)): influences of location on tidal flat and of nearby presence of mussel beds. Neth J Sea Res 31: 71-81

Kamermans P (1994a) Nutritional value of solitary cells and colonies of *Phaeocystis* sp. for the bivalve *Macoma balthica* (L.). Ophelia 39:35-44

Kamermans P (1994b) Similarity in food source and timing of feeding in deposit- and suspension-feeding bivalves. Mar Ecol Prog Ser 104:63-75

Kamermans P, Huitema HJ (1994) Shrimp (*Crangon crangon* L.) browsing upon siphon tips inhibits feeding and growth in the bivalve *Macoma balthica* (L.). J Exp Mar Biol Ecol 175:59-75

Kamermans P, Veer HW van der, Karczmarski L, Doeglas GW (1992) Competition in deposit- and suspension-feeding bivalves: experiments in controlled outdoor environments. J Exp Mar Biol Ecol 162:113-135

Kamermans P, Veer HW van der, Witte JIJ, Adriaans EJ (1999) Morphological differences in *Macoma balthica* (Bivalvia, Tellinacea) from a Dutch and three southeastern United States estuaries. J Sea Res 41:213-224

Kiørboe T, Møhlenberg F, Hamburger K (1985) Bioenergetics of the planktonic copepod *Acartia tonsa*: relation between feeding, egg production and respiration, and composition of specific dynamic action. Mar Ecol Prog Ser 26:85-97

Kuipers BR (1973) On the tidal migration of young plaice (*Pleuronectes platessa* L.) in the Wadden Sea. Neth J Sea Res 6:376–388

Kuipers RB (1977) On the ecology of juvenile plaice on a tidal flat in the Wadden Sea. Neth J Sea Res 11:56–91

Krüger F (1959) Zur Ernährungsphysiologie von *Arenicola marina*. Zool Anz 22: 115–120

Krüger F (1962) Experimentelle Untersuchungen zur ökologischen Physiologie von *Arenicola marina*. Kieler Meeresforsch 18:157–168

Krüger F (1964) Messungen der Pumptätigkeit von *Arenicola marina* L. im Watt. Helgoländer Wiss Meeresunters 18:70–91

Krüger F (1971) Bau und Leben des Wattwurmes *Arenicola marina*. Helgol Wiss Meeresunters 22:149–200

Larsen PS, Riisgård HU (1997) Biomixing generated by benthic filter feeders: a diffusion model for near-bottom phytoplankton depletion. J Sea Res 37:81–90

Levinton JS (1971) Control of Tellinacean (Mollusca: Bivalvia) feeding behavior by predation. Limnol Oceanogr 16:660–662

Levinton JS (1972) Stability and trophic structure in deposit-feeding and suspension-feeding communities. Am Nat 106:472–486

Levinton JS (1991) Variable feeding in three species of *Macoma* (Bivalvia: tellinacea) as a response to water flow and sediment transport. Mar Biol 110:375–383

Lin J, Hines AH (1994) Effects of suspended food availability on the feeding mode and burial depth of the Baltic clam, *Macoma balthica*. Oikos 69:28–36

Loo L-O, Jonsson PR, Sköld M, Karlsson Ö (1996) Passive suspension feeding in *Amphiura filiformis* (Echinodermata: Ophiuroidea): feeding behaviour in flume flow and potential feeding rate of filed populations. Mar Ecol Prog Ser 139:143–155

Meadows PS, Reid A (1966) The behavior of *Corophium volutator* (Crustacea: Amphipoda). J Zool 150:387–399

Meyhöfer E (1985) Comparative pumping rates in suspension-feeding bivalves. Mar Biol 85:137–142

Miller DC (1984) Mechanical post-capture particle selection by suspension- and deposit-feeding *Corophium*. J Exp Mar Biol Ecol 82:59–76

Miller DC, Bock MJ, Turner EJ (1992) Deposit and suspension feeding in oscillatory flows and sediment fluxes. J Mar Res 50:489–520

Mills EL (1967) The biology of an ampeliscid amphipod crustacean sibling species pair. Fish Res Bd Can 24:305–355

Møhlenberg F, Riisgård HU (1978) Efficiency of particle retention in 13 species of suspension feeding bivalves. Ophelia 17:239–246

Møhlenberg F, Riisgård HU (1979) Filtration rate, using a new indirect technique, in thirteen species of suspension-feeding bivalves. Mar Biol 54:143–148

Nielsen AM, Eriksen NT, Iversen JJL, Riisgård HU (1995) Feeding, growth and respiration in the polychaetes *Nereis diversicolor* (facultative filter-feeder) and *N. virens* (omnivorous) – a comparative study. Mar Ecol Prog Ser 125:149–158

Olafsson EB (1986) Density dependence in suspension-feeding and deposit-feeding populations of the bivalve *Macoma balthica*: a field experiment. J Anim Ecol 55:517–526

Olafsson EB, Persson LE (1986) The interaction between *Nereis diversicolor* O. F. Müller and *Corophium volutator* Pallas as structuring force in a shallow brackish sediment. J Exp Mar Biol Ecol 103:103–117

Petersen JK, Schou O, Thor P (1995) Growth and energetics in the ascidian *Ciona intestinalis* (L.). Mar Ecol Prog Ser 120:175–184

Peterson CH, Skilleter GA (1994) Control of foraging behaviour of individuals within an ecosystem context: the clam *Macoma balthica*, flow environment, and siphon-cropping fishes. Oecologia 100:256–267

Qian PY, Chia FS (1997) Structure of feeding palps and feeding behavior of the spionid polychaete *Polydora polybranchia*. Bull Mar Sci 60:502–511

Rasmussen E (1973) Systematics and ecology of the Isefjord marine fauna (Denmark). Ophelia 11:495

Reading CJ, McGrorty S (1978) Seasonal variations in the burying depth of *Macoma balthica* (L.) and its accessibility to wading birds. Est Coast Mar Sci 6:135–144

Reid RGB, Reid A (1969) Feeding process of members of the genus *Macoma* (Mollusca: Bivalvia). Can J Zool 47:649–657

Reiswig HM (1974) Water transport, respiration and energetics of three tropical marine sponges. J Exp Mar Biol Ecol 14:231–249

Riisgård HU (1988) Efficiency of particle retention and filtration rate in 6 species of Northeast American bivalves. Mar Ecol Prog Ser 45:217–223

Riisgård HU (1989) Properties and energy cost of the muscular piston pump in the suspension feeding polychaete *Chaetopterus variopedatus*. Mar Ecol Prog Ser 56:157–168

Riisgård HU (1991) Suspension feeding in the polychaete *Nereis diversicolor*. Mar Ecol Prog Ser 70:29–37

Riisgård HU (1998) Filter feeding and plankton dynamics in a Danish fjord: a review of the importance of flow, mixing and density-driven circulation. J Environ Man 53:195–207

Riisgård HU (2001) On measurement of filtration rates in bivalves – the stony road to reliable data, review and interpretation. Mar Ecol Prog Ser 211:275–291

Riisgård HU, Banta GT (1998) Irrigation and deposit feeding by the lugworm *Arenicola marina*, characteristics and secondary effects on the environment. A review of current knowledge. Vie Milieu 48:243–257

Riisgård HU, Ivarsson NM (1990) The crown-filament-pump of the suspension-feeding polychaete *Sabella penicillus*: filtration, effects of temperature, energy cost, and modelling. Mar Ecol Prog Ser 62:249–257

Riisgård HU, Larsen PS (1995) Filter-feeding in marine macro-invertebrates: pump characteristics, modelling and energy cost. Biol Rev 70:67–106

Riisgård HU, Larsen PS (2000) Comparative ecophysiology of active zoobenthic filter-feeding, essence of current knowledge. J Sea Res 44:169–193

Riisgård HU, Manríquez P (1997) Filter-feeding in fifteen marine ectoprocts (Bryozoa): particle capture and water pumping. Mar Ecol Prog Ser 154:223–239

Riisgård HU, Svane I (1999) Filter feeding in lancelets (amphioxus) *Brachiostoma lanceolatum*. Invert Biol 118:423–432

Riisgård HU, Randløv A, Kristensen PS (1980) Rates of water processing oxygen consumption and efficiency of particle retention in veligers and young post-metamorphic *Mytilus edulis*. Ophelia 19:37–47

Riisgård HU, Vedel A, Boye H, Larsen PS (1992) Filter-net structure and pumping activity in the polychaete *Nereis diversicolor*: effects of temperature and pump-modelling. Mar Ecol Prog Ser 83:79–89

Riisgård HU, Berntsen I, Tarp B (1996a) The lugworm *Arenicola marina* pump – characteristics, modelling and energy cost. Mar Ecol Prog Ser 138:149–156

Riisgård HU, Poulsen L, Larsen PS (1996b) Phytoplankton reduction in near-bottom water caused by filter-feeding *Nereis diversicolor* – implications for worm growth and population grazing impact. Mar Ecol Prog Ser 141:47–54

Rönn C, Bronsdorff E, Nelson WG (1988) Predation as a mechanism of interference within infauna in shallow brackish water soft bottoms; experiments with an infauna predator, *Nereis diversicolor* O. F. Müller. J Exp Mar Biol Ecol 116:143–157

Rosenberg R (1993) Suspension feeding in *Abra alba* (Mollusca). Sarsia 78:119–121

Self RFL, Jumars PA (1978) New resource axes for deposit feeders? J Mar Res 34:627–641

Shimeta J, Jumars PA (1991) Physical mechanisms and rates of particle capture by suspension-feeders. Oceanogr Mar Biol Annu Rev 29:191–257

Shimeta J, Koehl MAR (1997) Mechanisms of particle selection by tentaculate suspension feeders during encounter, retention, and handling. J Exp Mar Biol Ecol 209:47–73

Skilleter GA, Peterson CH (1994) Control of foraging behaviour of individuals within an ecosystem context: the clam *Macoma balthica* and interactions between competition and siphon cropping. Oecologia 100:268–278

Slattery M, McClintock JB, Bowser SS (1997) Deposit feeding: a novel model of nutrition in the Antarctic colonial soft coral *Germesia antarctica*. Mar Ecol Prog Ser 149: 299–304

Smith RI (1977) Physiological and reproductive adaptations of *Nereis diversicolor* to life in the Baltic Sea and adjacent waters. In: Reish, DJ, Fauchald K (eds) Essays on polychaeteous annelids. Univ Southern California Press, Los Angeles, pp 373–390

Specht DT, Lee H (1989) Direct measurement technique for determining ventilation rate in the deposit feeding clam *Macoma nasuta* (Bivalvia, Tellinaceae). Mar Biol 101: 211–218

Strickland JDH, Parsons TR (1972) A practical handbook of seawater analysis. Bull Fish Res Board Can 167:185–206

Taghon GL, Greene RR (1992) Utilization of deposited and suspended particulate matter by benthic 'interface' feeders. Limnol Oceanogr 37:1370–1391

Taghon GL, Nowell ARM, Jumars PA (1980) Induction of suspension feeding in spionid polychaetes by high particulate fluxes. Science 210:562–564

Theede HJ, Schaudinn J, Saffé F (1973) Ecophysiological studies on four *Nereis* species of the Kiel Bay. Oikos [Suppl] 15:246–252

Thomassen S, Riisgård HU (1995) Growth and energetics of the sponge *Halichondria panicea*. Mar Ecol Prog Ser 128:239–246

Thompson JK, Nichols FH (1988) Food availability controls seasonal cycle of growth in *Macoma balthica* (L.) in San Francisco Bay, California. J Exp Mar Biol Ecol 116: 43–61

Trevallion A (1971) Studies on *Tellina tenuis* da Costa, III. Aspects of general biology and energy flow. J Exp Mar Biol Ecol 7:95–112

Vedel A (1998) Phytoplankton depletion in the benthic boundary layer caused by suspension-feeding *Nereis diversicolor* (Polychaeta): grazing impact and effect of temperature. Mar Ecol Prog Ser 163:125–132

Vedel A, Riisgård HU (1993) Filter-feeding in the polychaete *Nereis diversicolor*: growth and bioenergetics. Mar Ecol Prog Ser 100:145–152

Vedel A, Andersen BB, Riisgård HU (1994) Field investigations of pumping activity of the facultatively filter-feeding polychaete *Nereis diversicolor* using an improved infrared phototransducer system. Mar Ecol Prog Ser 103:91–101

Wells GP, Dales RP (1951) Spontaneous activity patterns in animal behaviour: the irrigation of the burrow in the polychaetes *Chaetopterus variopedatus* Renier and *Nereis diversicolor* O. F. Müller. J Mar Biol Assoc UK 29:661–680

Williams JD, Mcdermott JJ (1997) Feeding behavior of *Dipolydora commensalis* (Polychaeta: Spionidae): particle capture, transport, and selection. Inv Biol 116:115–123

Wilson JG (1990) Gill and palp morphology of *Tellina tenuis* and *T. fabula* in relation to feeding. In: Morton B (ed) The bivalvia. Proceedings of a memorial symposium in honour of Sir Charles Maurice Yonge, Edinburgh, 1986. Hong Kong Univ Press, Hong Kong, pp 141–150

Wolff WJ (1973) The estuary as a habitat. Zool Meded (Leiden) 126:242

Yonge CM (1949) On the structure and the adaptations of the Tellinacea, deposit-feeding Eulamellibranchia. Philos Trans R Soc (B) 234:29–76

Zwarts L (1986) Burying depth of the benthic bivalve *Scrobicularia plana* (da Costa) in relation to siphon-cropping. J Exp Mar Biol Ecol 101:25–39

Zwarts L, Wanink J (1989) Siphon size and burying depth in deposit- and suspension-feeding benthic bivalves. Mar Biol 100:227–240

Part II

Biogenic Stabilization and Disturbances in Coastal Sediments

5 Microphytobenthos in Contrasting Coastal Ecosystems: Biology and Dynamics

D.M. Paterson and S.E. Hagerthey

5.1 Contrasting Shores

The rocky intertidal foreshore has long been an area intensely studied by ecologists, and the contrasts found between relative heights on the shore and between shores of different aspects have developed into a series of classical studies that outline the nature of the forces (exposure, competition, grazing) that dominate rocky shore ecology (Raffaelli and Hawkins 1996). These hard shores have natural advantages for scientists interested in intertidal ecology: they are usually accessible; the sessile organisms are easily observed; and the shores are relatively stable. However, the contrasts between sandy (non-cohesive) and muddy (cohesive) shores are equally of interest, although the shore themselves present more of a logistic challenge and hence contrasting accounts of the shores are less generally available. Also, except in the specialised circumstance of sea grass meadows and mangals, the primary producers (microphytobenthos) that dominate depositional systems are far less obvious than the intertidal macroalgae found on hard substrata. The following chapter addresses some of the differences found between sandy and muddy shores with emphasis on the dominant microphytobenthos.

5.2 The Microphytobenthos

The grouping of organisms together under the title "microphytobenthos" reflects no coherent phylogenetic relationship and, although mainly protists, the assemblage includes both prokaryotic and eukaryotic forms (Fig. 5.1). The dominant forms are often diatoms (Bacillariophyceae) that are a large and heterogeneous group of eukaryotic autotrophs, classified under the Heterokontophyta (van den Hoek et al. 1995). Estimates of species numbers vary from tens to hundreds of thousands and there is also recent discussion over

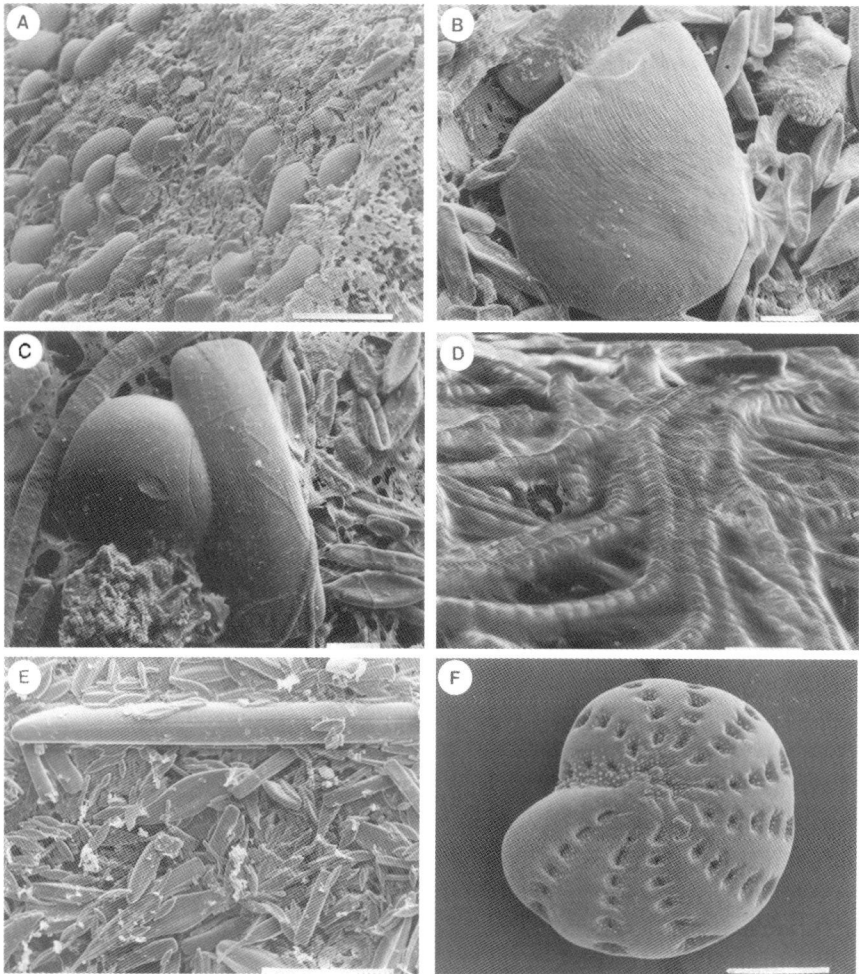

Fig. 5.1. A-E: Low-temperature scanning electron micrographs of sediment microphytobenthos. **A** Euglenids at the sediment surface during low tide. Bar marker=100 µm. **B** Individual euglenid on the sediment surface. Note characteristic pattern of the pellicle. *Bar*=10 µm. **C** Mixed assemblage of diatoms (single and chain forming), fungi (Chytrid [*F*]) and cyanobacteria (*CY*) on the sediment surface. *Bar*=10 µm. **D** Oblique view across sediment surface colonised by filamentous cyanobacteria. *Bar*=10 µm. **E** A mixed assemblage of epipelic diatoms ranging in size from 400 µm (*Gyrosigma balticum* [GB]) to small naviculoids <10 µm (*SN*). Twenty species are visible but more would be discerned after preparation of light microscopy. *Bar*=100 µm. Ambient SEM of a common benthic foram found among the microphytobenthos. Forams are often cyptic and overlooked in analysis. *Bar*=100 µm. Images were kindly supplied by E. Defew (**D**), M. Consalvey (**E**) and H. Austin (**F**)

the usefulness of the species concept for diatoms (Round et al. 1990; Mann 1999). Another major group are the cyanobacteria (Stal 1995) which are prokaryotic and members of the Eubacteria. Cyanobacteria are represented by far fewer species than the diatoms but provide considerable taxonomic difficulty due to their phenotypic plasticity (Palinska et al. 1996). Euglenoids are often a significant component of microphytobenthic assemblages (Paterson et al. 1998) and form a bright green carpet on the sediment surface when dominant (Fig. 5.1). In addition to the dominant forms, representatives of a variety of other protistan groups are found including dinoflagellates and flagellates (Patterson et al. 1989). The trophic position of many taxa is not always clear, for example, dinoflagellates may be photosynthetic, mixotrophic or heterotrophic depending on the species. Other mixotrophic taxa are more cryptic or rare so that they are often overlooked, as is their contribution to the productivity of the system. This group includes various forms of benthic foraminifera (Fig. 5.1) represented by both testate and agglutinating forms. The importance of mixotrophs to the general ecology of the system is poorly understood and they may well contribute significantly to ecosystem turnover through grazing and recycling of material at a microbial level. Ecological studies of benthic forms such as foraminifera are few and far between (cf. Moodley 1992) and it is clear that, even considering only the eukaryotic microorganisms that inhabit sediments, our current knowledge is patchy. The reason that we understand so little about the real nature and complexity of sediment systems can be explained by a number of factors:

1. Small forms, meiofauna and microphytobenthos, often dominate,
2. Many organisms are delicate and are lost or damaged on sampling,
3. The protista that dominate are difficult to identify, and
4. Techniques that measure at an appropriate scale for the determination of sediment properties have only recently been developed.

In addition, the diversity of form and the importance of the physical environment mean that an interdisciplinary effort is required to investigate the system.

5.3 Physical and Biological Sediment Properties

5.3.1 Sediment Types and Stability

Sediments are commonly described by reference to the Wentworth scale, which divides sediments into size classes (e.g., Allen 1984). Sandy sediments are described as comprising grains greater than 62.5 μm in diameter and are generally spherical, showing no significant inter-particle attraction (cohesive behaviour). Smaller sediments (below 62.5 μm) become increasingly more elongated or flat and have a greater surface-to-volume ratio. The consequence of this is that the charge distribution over the surface of the particle can cause significant attraction (cohesive behaviour) through Van der Waals forces. Thus cohesive particles do not act independently while non-cohesive sandy sediments are made up of particles behaving as individual units. Thus, the erosional behaviour of sandy sediment has been empirically determined and can now be mathematically described (Soulsby 1997).

$$\theta cr = 0.30/(1+1.2 D_*) + 0.055[1-\exp(-0.02 D_*)]$$

where θcr is the dimensionless threshold Shields parameter and D_* is the dimensionless grain size.

The understanding of cohesive sediment erosion is more complex and even the description and classification of the cohesive erosion process is a matter of debate (Amos et al. 1997; Paterson and Black 2000; Tolhurst et al. 2000). It is clear, however, that no single mathematical formulation exists to describe cohesive sediment erosion and that the modelling of cohesive sediment behaviour is much more complex (Parker 1997; Teisson 1997). However, much of the sedimentological discussion over equations and processes still neglects the biology of natural systems although this is changing rapidly. The cohesion, and therefore erosion threshold, of clay minerals is affected by the organic content of the sediment (Blanchard et al. 2000) and much of this organic material is produced by the activity of microphytobenthos. The organic material secreted by microbes is given the generic term "Extracellular Polymeric Substances" (EPS; Underwood et al. 1995; De Winder et al. 1999) and there is a strong relationship between microphytobenthos biomass and operational fractions of the organic material (Underwood and Smith 1998). This influence is well known for cohesive sediment but less so for sand. However, under field conditions, various forms of microbial development prevent sand grains behaving as individual particles and instead they clearly act in a cohesive manner (Yallop et al. 1994). The binding of sand particles by cyanobacteria is well known, while the influence of EPS has been shown in laboratory (Dade et al. 1990) and field experiments (Yallop et al. 1994). Thus, the idea of defining cohesion on the basis of particle size is limited to the sedimentological behaviour of the particles devoid of organic materials. In nature, cohesive or non-cohesive behaviour is dependent on the extent of microbial development and varies in response to biological patchiness, zonation and seasonality.

5.3.2 Physical Dynamics

Broadly, the nature of the sediment reflects the hydrodynamic forces that control sediment deposition and erosion. High-energy conditions prevent fine sediments being deposited and therefore coarse sediments predominate while, under low-energy conditions, only fine sediments are carried to the site and deposited. However, the forces of tides and winds behave in an episodic manner bringing in sediments of varied nature depending on the conditions. Under these circumstances, mixed flats can develop where both fine particles and coarse sediment are deposited resulting in poorly sorted sediments (Yallop et al. 1994). In fact, most natural intertidal sediments consist of a mixture of various sediment types. Biological processes have a role in developing and maintaining intermediate sediment beds in two ways. Firstly, biogenic products may add to the sediment matrix, for example, shell fragments, skeletal components (spicules), and cell walls (diatom frustules). In some situations, the sediment matrix is entirely organic in origin (e.g., coral sands and diatom ooze). Secondly, organic secretions add to the cohesion of the sediments (Sutherland et al. 1998a; Paterson and Black 1999). Thus, fine sediments can be trapped and retained despite current velocities that would normally lead to their resuspension (Faas et al. 1992; Underwood and Paterson 1993).

The sediment bed responds to hydrodynamic forcing and the flow over sediments creates a boundary layer (Paterson and Black 1999). Flow within the boundary layer over the bed can be described as laminar, smooth turbulent or rough turbulent (Fig. 5.2; Paterson and Black 1999). Laminar flow is rare under natural conditions in coastal waters (Brown et al. 1999); therefore, smooth or rough turbulent flow predominates. The transition between smooth and rough turbulent conditions is influenced by the velocity of the free stream flow and by the roughness of the bed. For smooth turbulent conditions, a small region near the bed experiences laminar flow (the viscous sub-layer); this layer breaks down on transition to rough turbulent flow. For a given flow, a smooth surface may help to maintain smooth turbulent flow while a rough surface promotes a transition to rough turbulent conditions (Fig. 5.2; Vogel 1994).

Organisms also influence flow (Paterson 1997) and, even at a microbial level, the effects are significant (Grant and Gust 1987; Dade et al. 1990). The interactions are complex; it is not simply that the bed resists flow more strongly because of the organic secretion or network effects, but also that the stress experienced on the bed as a result of water flow varies depending on the nature of the bed (Fig. 5.2). A rough bed will experience greater stress for the same overall flow than a smooth bed. This has important consequences for biota at the sediment surface. Under rough turbulent conditions, turbulent eddies impact the bed and the likelihood of sediment (or organism!) erosion is enhanced. No viscous sub-layer can be retained and the flux of material is

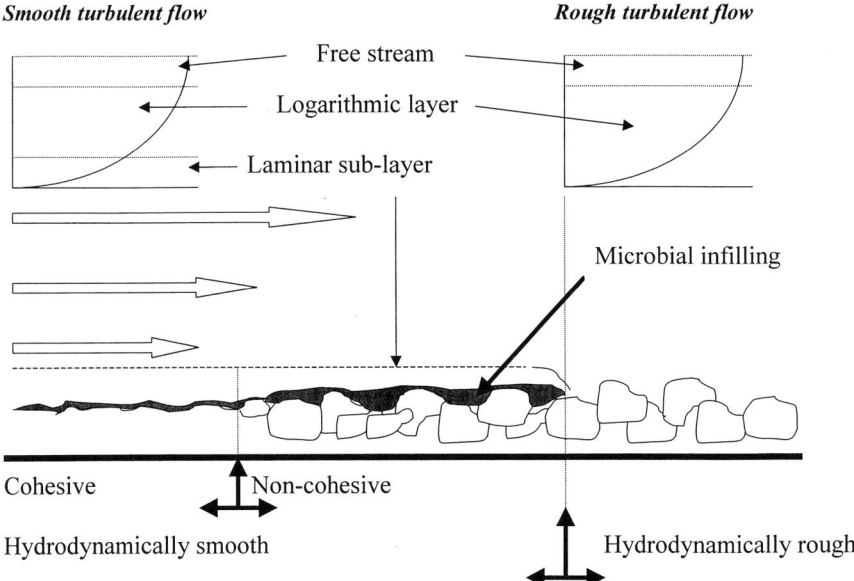

Fig. 5.2. Variation in flow condition above the bed. The infilling of spaces between particles reduces the "roughness" of the bed and can lead to smooth turbulent flow being retained. Over a rough bed, the transition to rough turbulent conditions occurs at a lower free stream velocity

dependent on the eddy diffusion coefficient rather than on the molecular diffusion coefficient, which applies within the viscous sub-layer. The viscous sub-layer therefore acts as a hydrodynamic and molecular buffer zone between the bed and the flow. This can significantly influence the flux of material (e.g., essential nutrients) to and from the bed. Vogel (1994) gives the example of sucrose in water, which has a molecular diffusion coefficient of 5×10^{-6} cm^2 s^{-1} while, under flow sufficient to induce rough turbulent conditions, eddy diffusion is 20 million times greater. Thus, under rough turbulent conditions, organisms must endure more stress and lose materials rapidly. This may be an advantage in terms of nutrient flux and waste removal but has a considerable cost in terms of energy and adaptation.

Roughness of a flat bed is related to the particle size and packing. The large particles in pure sandy sediment do not pack well and void spaces are left between particles that can be air- or water-filled. This is expressed as the porosity of the sediment (volume of voids/total volume occupied). Porosity varies with sediment packing and declines with sediment compaction. The character of the pore space is also important, large voids between particles allow for more efficient transport of fluid through the sediment whereas small

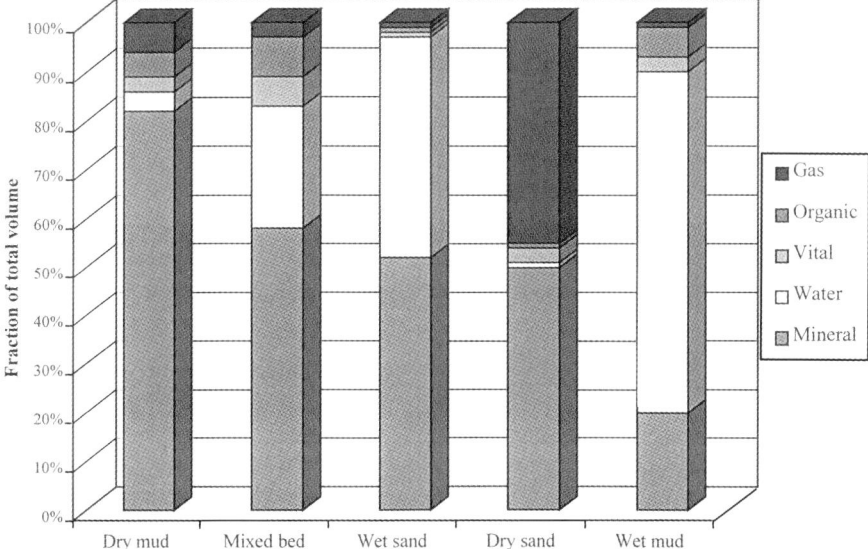

Fig. 5.3. Phase diagram of the composition of natural sediments. Representative sediment have been partitioned by fraction of total volume for five phases (Mineral, Water, Vital, Organic and Gas) and ranked (*left to right*) by mineral content. The variation in the proportional representation of phases between saturated and exposed sediments is clear. Intertidal sediments undergo such phase shifts during each exposure. Values above are only representative, not absolute (sources include Leeder 1982; Allen 1984; Yallop et al. 1994)

voids create resistance to flow. Thus, while porosity can be similar for different sediments, sediment permeability varies greatly for a given porosity.

Porosity and permeability are affected by sediment-dwelling biota. While many sedimentologists may consider anything that is not a particle to be a void, this is not actually the case in natural sediments. For example, calculations of porosity based on water content and derived by drying sediments exclude the potential for bound water (organic material) to fill void space. Cohesive sediments can be associated with a high organic content (Fig. 5.3) that includes extracellular polymeric substances (EPS) produced by microbes and infauna. EPS can have a variety of forms and properties (Decho 1994) but, despite being highly hydrated, EPS can retard the flow of solutes through the sediment reducing the permeability of the matrix. In extreme cases, structures known as blister mats develop where EPS forms a continuous film at the sediment surface. This surface becomes blistered as gas accumulates in the matrix (Yallop et al. 1994). Permeability approaches zero yet the porosity of the bed is unaffected. A subtidal equivalent has been found in biofilms that, although inherently buoyant due to gas accumulation, remain attached to the bed and stabilise the sediment (Sutherland et al. 1998a).

The influence of bed particles on the flow becomes critical when they are sufficiently large to penetrate the viscous sub-layer. If particles protrude into less than 1/3 of the sub-layer then the flow is unaffected (Brown et al. 1999). Particles protruding beyond this interfere with flow and the viscous sub-layer breaks down in the transition into rough turbulent flow. The size of the viscous sub-layer varies with flow but is usually less than 1 mm. For example, at a flow of 1 m s^{-1}, the laminar sub-layer is about 60 microns in depth. Thus, in theory, particles >20 µm can influence the transition between smooth and rough turbulent flow. This is over-simplified, however, since the shape of the particles is important. Plate-like or needle-shaped particles may orient in such as way that no obstruction to flow is caused (cohesive beds); sand grains are more spherical (equi-dimensional) and have the potential to influence flow transition. Under conditions where organisms infill the gaps between larger particles (by mesh work or polymer gels) smooth turbulent flow may be retained while, alternatively, organisms which create micro-relief features that penetrate the viscous sub-layer may enhance the transition to rough turbulent conditions.

5.4 Redefining Intertidal Sediments – The Five Phases of Depositional Environments

There are many classic texts which describe the physical nature of depositional systems, both mud and sand (e.g. Allen 1984), from a sedimentological approach. There are fewer sources detailing the nature of the microbiology of the sediments at more than a basic level, although this is beginning to change (Jickells and Rae 1997; Riding and Awramik 2000). However, there is still a tendency to separate the biological and physical components of the system, which may inhibit our understanding of natural processes in depositional environments. The natural deposition system can be consider to have five phases:

1. The mineral phase (sediment particles)
2. The vital phase (the organisms)
3. Non-living organic phase (products of secretion, detritus)
4. Free aqueous phase
5. Gas phase

The separation into five phases is, of course, arguable but serves as a device to understand the system complexity and to ensure the inclusion of living organisms in a conceptual framework (Fig. 5.3). Organic material can be considered as part of living biota but there are reasons to consider them separately. The influence of organisms can be "felt" long after their death or relocation since the structures and secretions they have made may remain. This is not confined to macrofauna but is also relevant to the microbiota whose most famous relics are the laminated microbial mounds or stromatolites found in the fossil record (Riding and Awramik 2000). Far more subtle remnants include polymeric material (Taylor et al. 1999), enzymes (Chrost 1991) and cellular remains. These phases do not only exist in isolation but may overlap allowing for material transformations and transfers to occur between phases. For example, water can contribute to three phases: as structural water with mineral particles (tightly bound and unavailable for chemical or biological transformation); as free water; and as biologically bound water where the molecules are associated with organic molecules such as cell walls, membranes or with a wide variety of polymers (Decho 1990). The balance between phases varies considerably between sediment types (Fig. 5.3).

Moreover, the biological influence on sediment characteristics and behaviour varies in space and time. The development of microbial assemblages is patchy (Saburova et al. 1995; Serôdio et al. 1997; Paterson et al. 1998) but changes can be rapid with biofilms developing in days (Underwood and Paterson 1993) with concomitant effects on sediment characteristics. The rapid growth rate and ability to exploit conditions is a feature of the protista and this has traditionally been exploited to examine ecological theory (Gause 1934; Petchey et al. 1999). Thus, the sediment matrix is a rapidly changing and heterogeneous environment made more variable by the activity of the organisms themselves (Decho 2000).

5.5 Comparative Biodiversity

Although no direct comparison has been made between the diversity of microphytobenthos on intertidal shores, assemblages on a sandy substratum are usually much more sparse and distributed over a much greater depth than for cohesive sediment (Round 1979; Yallop and Paterson 1994). The spatial distribution of microalgae on the mesoscale depends mostly on the granulometric composition of sediments and on factors influencing particle size and density (Oppenheim 1988; Saburova et al. 1995; Zong and Horton 1999). The life style of the microphytobenthos varies between two extremes: cells are attached to the grains or move through the interstitial spaces. The attached forms are termed "epipsammon" while the motile forms are termed "epipelon" (Round 1981). While few epipsammic forms occur in cohesive sediment because small grains provide limited surface area for attachment, both forms occur in mixed or sandy sediment, although epipelic forms are less common as the energy intensifies.

5.5.1 Non-cohesive Sediments

In regions of high hydrodynamic stress, benthic assemblages consist of organisms capable of firmly attaching to sediment particles and resistant to abrasion. Extreme environmental conditions can physically remove and damage epipsammic algae (Delgado et al. 1991). Under very extreme conditions few sediment grains may be colonised at all. Typical assemblages consist of epipsammic monoraphid diatoms, such as *Cocconeis* and *Achnanthes* (Round 1979; Yallop and Paterson 1994; Yallop et al. 1994), euglenoids (Kingston 1999), and coccoid cyanobacteria, such as *Merismopedia* (Wachendörfer et al. 1994; Noffke and Krumbein 1999). Attachment to sediment particles allows these microalgae to survive the hazards of physical transport, mainly particle collisions and burial. The density and composition of epipsammic algal assemblages are influenced by physical exposure and the duration, frequency, and timing of disturbance events that directly mediate key variables, such as mean grain size (Oh and Koh 1995) and organic content (Peletier 1996). The diversity of the microbial assemblages tends to increase as the magnitude of disturbance decreases. This is a statement of "The Intermediate Disturbance Hypothesis" (IDH, Connell 1978). Thus, the composition and structure of epipsammic assemblages varies depending on the frequency and extent of disturbance. This can be considered as a "disturbance continuum" which influences system diversity through the capabilities of the individual taxa in relation to their ability to adhere to particles (Wetherbee et al. 1998), their motility and rapidity of colonisation (Noffke and Krumbein 1999). At lower magnitudes and frequencies of disturbance, benthic diversity and biomass increase. Monoraphid diatom cell density can increase and araphid diatoms (e.g. *Fragilaria, Raphoneis*) appear. In regions of moderate hydrodynamic stress, Noffke and Krumbein (1999) found that a filamentous cyanobacterium, *Oscillatoria limosa*, could stabilise sandy surfaces because of its high mobility, rapid colonisation and binding efficiency.

As the effects of disturbance become even less, diversity may be affected in three ways. Firstly, for epipelic assemblages, diversity decreases as interspecific competition increases. Secondly, epipelic diatoms (e.g., *Navicula* and *Gyrosigma*), more familiar from muddy assemblages, colonise the non-cohesive sediments. This combination of epipsammic and epipelic forms further increases diversity. Examples of such mixed assemblages are reported in the literature. For example, epipsammic taxa dominated sediments in the Westerschelde Estuary and accounted for 90% of the cells counted from 74 species (Sabbe 1993). Yet, in terms of diversity, only 32 of the 74 taxa were true epipsammic forms belonging to the genera *Achnanthes, Amphora, Catenula, Cocconeis, Fragilaria*, and *Opephora*. Thus, although epipelic diatoms contributed very little to the overall biomass (10%) in the Westerschelde Estuary, epipelic species (primarily *Navicula*) accounted for the remaining 42 species and greatly increased the diversity of the assemblage. Thirdly, dense mats of cyanobacteria often cover non-cohesive sediments (Wachendörfer et al. 1994; Yallop and Paterson 1994; Yallop et al. 1994; Noffke and Krumbein 1999). Complex mixtures of cyanobacteria and diatoms have been described for both freshly colonised sediments and mature mat systems (Riege and Villbrandt 1994; Yallop and Paterson 1994; Yallop et al. 1994). Typically, these mats are dominated by filamentous cyanobacteria such as *Microcoleus chthonoplastes*, *Oscillatoria* spp., *Spirulina* sp., *Phormidium* sp. and the coccoids *Gloecapsa* sp., and *Merismopedia* sp.

The microscale distribution of particles of varying sizes can also directly influence diversity. Saburova et al. (1995) suggested that the spatial distribution of microalgae on the mesoscale depends mostly on the composition of sediments (and on factors influencing particle size and density). They found that the maximum numbers of microalgae coincided with the crests of sandy mounds presumably due to sediment texture, as the depressions had a higher percentage of silt. The distribution and abundance of crevices in particles may also influence microphytobenthic diversity (Bergey 1999). Small crevices contain young stages or small taxa. Moderate crevices are effective refugia and sustain greater diversity. Very large crevices do not enhance diversity because there is less protection from disturbances (e.g., predators and particle collisions). Other environmental variables can also influence diversity on non-cohesive sediments. Growth rates for cyanobacteria were higher on coarse sand (>63 µm) at 15°C, whereas *Nitzschia* sp. was the competitive dominant for all sediment types at 10°C. At 25°C, the filamentous cyanobacterium *Microcoleus chthonoplastes* dominated (Watermann et al. 1999).

5.5.2 Cohesive Sediments

Hydrodynamic stress is less influential in determining diversity on cohesive sediments. However, while surveys of microphytobenthic assemblages are common (e.g., Oppenheim 1991; Underwood et al. 1998), few studies have directly compared diversity between non-cohesive and cohesive sediments or reported data on sediment characteristics (cf. Underwood 1994; Yallop et al. 1994). Whereas epipsammic algae dominate non-cohesive sediments, epipelic algae dominate cohesive sediments. Epipelic diatoms have adaptations to cope with the heterogeneity in physicochemical conditions (Admiraal 1984) and appear to be more competitive on finer grained sediments. For example, diatoms *Phaeodactylum tricornutum* and *Nitzschia* sp. have higher growth rates at 10°C and 15°C on mixed sediment (50:50 fine sand [<63 µm]/mud) than on fine sand alone and under these conditions outcompete cyanobacteria (Watermann et al. 1999).

In a study of the epipelic diatoms of the Severn Estuary, UK, Underwood (1994) found that diversity on the low shore was higher than mid and high shore but did not differ between the three sampling sites (Severn Estuary: Sand Bay; Portishead; and Aust). On average, Shannon index values for each site and number of taxa were less than 1.78 and 16, respectively. However, the taxonomic composition for the mid-shore site was different from the low and high shore. Sediment shear strength, bulk carbohydrate, and colloidal carbohydrate were all greater for the mid-shore site than for the low- and high-shore sites. *Navicula phyllepta, Cylindrotheca signata, Gyrosigma limosum* (formerly *spencerii*), *Entomoneis paludosa*, and *Nitzschia epithemioides* were positively correlated with sediment bulk carbohydrate and colloidal carbohydrate and had higher relative abundances for the mid-shore site, suggesting that EPS production was higher for these taxa and directly increased sediment stability. The mid-shore site also had high relative abundance of euglenoids in mid summer. Taxa associated with both the high- and low-shore sites, *Navicula pargemina, Navicula flanatica, Raphoneis minutissima, Coscinodiscus* sp. 1, and *Cymatosira belgica*, were negatively correlated with colloidal carbohydrate indicating that these species do not affect sediment stability as much.

5.5.3 Niche Diversity

The diversity of microphytobenthos therefore varies between sandy and muddy sites. Maximum diversity appears to occur at the interface between clean sands and pure mud where epipsammic and epipelic forms can coexist. This may reflect the availability of sites and resources for the cells and the variety of niches available. The interfaces between the sediment and overlying media are areas of intense vertical gradients (Stal 1995; Taylor and Paterson 1998; Paterson and Black 2000). These gradients are created and maintained by the activity of organisms at the sediment surface (Jickells and Rae 1997). The severity of the gradients is related to the nature of the sediment (porosity, permeability) and the intensity of the biological activity. In general terms, gradients are more severe where biomass is high and where sediment particles are small. Thus, cohesive sediments of reasonable nutrient status maintain the sharpest gradients of change. This implies very restricted spatial niche separation along the gradients in cohesive sediments. Thus, within the aerobic region, spatial zonation cannot be established easily but there is some evidence indicating that temporal displacement occurs. Early observations of diatom migration noted that cells came to the surface and were replaced by different forms at various stages of the tidal cycle (Round 1981). This suggestion has been given support by electron microscopy of intertidal sediments showing euglenids replacing diatoms at the surface (Paterson et al.

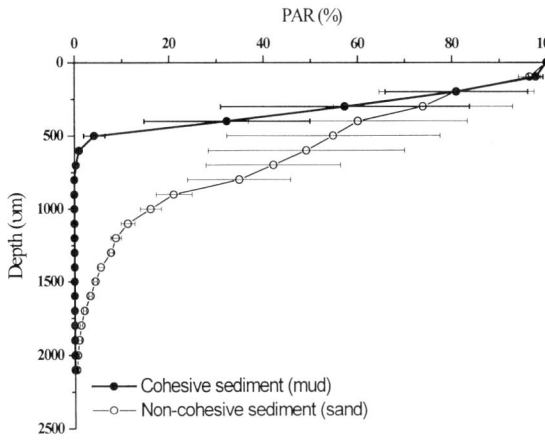

Fig. 5.4. Depth of photosynthetic active radiation (*PAR*) penetration in cohesive and non-cohesive intertidal sediments ($n=3$, *bars*=SE)

1998). This mechanism of migration between levels has also been shown for the sulphur bacteria *Thioploca* spp. (Huettel et al. 1996) as a mechanism for overcoming diffusion limitations to cell metabolism.

A similar situation may be in operation for autotrophs where light, carbon and nutrient availability must be balanced. This may provide the evolutionary drive that has enhanced the motility mechanism in microphytobenthos. Light availability is central to niche separation and light penetrates non-cohesive sediments to greater depth (Fig. 5.4) than for cohesive sediments where it is often attenuated within 1 mm or less (Paterson et al. 1998). Thus, motile species have a competitive advantage because they can exploit variation in the physicochemical environment with depth (Admiraal 1984). Vertical migration occurs among biraphid diatoms (Hay et al. 1993; Paterson et al. 1998), euglenoids (Kingston 1999) and filamentous cyanobacteria (Noffke and Krumbein 1999). The diatom *Hantzschia virgata* var. *intermedia* migrated from just above the anaerobic zone (depth ~3 mm) to the surface (Round 1981), while several motile diatoms, *Navicula cancellata*, *Nitzschia spathulata*, and *Amphora cymbifera*, were found living 4 mm below the surface but did not migrate.

Where light penetrates the sediment (non-cohesive), organisms may be found in discrete zones along a depth profile. This vertical separation increases microenvironmental variability and the number of available niches and therefore the potential diversity of non-cohesive assemblages. These stratified assemblages reach their maximum expression in the versicolour sand or "Farbenstreifen-Sandwattes" (Gerdes and Krumbein 1987), as exemplified by well-known sites on the German island of Mellum (Gerdes et al. 1987). Thus, muddy sites support intense gradients and are spatially restricted but have the advantage that nutrients are sequestered onto fine particles and are available for epipelic algae. Sandy sites are inherently dynamic and physical stress is high while nutrient levels in the sediments tend to be low leading to low diversity. The mixed assemblage site combines advantages from both systems (nutrients, attachment sites on particles, light penetration) and supports high diversity.

5.6 Sediment Stability

From the above discussion it is clear that grain size has a fundamental role to play in the nature of the microphytobenthos assemblages that colonise sediments. However, it is now recognised that the activity of organisms also influences the nature, behaviour and distribution of sediments (Noffke and Krumbein 1999; Blanchard et al. 2000). The biota stabilising non-cohesive sediments consists of bacteria (Grant and Gust 1987; Dade et al. 1990), algae (Neumann et al. 1970), diatoms (Holland et al. 1974; Vos et al. 1988; Madsen et al. 1993), and cyanobacteria (Neumann et al. 1970; Grant and Gust 1987; Madsen et al. 1993; Noffke and Krumbein 1999). The most pervasive effect is probably through the secretion of organic substances (extracellular polymeric substances, EPS) present in the sediment matrix (Paterson 1997). Whilst biogenic effects cannot protect sediments against extreme events, the influence of biogenic mediation on the day-to-day fluxes of material from the bed can be highly significant (Tolhurst et al. 2000). Most stability studies have concentrated on the carbohydrate (bulk or colloidal) fraction of the EPS found in sediments (Underwood et al. 1995; De Winder et al. 1999; Yallop et al. 2000). Some discrimination between species is now possible. *Gyrosigma limosum* and *Nitzschia sigma* were associated with high total carbohydrate concentrations in the Colne Point saltmarsh, Essex, UK (Underwood et al. 1998). In contrast, for the diatoms *Bacillaria paxillifer*, *Achnanthes longipes*, and *Navicula pelliculosa*, mucilage production was low and did not enhance sediment stability (Holland et al. 1974). Smith and Underwood (1998) found that the epipelic diatoms *Cylindrotheca closterium*, *Navicula perminuta*, and *Nitzschia sigma* produced more EPS when migrating. These studies suggest that the taxonomic composition of epipelic diatom assemblages may determine the stability of cohesive sediments. This possibility is supported by the recent work of Riethmuller et al. (2000) using remote sensing of algal distribution linked to ground erosion measurements (Riethmuller et al. 1998). Assemblages containing species capable of high EPS production reduce sediment erosion more than assemblages producing less EPS. Epipelic forms seem to be better stabilisers than epipsammic forms. Thus, biogenic stabilisation via organic secretions increases as the proportion of epipelic forms increases.

5.7 Conceptual Model

The stability of intertidal sediments is a function both of hydrodynamics and biology (Fig. 5.5). Here we present a conceptual model that provides an overview of the complex interactions that regulate microphytobenthos assemblage structure and consequently the biostabilisation of intertidal sediments. Along the continuum from non-cohesive to cohesive sediments, the importance of hydrodynamic processes declines. For non-cohesive sediments, the frequency and magnitude of disturbance dictates diversity and limits sediment biostabilisation by regulating the structure of microphytobenthic assemblages (IDH). However, as the frequency and magnitude of disturbance declines toward the cohesive/non-cohesive boundary, the diversity of the system depends on the colonisation pattern. The ratio of epipsammic to epipelic algae decreases as cohesive sediment accumulates in response to lower hydrodynamic stress. Sediment stability for these mixed assemblages and sediments is expected to be relatively high because of EPS production by epipelic diatoms and the formation of mats and biofilms (Yallop et al. 1994).

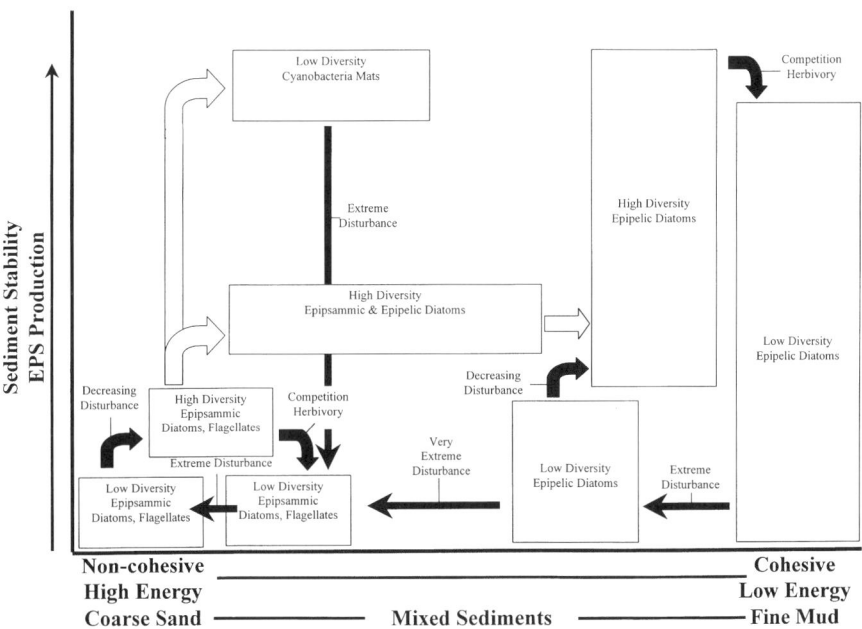

Fig. 5.5. Conceptual model of microphytobenthic assemblages associated with sediment type and sediment stability. *Black arrows* represent the physical and biological processes that can influence microphytobenthic diversity and sediment stability. *White arrows* represent alternative assemblages that may develop as a result of local processes

However, where cyanobacteria become dominant, they not only dramatically affect sediment stability but also reduce diversity since the mats tend to be dominated by few species. Thus, beginning with a coarse high-energy sandy system, diversity is expected to increase as sediment characteristics change (non-cohesive >> mixed flat) then decline again (mixed flat >> cohesive) as sediment grain size decreases.

Superimposed on the relationships between microphytobenthic diversity, EPS production, sediment stability, and sediment characteristics are three important factors. First, microphytobenthic production is dependent on nutrients. The supply rate of resources (e.g., nitrogen and phosphorus) affects microphytobenthic biomass and species composition (Tilman 1982; Grover 1997). Here low nutrient supply rates and interspecific competition limit diversity. Second, sediment dwelling fauna can affect sediment stability in three ways:

1. Regulation of microphytobenthic biomass and species composition through direct consumption,
2. Bioturbation of the sediments, and
3. Secretion of EPS or mucilage

In natural environments, nutrient supply rates and herbivory work in concert to structure microphytobenthic algal assemblages. Whereas grazers increase diversity and perhaps decrease sediment stability under nutrient-limited conditions, it is unclear what influence grazers have on microbial assemblages at intermediate and high resource supply rates (Proulx and Mazumder 1998). Third, the development of a microbial mat establishes a feedback mechanism by which EPS traps finer particles, thus changing the nature of the sediment and potentially the diversity (Daborn et al. 1993; Underwood and Paterson 1993).

The complex interactions between the physical and biological properties that regulate and maintain microbial assemblages are just beginning to be understood. However, the primary factors controlling diversity are not well known and areas requiring study are the influence of mixtures of sediment types; successional changes, interaction between nutrients and grazing pressure; biostabilisation, and bioturbation. The nature of trophic interactions and how infauna affect microphytobenthic structure and function are particularly interesting in terms of the ecosystem dynamics, Although these have still to be elucidated, some initial studies have described trophic cascades with direct physical effects (Daborn et al. 1993). While a great deal of work is required in these areas, conversely, intertidal sediments may provide a valuable model and an experimental system in which to investigate ecological theory (see Chap. 16).

5.8 Conclusions

Hydrodynamic forces regulate the distribution of sediments within coastal depositional ecosystems and hence the structure and function of microphytobenthic assemblages. At the extremes, non-cohesive sediments are typically associated with high-energy environments, and microphytobenthos is sparse and dominated by microbes that attach directly to individual grains. In contrast, motile microbes dominate the cohesive sediments of low-disturbance environments. However, microphytobenthos are not passive inhabitants and can directly affect sediment stability through the production and secretion of organic polymers, known as extracellular polymeric substances (EPS), and network formation (cf. cyanobacteria). The biostabilisation of sediments thus serves as a feedback mechanism regulating sediment erosion and sediment characteristics. The degree of biostabilisation also differs dramatically between non-cohesive and cohesive sediments. In this chapter we have outlined some of the physical and biological differences between non-cohesive and cohesive sediments. The physical differences include the hydrodynamic processes that control sediment distribution and erosion. The biological differences include adaptation by the microphytobenthos, diversity, and the complex factors that regulate sediment biostabilisation and assemblage structure.

Acknowledgements. This review was supported by contributions from the EU through the BIOPTIS (MAS3-CT97–0158) and CLIMEROD (MAS3-CT98-0166) programmes and from work and discussions with Dr. T. Tolhurst, Heather Austin (University of St. Andrews), Emma Defew, Catherine Biles (NERC GR3/12370) and Mireille Consalvey (NERC GR3/11782). ED, CB and MC also supplied negatives for Fig. 5.1. Irvine Davidson completed the SEM studies. This help is gratefully acknowledged.

References

Admiraal W (1984) The ecology of estuarine sediment-inhabiting diatoms. In: Round FE, Chapman DJ (eds) Progress in phycological research, vol 3. Biopress, Bristol, UK pp 269–322

Allen JRL (1984) Sedimentary structures: their character and physical basis. Elsevier, Oxford

Amos CL, Feeney T, Sutherland TF, Luternauer JL (1997) The stability and erodibility of fine-grained sediments from the Fraser River delta foreshore and upper foreslope. Est Coastal Shelf Sci 45:507–524

Bergey E (1999) Crevices as refugia for stream diatoms: effect of crevice size on abraded substrates. Limnol Oceanogr 44:1522–1529

Blanchard GF, Paterson DM, Stal L, Richard P, Galois R, Huet V, Kelly J, Honeywill C, de Brouwer J, Dyer K, Christie M, Seguignes M (2000) The effect of geomorphological structures on potential biostabilisation by microphytobenthos on intertidal mudflats. Continental Shelf Res (special issue) 20(10/11):1243–1256

Brown E, Colling A, Park D, Phillips J, Rothery D, Wright J (1999) Waves, tides and shallow-water processes, 2nd edn. Open Univ, Milton Keynes

Chrost RJ (1991) Microbial enzymes in aquatic environments. Springer, Berlin Heidelberg New York

Connell JH (1978) Diversity in tropical rainforests and coral reefs. Science 199: 1302–1310

Daborn GR, Amos CL, Berlinsky M, Christian H, Drapeau G, Faas RW, Grant J, Long B, Paterson DM, Perillo GME, Piccolo MC (1993) An ecological "cascade" effect: migratory birds affect stability of intertidal sediments. Limnol Oceanogr 38:225–231

Dade WB, Davies JD, Nichols PD, Nowell ARM, Thistle D, Trexler MB, White DC (1990) Effects of bacterial exopolymer adhesion on the entrainment of sand. Geomicrobiol J 8:1–16

Decho AW (1990) Microbial exopolymer secretions in ocean environments: their role(s) in food webs and marine processes. Oceanogr Mar Biol Annu Rev 28:73–153

Decho AW (1994) Molecular-scale events influencing the macro-scale cohesiveness of exopolymers. In: Krumbein WE, Paterson DM, Stal LJ (eds) Biostabilization of sediment. BIS-Verlag, Oldenburg, pp 135–148

Decho AW (2000) Exopolymer microdomains as structuring agents for heterogeneity within microbial biofilms. In: Riding RE, Awramik SM (eds) Microbial sediments. Springer, Berlin Heidelberg New York, pp 217–225

Delgado M, de Jonge VN, Peletier H (1991) Effect of sand movement on the growth of benthic diatoms. J Exp Mar Biol 145:221–231

De Winder B, Staats N, Stal LJ, Paterson DM (1999) Carbohydrate secretion by phototrophic communities in tidal sediments. J Sea Res 42:131–146

Faas RW, Christian HA, Daborn GR (1992) Biological control of mass properties of surficial sediments: an example from Starr's Point tidal flat, Minas Basin, Bay of Fundy. Nearshore and estuarine cohesive sediment dynamics. American Geophysical Union, vol 42. Springer, Berlin Heidelberg New York, pp 360–377

Gause GF (1934) Experimental demonstration of Volterra's periodic oscillation in numbers of animals. J Exp Biol 12:44–48

Gerdes G, Krumbein WE (1987) Biolaminated deposits. Lecture notes in earth sciences, vol 9. Springer, Berlin Heidelberg New York

Gerdes G, Krumbein WE, Reineck HE (1987) Mellum, Portrait einer Insel. Verlag Waldemar Kramer, Frankfurt/Main

Grant J, Gust G (1987) Prediction of coastal sediment stability from photopigment content from mats of purple sulphur bacteria. Nature 330:244–246

Grover JP (1997) Resource competition. Chapman Hall, London

Hay SI, Maitland TC, Paterson DM (1993) The speed of diatom migration through natural and artificial substrata. Diatom Res 8:371–384

Holland AF, Zingmark RG, Dean JM (1974) Quantitative evidence concerning the stabilization of sediments by marine benthic diatoms. Mar Biol 27:191–196

Huettel M, Forster S, Kloser S, Fossing H (1996) Vertical migration in the sediment-dwelling sulfur bacteria *Thioploca* spp. in overcoming diffusion limitations. App Environ Microbiol 62:1863–1872

Jickells TD, Rae JE (1997) Biogeochemistry of intertidal sediments. CUP, Cambridge

Kingston MB (1999) Effect of light on vertical migration and photosynthesis of *Euglena proxima* (Euglenophyta). J Phycol 35:245–253

Leeder MR (1982) Sedimentology: Process and product. Harper Collins, London
Madsen NP, Nillson P, Sundback K (1993) The influence of benthic microalgae and the stabilisation of a subtidal sediment. J Exp Mar Biol Ecol 170:159–178
Mann DG (1999) The species concept in diatoms. Phycologia 38:437–495
Moodley LM (1992) Experimental ecology of benthic foraminifera in soft sediments and its (paleo) environmental significance. PhD thesis, Vrijie universiteit te Amsterdam, Netherlands
Neumann AC, Gebelein CD, Scoffin TP (1970) The composition, structure, and erodibility of subtidal mats, Abaco, Bahamas. J Sed Petrol 40:274–297
Noffke N, Krumbein WE (1999) A quantitative approach to sedimentary surface structures contoured by the interplay of microbial colonization and physical dynamics. Sedimentology 46:417–426
Oh SH, Koh CH (1995) Distribution of diatoms in the surficial sediments of the Mangyung-Dongjin tidal flat, west coast Korea (Eastern Yellow Sea). Mar Biol 122:487–496
Oppenheim DR (1988) The distribution of epipelic diatoms along an intertidal shore in relation to principal physical gradients. Bot Mar 31:65–72
Oppenheim DR (1991) Seasonal changes in epipelic diatoms along an intertidal shore, Berrow Flats, Somerset. Mar Biol Assoc UK 71:579–596
Palinska KA, Liesack W, Rhiel E, Krumbein WE (1996) Phenotype variability of identical genotypes: the need for a combined approach in cyanobacterial taxonomy demonstrated on *Merismopedia*-like isolates. Arch Microbiol 166:224–233
Parker WR (1997) On the characterisation of cohesive sediments for transport modelling. In: Black KS, Paterson DM, Cramp A (eds) Sedimentary processes in the intertidal zone. Geological Society special publication 139. Geological Society, London, pp 3–14
Paterson DM (1997) Biological mediation of sediment erodibility: ecology and physical dynamics. In: Burt N, Parker R, Watts J (eds) Cohesive sediments. Wiley, Chichester, pp 215–229
Paterson DM, Black KS (1999) Water flow, sediment dynamics, and benthic biology. In: Raffaelii D, Nedwell D (eds) Advances in ecological research. Academic Press, London, pp 155–193
Paterson DM, Black KS (2000) Siliclastic intertidal microbial sediments. In: Riding RE, Awramik SM (eds) Microbial sediments. Springer, Berlin Heidelberg New York, pp 217–225
Paterson DM, Yates MG, Wiltshire KH, McGrorty S, Miles A, Eastwood JEA, Blackburn J, Davidson I (1998) Microbiological mediation of spectral reflectance from intertidal cohesive sediments. Limnol Oceanogr 43:1207–1221
Patterson DJ, Larsen J, Corliss JO (1989) The ecology of heterotrophic flagellates and ciliates living in marine sediments. Prog Protistol 3:185–277
Peletier H (1996) Long-term changes in intertidal estuarine diatom assemblages related to reduced input of organic waste. Mar Ecol Prog Ser 137:265–271
Petchey OL, McPhearson PT, Casey TM, Morin PJ (1999) Environmental warming alters food-web structure and ecosystem function. Nature 402:69–72
Proulx M, Mazumder A (1998) Reversal of grazing impact on plant species richness in nutrient-poor vs. nutrient-rich ecosystems. Ecology 79:2581–2592
Raffaelli D, Hawkins S (1996) Intertidal ecology. Chapman Hall, London
Riethmuller R, Hakvoort JHM, Heinke M, Heymann K, Khul H, Witte G (1998) Relating erosion shear stress to tidal flat surface colour. In: Black KS, Paterson DM, Cramp A (eds) Sedimentary processes in the intertidal zone. Geological Society special publication 139. Geological Society, London, pp 1–10

Riethmuller R, Heinke M, Kuhl H, Keuke-Rudiger R (2000) Chlorophyll *a* concentration as a potential index of sediment surface stabilisation by microphytobenthos. Cont Shelf Res (special issue) 20(10/11):1351–1372

Riding RE, Awramik SM (2000) Microbial sediments. Springer, Berlin Heidelberg New York

Riege H, Villbrandt M (1994) Norderney survey. In: Krumbein WE, Paterson DM, Stal LJ (eds) Biostabilization of sediment. BIS-Verlag, Oldenburg, pp 339–360

Round FE (1979) A diatom assemblage living below the surface of intertidal sand flats. Mar Biol 54:219–223

Round FE (1981) The ecology of the algae. CUP, Cambridge

Round FE, Crawford RM, Mann DG (1990) The diatoms. CUP, Cambridge

Sabbe K (1993) Short-term fluctuations in benthic diatom numbers on an intertidal sandflat in the Westerschelde Estuary (Zeeland, The Netherlands). Hydrobiologia 269/270:275–284

Saburova MA, Polikarpov IG, Burkovsky IV (1995) Spatial structure of an intertidal sandflat microphytobenthic community as related to different spatial scales. Mar Ecol Prog Ser 129:229–239

Serôdio J, da Silva JM, Catarino F (1997) Non-destructive tracing of migratory rhythms of intertidal benthic microalage using *in vivo* chlorophyll *a* fluorescence. J Phycol 33:542–553

Smith DJ, Underwood GJC (1998) Exopolymer production by intertidal epipelic diatoms. Limnol Oceanogr 43:1578–1591

Soulsby S (1997) Dynamics of marine sands. Thomas Telford, London

Stal LJ (1995) Physiological ecology of cyanobacteria in microbial mats and other communities. New Phytol 131:1–32

Sutherland TF, Amos CL, Grant J (1998a) The effect of buoyant biofilms on the erodibility of sublittoral sediments of a temperate microtidal estuary. Limnol Oceanogr 43:225–235

Sutherland TF, Grant J, Amos CL (1998b) The effect of carbohydrate production by the diatom *Nitzschia curvilineata* on the erodibility of sediment. Limnol Oceanogr 43:65–72

Taylor I, Paterson DM (1998) Microspatial variation in carbohydrate concentrations with depth in the upper millimetres of intertidal cohesive sediments. Est Coastal Shelf Sci 46:359–370

Taylor I, Paterson DM, Mehlert A (1999) The quantitative variability and monosaccharide composition of sediment carbohydrates associated with intertidal diatom assemblages. Biogeochemistry 45:303–327

Teisson C (1997) A review of cohesive sediment transport models. In: Black KS, Paterson DM, Cramp A (eds) Sedimentary processes in the intertidal zone. Geological Society special publication 139. Geological Society, London, pp 367–381

Tilman D (1982) Resource competition and community structure. Princeton Univ Press, Princeton

Tolhurst TJ, Riethmüller R, Paterson DM (2000) *In situ* versus laboratory analysis of sediment stability from intertidal mudflats. Continental Shelf Research (special issue) 20(10/11):1317–1334

Underwood GJC (1994) Seasonal and spatial variation in epipelic diatom assemblages in the Severn Estuary. Diatom Res 9:451–472

Underwood GJC, Paterson DM (1993) Recovery of intertidal benthic diatoms after biocide treatment and associated sediment dynamics. J Mar Biol Assoc UK 73:24–45

Underwood GJC, Smith DJ (1998) Predicting epipelic diatom exopolymer concentrations in intertidal sediments from sediment chlorophyll *a*. Microbiol Ecol 35:116-125

Underwood GJC, Paterson DM, Parkes RJ (1995) The measurement of microbial carbohydrate exoploymers from intertidal sediments. Limnol Oceanogr 40:1243-1253

Underwood GJC, Phillips J, Saunders K (1998) Distribution of estuarine benthic diatom species along salinity and nutrient gradients. Eur J Phycol 33:173-183

van den Hoek C, Mann DG, Jahns HM (1995) Algae: an introduction to phycology. CUP, Cambridge

Vogel S (1994) Life in moving fluids, 2nd edn. Academic Press, London

Vos PC, de Boer PL, Misdrop R (1988) Sediment stabilization by benthic diatoms in intertidal sandy shoals; qualitative and quantitative observations. In: de Boer PL, van Gelder A, Nio SD (eds) Tide-influenced sedimentary environments and facies. Reidel, The Netherlands, pp 511-526

Wachendörfer V, Riege H, Krumbein WE (1994) Parahistological sediment thin sections. In: Krumbein WE, Paterson DM, Stal LJ (eds) Biostabilization of sediment. BIS-Verlag. Oldenburg. pp 257-278

Watermann F, Hillebrand H, Gerdes G, Krumbein WE, Sommer U (1999) Competition between benthic cyanobacteria and diatoms as influenced by different grain sizes and temperatures. Mar Ecol Prog Ser 187:77-87

Wetherbee R, Lind JL, Burke J, Quatrano RS (1998) The first kiss: establishment and control of initial adhesion by raphid diatoms. J Phycol 34:9-15

Yallop ML, de Winder B, Paterson DM, Stal LJ (1994) Comparative structure, primary production, and biogenic stabilization of cohesive and non-cohesive marine sediments inhabited by microphytobenthos. Est Coastal Shelf Sci 39:565-582

Yallop ML, Paterson DM (1994) Seasonal field studies: survey of Severn Estuary. In: Krumbein WE, Paterson DM, Stal LJ (eds) Biostabilization of sediment. BIS-Verlag, Oldenburg, pp 280-325

Yallop ML, Paterson DM, Wellsbury P (2000) Interrelationships between rates of microbial production, exopolymer production, microbial biomass and sediment stability in biofilms of intertidal sediments. Microbial Ecol 39:116-127

Zong Y, Horton BJ (1999) Diatom zones across intertidal mudflats and coastal saltmarshes in Britain. Diatom Res 13(2):375-394

6 Sediment Dynamics by Bioturbating Organisms

G.C. CADÉE

6.1 Introduction

Judging from the few pages dealing with bioturbation in books on coastal sediment dynamics (e.g. Hails and Carr 1975; Dyer 1986), one might conclude bioturbation to be unimportant in sediment dynamics: waves and (tidal) currents have a predominant role in sediment transport and erosion. Is this conclusion wrong? Probably not: For the Lister Basin near Sylt, Bayerl et al. (1998) conclude that reworking of sediments is mainly physical, certainly in exposed parts hardly influenced by bioturbation. With an increase in current strength and/or wave exposure (energy in the environment), sediment transport will increase. This is not the case with bioturbation. In very high-energy environments, sediment transport is too high for benthic organisms to live, so bioturbation is also absent. Under absence of water movement, benthic organisms also will not thrive. If oxygen deficiency also occurs they may be even completely absent. In between these extremes of environmental energy, benthic organisms and, therefore, bioturbation play a role. Sediment transport by bioturbation is mainly vertical, within the sediment column, whereas currents and waves transport sediment along the bottom. However, both are linked, e.g. sediment brought to the surface by organisms may be taken up and dispersed by currents. We can visualise sediment transport in a very simple model (Fig. 6.1) in which energy is on the X-axis, sediment dynamics on the Y-axis. Sediment dynamics (sediment movement) due to physical forces increases continuously with increasing energy. Sediment dynamics by bioturbating organisms is absent at very low- and very high-energy levels and is important in between at medium-energy levels. It is at these medium-energy levels that bioturbation adds to sediment dynamics in coastal areas. In high-energy environments – those most interesting for students of coastal sediment dynamics – the contribution of bioturbation to sediment dynamics is relatively less important.

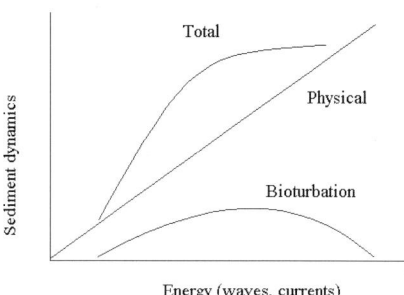

Fig. 6.1. Simplified model showing physical (mainly horizontal) and biological (mainly vertical) sediment reworking (both separate and combined=total) on the Y-axis plotted against energy (current-strength, wave exposure) on the X-axis. Bioturbation is mainly active at intermediate energy levels

Not all coastal areas, however, are high-energy environments. In an interesting paper, Reineck (1977) used the amount of bioturbation visible in cores of tidal flat sediments to estimate the relative energy level at stations distributed over a large tidal flat area in the German Wadden Sea. Mainly cross-bedded sediments without traces of bioturbation indicated the highest energy areas, exposed to higher waves and stronger tidal currents. Totally bioturbated sediments occurred in areas with relatively low-energy levels. Cores with the top centimetres cross-bedded, but the lower part bioturbated, indicated medium-energy levels: only the top of the sediment was regularly physically disturbed (tidal flats with ripple marks). Also in an earlier paper, Reineck (1976) used increase in bioturbation structures with water depth in a transect from 0 to 20 m depth off the island of Norderney, North Sea, to illustrate decrease of sediment transport with water depth.

Reineck (1967) produced a more complicated model than the one I present here, in which he relates bioturbation to both erosion and sedimentation (slightly modified in Reineck and Singh 1975). Schäfer (1962, 1963) nicely illustrated the biofacies (sedimentary structures and organic remains) that form under the different energy regimes. I have sorted his biofacies types in Fig. 6.2 according to the same energy axis of my model in Fig. 6.1, which Schäfer did not. Probably the awkward terminology he used, such as 'letal isostrate' and 'vital heterostrate', made his classification unpopular. 'Letal' stands for without benthos (the extremes left and right) as opposed to 'vital' (with benthos, middle part). The layering is called isostrate (in 1963 pantostrate) when it is complete and continuous (left side). With increasing energy, it changes to heterostrate (lipostrate in 1963), i.e. incomplete, non-continuous, with many erosional unconformities.

In this chapter I will briefly deal with some aspects of bioturbation and sediment dynamics based on a personal and eclectic choice of papers from the ever increasing bioturbation literature. I will provide some history indicating the prominent role the Wadden Sea played in early sedimentology and bioturbation research, thanks to, among others, Häntzschel, Hertweck, Linke, Reineck, Richter, Schäfer, Van Straaten, and Wohlenberg. I will indicate

Sediment Dynamics by Bioturbating Organisms

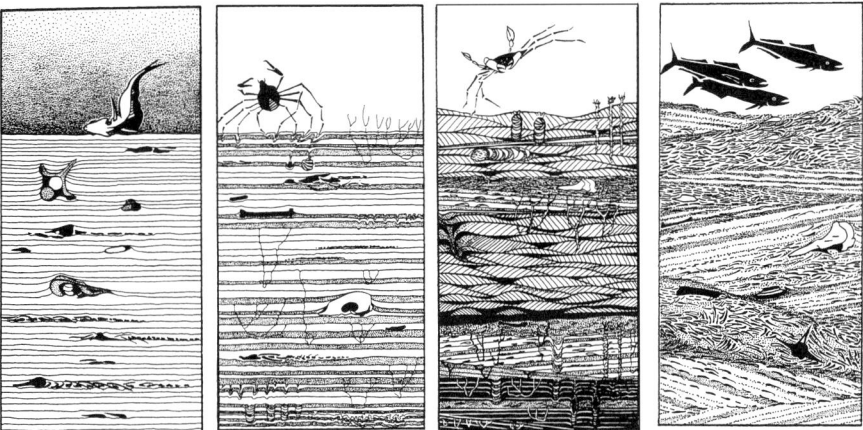

Fig. 6.2. Schäfer's (1962, 1963) biofacies types plotted along an energy axis as in Fig. 6.1. Bioturbation structures in the sediment are only preserved at intermediate energy levels

that bioturbation is more than sediment reworking by deposit feeders, as might be thought simply by reading, e.g., Levinton (1995) or Boudreau (1998). Finally, I will briefly deal with latitudinal variation, short-term seasonal variation, and changes in bioturbation in historical times. Measurements of bioturbation and the units in which they are presented in the literature are not uniform. For comparisons I will express data as the layer of sediment that bioturbators rework annually, making such data also comparable to sedimentation data. Keep in mind that 10 cm may imply 10 times reworking the top centimeter or one time reworking the top 10 cm per year.

Reviews have been produced in the past that include bioturbation, biodeposition and bioresuspension to which I refer for a more general introduction (e.g. Rhoads 1974; Lee and Swartz 1980; Rhoads and Boyer 1982; Meadows and Meadows 1991; Levinton 1995; Graf and Rosenberg 1997; Cadée 1998). Black et al. (1998) published a book on research related to LISP (Littoral Investigations of Sediment Properties) in which biota play a dominant role in many of the papers, indicating a changed attitude of the study of sediment dynamics in the 1990s, also visible in a recent overview by Paterson and Black (1999). The study of trace fossils shows many overlaps with bioturbation, has a long history, and is also an expanding field (see Abel 1935; Seilacher 1964; Frey 1975; Ekdale et al. 1984; Bromley 1990); it even has its own journal, *Ichnos* (started in 1990).

6.2 History of Bioturbation Research

The word bioturbation was first used by Richter (1952), but studies in sediment reworking by organisms had started much earlier. Darwin was the first. He showed a lifelong interest in bioturbation by earthworms. His first paper on this subject (Darwin 1837) was inspired by his uncle and future father-in-law, Josiah Wedgwood, who suggested that the fact that fragments of burnt marl and cinders strewed over the surface of several meadows were found after 10 years as a layer sunken to a depth of 2–3 inches below the turf, was due to the large quantities of fine earth continually brought to the surface by worms in the form of castings. In December 1842, just after settling in Down House, Darwin spread a quantity of broken chalk over part of a field near his house "for the sake of observing at some future period at what depth it would become buried". This future period arrived in 1871 after 29 years (!) and the chalk was observed at a depth of 7 inches, resulting in an average sediment reworking rate of 0.22 inches per year. The last book Darwin wrote (1881) is devoted to the activity of worms. It was a best-seller in his time and still inspiring to read (see e.g. Pemberton and Frey 1990). Keith (1942), restudying the same fields in Downe, was able to recognise some of the material Darwin had used. Darwin also collected data on the worm castings brought to the surface within a given time and area. What is so interesting in Darwin's experiments in my opinion is that the same methods to estimate bioturbation are still in use. Castings are collected to estimate deposit-feeder activities, and the rate at which particles sink below the surface is also used nowadays to estimate bioturbation rates. Glass beads, fluorescent sand grains, or radioactive materials have long since replaced the cinders and burnt marl; and one other difference is that we want our results within days or weeks, not years.

Davison (1891), inspired by Darwin's book, carried out some measurements on the castings of the polychaete *Arenicola marina* on a tidal flat off Northumberland in August 1891. He assumed the same number and weight of castings to be produced every tide and so calculated the annual production of castings. In earlier work (Cadée 1976), I compared annual estimates of sediment reworking by *Arenicola marina* of Davison and later authors. They usually based their estimates on short-term observations and were unable to deal adequately with the seasonal trend in *Arenicola* activity. They estimated *Arenicola* to be able to rework an amount of sediment equalling a sediment layer of 10–60 cm per year depending on the density of animals. Somewhat lower values were obtained in measurements over the whole year (Cadée 1976; Riisgard and Banta 1998).

A few early quantitative studies dealt with deposit feeders outside Europe, e.g. Crozier (1918) collected data on a holothurian (*Stichopus*) in Bermuda and Gardiner (1931) on a holothurian in the Maldive Islands. McGinitie (1934) came to extremely high estimates of over 1 m sediment reworking per year for a burrowing crustacean *Callianassa californiensis* in California. Dapples (1938, 1942) summarised the work up to that period and informed sedimentologists of the importance of bioturbation. Rhoads (Rhoads 1963, 1967, 1974; Rhoads and Young 1970, 1971) in particular was influential in the upsurge of (quantitative) bioturbation research that has taken place since the 1960s.

The tidal flats of the Wadden Sea were the study object of German researchers particularly after Richter had strongly advocated the foundation of the institute "Senckenberg am Meer" in Wilhelmshaven (1928), an institute especially for "Aktuopaläontologie" (Richter 1928) and the study of recent sediments, actuogeology. Schäfer (1962) gives an excellent review of his own work carried out in the Wadden Sea and the North Sea and that of others in Germany such as Wohlenberg (1937) on Sylt and Linke (1939) in the Jade Bight. About 200 pages of his book deal with sediment disturbance by organisms in relation to their different activities (food gathering, hole- and tube-building, locomotion, resting and escape): a wealth of information timely translated into English in 1972. This Senckenberg school of research of recent sediments not only continued in the Wadden Sea and North Sea (e.g. Hertweck 1970, 1994), but also extended its area of research outside the North Sea, often in co-operation with local institutes, to Italy (e.g. Reineck and Singh 1971), Georgia, USA (e.g. Howard and Reineck 1972), and Taiwan (e.g. Reineck and Cheng 1978). A successful textbook by Reineck and Singh (1975) further helped to propagate the knowledge gained.

Boudreau (1998), reviewing literature on tracer-identified mixed-layer thickness of marine sediments – which is the combined effect of mixing by bioturbation and physical processes – observed a world-wide mean value ($\pm$SD) of 9.8$\pm$4.5 cm from the intertidal to the deep sea. For modellers this is an interesting outcome, and we may ask why continue with bioturbation studies: this mixing of sediments is for the large part due to bioturbation. However, a search for recent literature using Current Contents (January 1997 to May 1999) and bioturbation as a keyword, gave some 100 papers, indicating bioturbation research is still very much alive. Of these papers, some 70 % dealt with bioturbation in marine environments, and some 50 % with the role of bioturbation in mineralisation or burrowing and release of pollutants. Rate measurements of bioturbation are still popular (25 %), the rest deal with effects on sediment properties, biological interactions via bioturbation, autecology of bioturbators, and modelling.

6.3 Types of Bioturbation

6.3.1 Crawling and Dwelling Traces

Birds that walk over a tidal flat, or meiofauna that moves between the sand grains, rework surface sediments. Schäfer (1956) carried out some excellent experiments to study animals moving on the sediment or resting in the surface layer (see some examples in Fig. 6.3). He provided a layered sediment in aquariums and studied disturbance of the layering, either by adding new layered sediment during the experiment, studying how the animals worked upwards (escape traces), or introducing the animals on top of the layered sediment. These experiments were very nicely illustrated. The study of the traces formed is now a separate science: ichnology. Trace fossils are keys to paleo-environmental studies of sediments where traces may be the only fossils found. Unlike body fossils, they preserve fossil behaviour (Seilacher 1967).

Organisms may only find shelter in the sediment (dwelling traces in Seilacher's 1964 terminology). This may vary from shrimps and flatfish hiding themselves temporarily in surface sediments to large permanent burrows formed by tilefish: large specimens of 1 m length produced crater-like holes of 4–5 m in diameter and at least 2–3 m deep on the shelf off Long Island (Able et al. 1982, 1987). Even deeper burrows (5–9 m, diameter 0.5 cm) were described by Bromley et al. (1975), from the Upper Cretaceous and Danian of NW, the dwelling place of an unknown organism. Temperate areas lack the species-rich communities of crustaceans (ghost crabs *Ocypode*, fiddler crabs *Uca*) so abundant in (sub)tropical intertidal areas (Verwey 1930; Krejci-Graf 1935; Frey and Mayou 1971). Katz (1980) was the first to estimate the sediment-reworking activity of fiddler crabs related to digging of their dwelling burrows; they leave their burrows to feed on algae which they scrape from the sediment surface.

The burrowing callianassid crustaceans belong to the most important sediment reworkers with regards to the amount reworked as well as to the depth to which they can burrow, in coastal areas as well as subtidal. Their deep burrows have a good fossilisation potential (Frey et al. 1978). Bromley (1990) reviews both recent and fossil examples. *Callianassa major* lives in the intertidal of the Atlantic coast of southeast USA, in a communal burrow system formed in a horizontal plane 3–5 m below the surface. With long vertical shafts the animals remain in contact with the overlying water from which they filter their food. Some species are gardeners, bringing seagrass leaves into their burrows and feeding on the bacteria cultures, such as *Callianassa longiventris* and *C. acanthochirus* (Suchanek 1983). Other callianassids are deposit feeders, e.g. *Callianassa japonica* on intertidal sand-

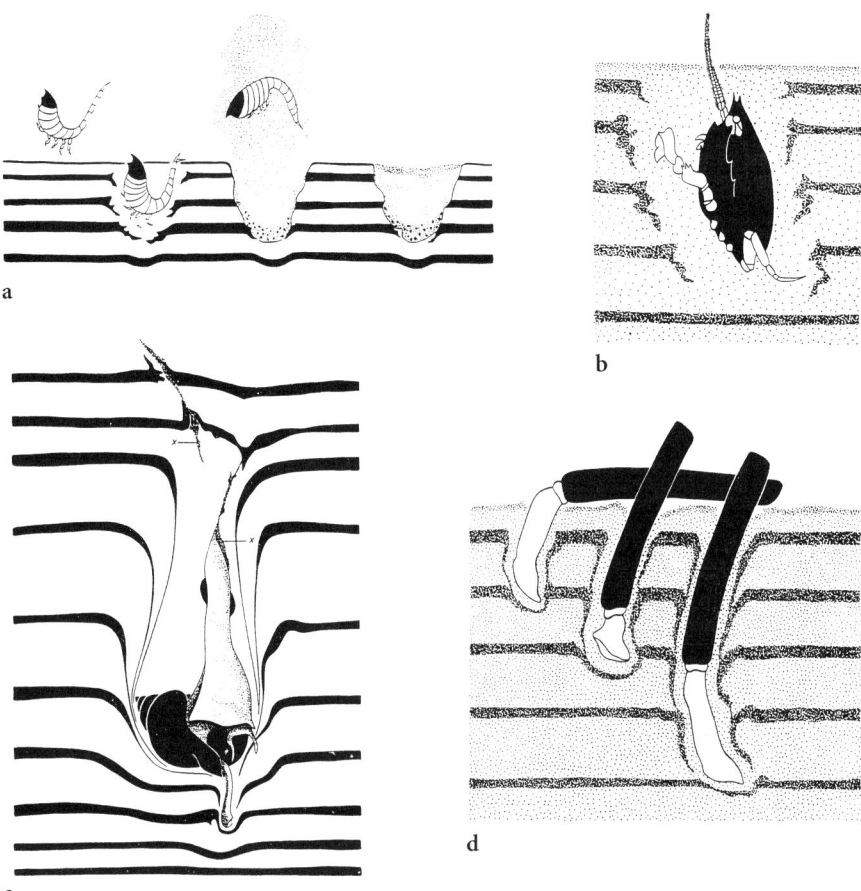

Fig. 6.3a–d. Experimental disturbance of layered sediments in aquariums by different organisms. **a** Temporarily resting Cumacea; **b** burying crab *Corystes*; **c** *Buccinum* moving towards surface; and **d** digging *Ensis*. (Schäfer 1956)

flats in Japan (Tamaki 1988); *Callianassa filholi* in New Zealand tidal flats (Berkenbusch and Rowden 1999); and *Callianassa subterranea* in the subtidal North Sea (Witbaard and Duineveld 1989; Rowden et al. 1998).

6.3.2 Deposit Feeders

Deposit feeders are often mentioned as the main bioturbators (Levinton 1995; Boudreau 1998). They may transport sediment in a vertical direction, either bringing sediment from deeper layers to the surface (conveyer-belt species, Rhoads 1974, e.g. the polychaete *Heteromastus* and some callianassid crustaceans), or surface sediments to deeper layers (inverted conveyer-belt species, e.g. the polychaete *Scolelepis* see Wohlenberg 1937). They may also mix the surface sediment layer by surface deposit feeding (*Yoldia limatula*, see Rhoads 1963) or grazing (the gastropod *Hydrobia ulvae*). Deep layers may be mixed by deposit feeders that bulldoze horizontally through subsurface sediments (e.g. heart urchins, Howard et al. 1974). Funnel feeders feed at depth and surface sediment sinks to their feeding depth via a funnel. This works best in loose sediments. One of the best-known examples is *Arenicola marina* (Wells 1966; Riisgard and Banta 1998). This is also a good example that deposit feeders do not always keep to one mode of feeding. The funnel feeder *Arenicola* may also feed on deep sediments (Richter 1924; Rijken 1979). Gardening (stimulation of microbial growth by irrigation or bioturbation) has been proposed to enrich the sediment eaten (Hylleberg 1975). Riisgard and Banta (1998) demonstrated that *Arenicola* cannot make a living as true filter feeders, as proposed by Krüger (1959). For other examples of versatile deposit feeders, see Cadée (1984) and Riisgard and Kamermans (Chap. 4).

Callianassid crustaceans rework more sediment per surface area than other bioturbators (see compilations in Cadée 1976; Lee and Swartz 1980). McGinitie (1934) estimated that a population of *Callianassa californiensis* reworked annually an amount of sediment equalling a layer of over 1 m. Myrick and Flessa (1996) observed the same callianassid to burrow to a depth of 1.15 m on the tidal flat of Bahia la Choya, northern Gulf of California, Mexico, where the population reworked a layer of 56 cm annually.

Whereas most sediment transport by bioturbators is in a vertical direction, their activity may also increase horizontal sediment transport by waves and currents: In the Danish Wadden Sea, Wesenberg-Lund (1905) already observed erosion by the incoming tide of *Arenicola* faecal casts deposited on the tidal flat during low tide, bringing fine particles into suspension. Much later Rhoads and Young (1970), in a now 'classical' paper, noted that intensive bioturbation by deposit feeders in Buzzards Bay produced a fluid sediment surface that was easily resuspended by low-velocity tidal currents, thereby preventing establishment of suspension feeders.

6.3.3 Larger Predators and Grazers

Recently, more attention is being paid to feeding holes made by predators on benthic macrofauna or by grazers on plant rhizomes. It is now realised that they may also produce considerable sediment disturbance. Side-scan sonar revealed up to 4 m long, 2 m wide and 0.4 m deep holes in the surface sediments of the Bering Sea (Johnson and Nelson 1984). They could be related to Gray whales (*Eschrichtius robustus*) that feed there on benthic tube-building ampeliscid amphipods. Abundant amphipod tubes commonly coalesce to form a mat that efficiently fixes the sediment surface and protects it from scouring by currents. Current scouring enlarges whale-feeding pits. Pits also act as loci of detritus accumulation during summer quiescence. Commonly, erosion of areas between pits will leave a single large pit that can be as large as 8×20 m. Whales thus profoundly disturb the substrate. They trigger substantial current scouring. The volume of sediment injected into the water column by feeding whales was estimated to be at least 1.2×10^9 m^3 per year (in an area of 22.000 km^2). Fresh pits occupied 5.6% of the feeding area. From this we can calculate that they rework annually an amount of sediment equalling a layer of 5 cm over the whole area. Gray whales also feed in intertidal areas (Weitkamp et al. 1992): in Puget Sound they were feeding on ghost shrimps (*Callianassa californiensis*), making holes of ca. 10 cm deep and ca. 2×3 m in size.

During the same side-sonar studies, long linear traces were observed on average 47 m long and 0.4 m deep, which were related to walrus feeding (Nelson et al. 1987). Such traces had already been studied by scuba divers in the same area (Oliver et al. 1983): they suggested that tusks were not used to excavate prey. In a series of papers Kastelein and Mosterd (1989) and Kastelein et al. (1991, 1994) describe how walrus feed. They first root into the substrate using the upper edge of their snout, and then stir up the sand by water jetting with their mouth; when they have reached the bivalve they are searching for (i.a. *Mya truncata*), they use strong oral suction to extract the meat (Fig. 6.4).

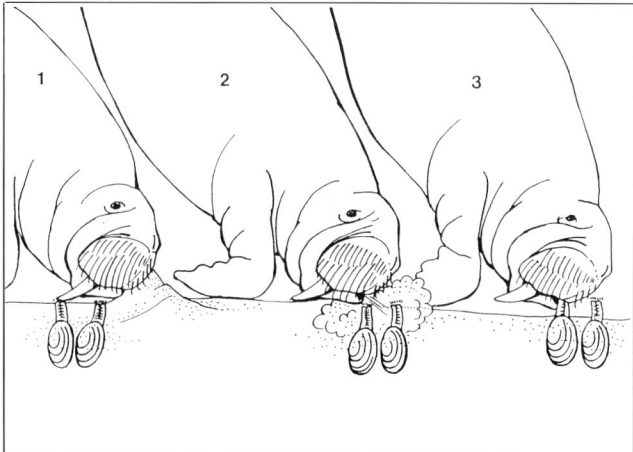

Fig. 6.4. In search for food walrus plough the sediment, use a water-jet to dig out bivalves encountered, and suck the meat from the bivalves leaving shells intact. (Kastelein and Mosterd 1989)

Water jets are also used by the New Zealand eagle ray *Myliobatis tenuicaudatus* to dig out bivalves they are feeding on (Gregory et al. 1979); others claim *Myliobatis* dig holes by flapping their pectoral fins (McGinitie and McGinitie 1968). The holes produced by three species of stingrays (*Dasyatis*) reach sizes of up to 1 m in diameter and up to 30 cm in depth (Howard et al. 1977). VanBlaricom (1982) observed somewhat smaller holes that had been produced by the round stingray *Urolophus halleri* and the bat ray *Myliobatis californica* off La Jolla, California. He observed a strong seasonal variation and an average daily disturbance rate of 0.84% of the transect. Taking 5 cm as the average depth of the entire excavation made, I calculate that rays rework annually an amount of sediment equalling a layer of 15 cm (365 × 0.0084 × 5 cm = 15 cm). Myrick and Flessa (1996) estimated the same two species of rays to rework as much as 1 m of sediment annually on the lower tidal flats of Bahia la Choya, northern Gulf of California, Mexico.

Very comparable holes to those produced by rays were reported by Cadée (1990) from the Dutch Wadden Sea (Fig. 6.5). These were, however, made by shelducks, *Tadorna tadorna*. Other ducks, such as the eider *Somateria mollissima*, may make similar holes during food collecting (Wietfield 1980). Black-headed gulls *Larus ridibundus* produced up to 3 m long, 15 cm wide and ca. 5 cm deep feeding traces by trampling and walking backwards in shallow water left on tidal flats (Cadée 1990). By trampling, they fluidise the sediment in which worms and bivalves then float and thus become an easy prey. These traces have wrongly been explained as resting traces (e.g. Hertweck 1970; see Cadée 1990 for discussion). Sediment reworking by these birds showed strong seasonal fluctuation and equalled a sediment layer of 2.5 cm per year or ca. 10% of that reworked by deposit feeders on the same tidal flat. Flamingoes may produce very characteristic feeding holes when feeding on tidal flats: circular holes, about 1 m in diameter, with a hill in the middle (Fig. 6.6; photograph taken by H. Michaelis on the Banc d'Arguin, Mauritania, and kindly made available for reproduction here). Treading around and around on mud, they sieve the sediment with their very characteristically structured bill for small organisms or even microscopic algae; discarded sediment piles up in the middle of the ring (Jenkin 1957).

Digging for prey is described for the sea otter *Enhydra lutris* in Alaskan waters (Calkins 1977). In shallow water they make holes 15–45 cm across and up to 50 cm deep, to dig out bivalves. In Monterey Bay, California, Hines and Loughlin (1980) observed sea otters digging holes even larger (150 × 50 cm large and up to 50 cm deep) made during successive dives by the same animal. The crab *Cancer pagurus* digs holes in subtidal sands in search of food of up to 50 cm in diameter (Hall et al. 1991). Woodin (1978) describes feeding pits of horse-shoe crabs (*Limulus polyphemus*) and blue crabs (*Callinectes sapidus*) on a tidal flat in Virginia.

Sediment Dynamics by Bioturbating Organisms

Fig. 6.5. Trampling holes made by Shelducks in search of food on tidal flats in the Dutch Wadden Sea. Sand from the round to oval holes deposited on one side of the hole, average diameter ~40 cm, depth 10 cm (size ruler 25 cm, for more information, see Cadée 1990)

Fig. 6.6. Circular holes with central mound produced by food-searching flamingoes (photographed at the Banc d'Arguin, Mauritania, 1988 by Hermann Michaelis and kindly made available)

Herbivores, particularly those feeding on rhizomes below the sediment surface, also make holes to reach their food. This is described for the Greater Snow Geese (*Anser caerulescens atlanticus*) by Dionne (1985): in search of rhizomes of *Scirpus americanus* geese dig many thousands of small holes in the substrate causing on average 10 cm lowering of the lower tidal marsh of the St. Lawrence Estuary. Reise (1985; Fig. 7.3) illustrates pits dug by ducks (*Anas penelope*) and geese (*Branta bernicla*) in search of rhizomes of intertidal seagrass *Zostera noltii* in Königshaven, Sylt. Dugongs (*Dugong dugong*), herbivorous sea-mammals, are reported to produce unvegetated feeding trails in seagrass fields (Nakaoka and Aioi 1999) or even to change seagrass beds into bare sand flats (Preen 1995).

These examples may illustrate how important other organisms – in particular large herbivores and predators – are in sediment reworking next to deposit feeders. Their under-water hole digging may produce large clouds of suspended sediment in the water column, thereby also causing lateral transport of sediment. Holes made in the vegetation by, e.g., dugongs (Preen 1995), or in the mats of tube-building amphipods the Gray whales use as food, enhance erosion (Johnson and Nelson 1984). They may influence sediment dynamics on a scale similar to that of deposit feeders.

6.4 Seasonal Variation

"Change is the norm within the estuarine system, whether related to individual tides, to lunar cycles of tides, to seasonal or annual variations, or even to controls with longer periodicities," wrote McManus (1998) and "some of the greatest dangers to present-day estuarine science lie in the need to respond to outside pressures: to provide rapid answers for administrators under pressure ... and to ensure rapid publication".

Bioturbation, like most biological processes, shows variations on different time scales from tidal, diurnal, seasonal to interannual. Quantification of sediment reworking therefore needs to be based on at least one year of measurements to give meaningful results. Seasonal variation in bioturbation is best studied in temperate areas; little is known about seasonal variation in tropical areas. A large seasonal variation in sediment reworking in temperate areas is apparent from studies of *Arenicola marina* (Cadée 1976; Riisgard and Banta 1998) and burrowing callianassid shrimps such as *Callianassa subterranea* in the North Sea (Rowden et al. 1998) and the intertidal *C. filholi* in New Zealand (Berkenbusch and Rowden 1999). Year-to-year variation may also be considerable and could be related to comparable variations in density in the polychaete *Heteromastus filiformis* on tidal flats (Cadée 1979). Hole-digging activity of predators such as birds shows clear seasonal variations (Cadée

1990). Geese feeding on *Scirpus* rhizomes are only active in the St. Lawrence estuary tidal marshes twice a year when they visit this area during their migration (Dionne 1985). Gray whales feed mainly in the summer season, so their sediment reworking by suction-sieving for their food crustaceans is seasonal (Johnson and Nelson 1984; Weitkamp et al. 1992). Rays off California showed a strong seasonality in hole digging (VanBlaricom 1982).

This strong seasonality influences sediment properties such as stability (erodability). Intertidal sediments are stabilised by exopolymers (mucopolysaccharides) produced by diatoms during locomotion between the sediment grains (e.g. Underwood and Paterson 1993; Chap. 5). Disturbing this diatom population by sediment reworking or feeding on these diatoms therefore might destabilise the sediment. This indeed has been observed in the Dollard estuary (Dutch Wadden Sea; Kornman and De Deckere 1998). Benthic diatom growth in April increased sediment erosion resistance as compared with the winter situation, but in May the increased population of the amphipod *Corophium volutator* caused diatoms to decrease and erodability to increase again, both by their feeding on diatoms and by bioturbation. Quite interestingly, Daborn et al. (1993) published a paper that may almost be seen as a sequel. On a tidal flat in the Bay of Funday they observed a decrease in bioturbation and grazing pressure on diatoms, due to the feeding of migratory birds on *Corophium volutator* in August: the diatoms could increase again and, with the mucopolysaccharides they produced, the stabilisation of the sediment increased. We might complete the seasonal cycle of sediment stability by assuming again lower values in winter due to lower diatom densities (Underwood and Paterson 1993) and a higher frequency of sediment disturbance by strong winds.

To study the relative importance of physical forces and bioturbation on sediment transport, short-term *in situ* studies are insufficient (McManus 1998); they will miss this seasonal variation. However, in the sediments, effects of bioturbation and physical forces remain visible. The study of sediment cores from tidal flats and near-shore sediments gives us an insight into the relative importance of energy level and bioturbation in sediment dynamics, as shown by Reineck (1963, 1976, 1977). Howard (1975) indicates how bioturbation traces can be used to study depositional regimes in older sediments.

6.5 Latitudinal Variation

Bioturbation by the total community cannot be estimated in a simple way. Darwin's elegant method of measuring the sinking velocity of particles could work on land. In marine (and fresh water) environments it measures the combined effect of biological and physical mixing. The same holds for mixing depth estimated from the vertical distribution of isotopes (e.g. ^{210}Pb). Only if physical mixing is low will these methods estimate bioturbation, as for Discovery Bay, Jamaica, in the calm summer period, but not in the stormy winter (Aller and Dodge 1974). For more exposed coastal areas the best method seems to estimate bioturbation of the (main) bioturbators separately and than add these estimates for community bioturbation data. Only few studies give such information. Moreover, comparisons of literature data on bioturbation face the problem of differences in methods used and data provided.

The greatest diversity of coastal life occurs in the tropics (Ekman 1953; Briggs 1995), and this increase in species diversity from high to low latitudes occurs particularly in epifaunal species such as crabs and nudibranchs (Thorson 1957). Some of the infaunal groups that Thorson studied did not show such an increase. This might imply that the diversity of deposit-feeding polychaetes and bivalves does not show a strong latitudinal variation, but their density and bioturbation activity might still vary with latitude. As already mentioned, arctic and temperate coasts lack the rich diversity of hole-digging callianassid shrimps and crabs such as ghost crabs *Ocypode* and fiddler crabs *Uca* of (sub)tropical beaches and intertidal flats.

The best-studied coastal faunas are usually in temperate areas, which makes comparisons from (ant)arctic to tropical areas difficult. Coastal areas of high latitudes are too hostile for bioturbators due to low temperatures and the high amount of sediment disturbance by drift ice. As an example of subarctic intertidal flats I will use the study of Aitken et al. (1988) at Baffin Island, Canada, near the Arctic Circle. They found a boreal *Macoma balthica* community with bioturbators such as the bivalve *Macoma balthica* and the polychaete *Arenicola marina*, which are species that also occur on temperate tidal flats of i.a. the Wadden Sea. The authors state their sediment-reworking capacity to equal that of their counterparts in the Wadden Sea, but their density was lower as well as the period of the year they could bioturbate. For tidal flats in the temperate Dutch Wadden Sea, I arrived at an annual bioturbation activity equalling a sediment layer of almost 40 cm by adding sediment-reworking rates of the most important bioturbators (Cadée 1976, 1979, 1990); this is probably four times higher than on Baffin Island.

Myrick and Flessa (1996) studied bioturbation on a tropical tidal flat area in Bahia la Choya, northern Gulf of California, Mexico. They measured an annual sediment-reworking rate of 0.5 m by callianassid shrimps on high tidal flats and of 1 m by rays on low tidal flats. The study of Aller and Dodge (1974) in Discovery Bay, Jamaica, provides data for bioturbation in a tropical lagoon. They confined their measurements to the warm and calm summer period (April–October), which makes extrapolation to include the cooler stormy winter period difficult. Deposit-feeding callianassid crustaceans were the main bioturbators, reworking an amount of sediment equalling a layer of 6–7 cm per week. For the seven summer months this would mean a layer of ~2 m, to which an unknown amount of winter bioturbation has to be added.

The few data available suggest bioturbation activity to increase at least one order of magnitude from the subarctic to the equator mainly due to the activity of callianassid shrimps in tropical areas. Everywhere bioturbation is higher in sheltered than in exposed coastal areas.

6.6 Changes in Historical Times

How interesting it would be to travel back in time some hundreds of years and to study bioturbation in a pristine Wadden Sea! We would be able to observe Gray whales and rays digging holes for food. Gray whales are extinct now in the Atlantic Ocean, but once occurred in the Wadden Sea area (Van Deinse and Junge 1937; De Smet 1981). Rays have been decimated mainly in the last 100 years due to fisheries (Walker and Heessen 1996). Before the eelgrass *Zostera marina* disappeared from the Dutch Wadden Sea due to a still ill-understood "wasting disease" in the early 1930s (Polderman and Den Hartog 1975), large populations of Brent geese (*Branta bernicla*) grazed mainly on the rhizomes of this species during winter, causing undoubtedly a considerable bioturbation. Perhaps no tidal basin has been studied for such a long time as the area near Sylt, where Möbius started research as early as 1869 (see Reise 1982). Reise observed considerable changes in macrobenthos; sessile taxa suffered most (mainly from shrimp fisheries), and some new species arrived. Together with these changes, the Wadden Sea biota of mammals, fish, macrobenthic species and eelgrass vegetation, and bioturbation will also have changed.

Man's influence on the marine environment has increased with time, this also influences sediment dynamics. Fishing activity such as beam-trawling and dredging for *Arenicola* and shellfish are also sources of sediment disturbance. As early as the end of the sixteenth century, Dutch fishermen asked Prince William of Orange to place restrictions on the use of trawls, because this would make the grounds rough and probably lower future catches (de Groot 1984). Only recently has research clearly indicated the disastrous effects of fishing activity, increasingly effecting the sea bottom (Lindeboom and de Groot 1998), both the sediment and its macrobenthos, the so-called 'non-target' species. Shellfisheries have become more and more dangerous for benthic life. Piersma and Koolhaas (1997) warned or lasting changes due to shellfisheries in the Wadden Sea. Bivalve-rich silty sediments in the lee of a mussel-bank have been transformed by removal of the mussel-bank and successive mechanical harvesting of cockles into a permanently sandy, lower-lying and higher dynamic intertidal flat, where bivalves can no longer settle and maintain their populations. New developments in mechanical harvesting of bivalves also enable collecting of deep-living species such as *Ensis directus*. In the North Sea shell-fishery is now able to fish 1 ha per hour down to a maximum depth of 40 m, scraping and sieving the upper 30 cm of the bottom (Leopold 1999). This is not bioturbation but habitat destruction by man. What will be left of the North Sea ecosystem in the twenty-first century?

6.7 Conclusions

Sediment reworking by organisms occurs in almost all coastal areas. Under very high-energy conditions, where few infaunal organisms can survive, bioturbation is negligible. Estimates of sediment mixing depth can be obtained, e.g., from vertical ^{210}Pb distributions, but they give the combined effects of physical and biological sediment reworking and thus underestimate bioturbation. Structures in sediment cores can be used as an archive of the relative role of physical and biological sediment disturbance. It is not only deposit feeders and all organisms crawling on, or dwelling in, the sediment that cause bioturbation, but also predators (including birds, rays and the Gray whale) on infaunal macrobenthos, and herbivores such as geese and dugong feeding on sea grass rhizomes. Bioturbation in the coastal zone increases one order of magnitude from the sub-arctic towards the tropics, particularly due to the rich and diverse (sub)tropical crustacean fauna. In temperate areas bioturbation shows seasonal variation; for its quantification, year-round studies are necessary. Little is known of seasonal variation in tropical areas. In the Wadden Sea, bioturbation has considerably changed in historical times: Gray whales and rays have disappeared due to fishing activities; geese feeding on rhizomes of seagrass have disappeared with the disappearance of *Zostera* due to a "wasting disease". Human impact on the sediment has increased due to trawling and mechanical harvesting of shellfish and lugworms, now disturbing the top 30 cm of the sediment. This long-term and still-increasing human impact continues to change the Wadden Sea ecosystem, and such is the future of many over-fished coastal areas in the world.

Acknowledgements. The author is very grateful for critical remarks by referees Herman Michaelis and Akio Tamaki. This is NIOZ publication nr. 3486.

References

Abel O (1935) Vorzeitliche Lebenspuren. Fischer, Jena, 644 pp
Able KW, Grimes CB, Cooper RA, Uzmann JR (1982) Burrow construction and behavior of tilefish, *Lopholatilus chamaeleonticeps*, in Hudson submarine canyon. Env Biol Fish 7:199–205
Able KW, Twichell DC, Grimes CB, Jones RS (1987) Tilefish of the genus *Caulolatilus* construct burrows in the sea floor. Bull Mar Sci 40:1–10
Aitken AE, Risk MJ, Howard JD (1988) Animal-sediment relationships on a subarctic intertidal flat, Pangnirtung Fjord, Baffin Island, Canada. J Sed Petrol 58:969–978
Aller RC, Dodge RE (1974) Animal-sediment relations in a tropical lagoon Discovery Bay, Jamaica. J Mar Res 32:209–232
Bayerl K, Austen I, Köster R, Pejrup M, Witte G (1998) Dynamik der Sedimente im Lister Tidebecken. In: Gätje C, Reise K (eds) Ökosystem Wattenmeer – Austausch-, Transport- und Stoffumwandlungsprozesse. Springer-Verlag, Berlin Heidelberg New York, pp 127–159
Berkenbusch K, Rowden AA (1999) Factors influencing sediment turnover by the burrowing ghost shrimp *Callianassa filholi* (Decapoda: Thalassinidea). J Exp Mar Biol Ecol 238:283–292
Black KS, Paterson DM, Cramp A (eds) (1998) Sedimentary processes in the intertidal zone. Spec Publ Geol Soc Lond 139, pp 1–409
Boudreau BP (1998) Mean mixed depth of sediments: The wherefore and the why. Limnol Oceanogr 43:524–526
Briggs JC (1995) Global biogeography. Developments in palaeontology and stratigraphy, vol 14. Elsevier, Amsterdam
Bromley RG (1990) Trace fossils, biology and taphonomy. Special topics in paleontology. Unwin and Hyman, London (2nd edn 1996)
Bromley RG, Curran HA, Frey RW, Gutschick RC, Suttner LJ (1975) Problems in interpreting unusually large burrows. In: Frey RW (ed) The study of trace fossils. Springer-Verlag, Berlin Heidelberg New York, pp 351–376
Cadée GC (1976) Sediment reworking by *Arenicola marina* on tidal flats in the Dutch Wadden Sea. Neth J Sea Res 10:440–460
Cadée GC (1979) Sediment reworking by the polychaete *Heteromastus filiformis* on a tidal flat in the Dutch Wadden Sea. Neth J Sea Res 13:441–456
Cadée GC (1984) 'Opportunistic feeding', a serious pitfall in trophic structure analysis of (paleo)faunas. Lethaia 17:289–292
Cadée GC (1990) Feeding traces and bioturbation by birds on a tidal flat, Dutch Wadden Sea. Ichnos 1:23–30
Cadée GC (1998) The influence of benthic fauna and microflora. In: Eisma D (ed) Intertidal deposits, river mouths, tidal flats, and coastal lagoons. CRC Press, Boca Raton, pp 383–402
Calkins DG (1977) Feeding behavior and major prey species of the sea otter, *Enhydra lutris*, in Mantague Strait, Prince William Sound, Alaska. Fish Bull 6:125–131
Crozier JW (1918) The amount of bottom material ingested by holothurians (*Stichopus*). Contr Bermuda Biol Stn 88:379–389
Daborn GR, Amos CL, Brylinsky M, Christian H, Drapeau G, Faas RW, Grant J, Long B, Paterson DM, Perillo GME, Piccolo MC (1993) An ecological cascade effect: migratory birds affect stability of intertidal sediments. Limnol Oceanogr 38:225–231
Dapples EC (1938) The sedimentational effects of the work of marine scavengers. Am J Sc (5th series) 36:54–65

Dapples EC (1942) The effects of macro-organisms upon near-shore marine sediments. J Sed Petrol 12:118-126
Darwin C (1837) On the formation of mould. Trans Geol Soc Lond 5:505-509
Darwin C (1881) On the formation of vegetable mould through the action of worms with observations on their habits. Murray, London
Davison C (1891) On the amount of sand brought up by lobworms to the surface. Geol Mag 8:489-493
De Groot SJ (1984) The impact of bottom trawling on benthic fauna of the North Sea. Ocean Manag 9:177-190
De Smet WMA (1981) Evidence of whaling in the North Sea and English Channel during the middle ages. In: Mammals in the seas. FAO Fish Ser 5(III):301-309
Dionne C (1985) Tidal marsh erosion by geese, St. Lawrence estuary, Québec. Géogr Phys Quat 39:99-105
Dyer KR (1986) Coastal and estuarine sediment dynamics. John Wiley and Sons, Chichester
Ekdale AA, Bromley RG, Pemberton SG (1984) Ichnology, Trace fossils in sedimentology and stratigraphy. SEPM Short Course 15:1-317
Ekman S (1953) Zoogeography of the sea. Sidgwick and Jackson, London
Frey RW (ed) (1975) The study of trace fossils. Springer-Verlag, Berlin Heidelberg New York
Frey RW, Mayou TV (1971) Decapod burrows in Holocene barrier island beaches and washover fans. Senckenbergiana Marit 3:53-77
Frey RW, Howard JD, Pryor WA (1978) *Ophiomorpha*: its morphologic, taxonomic, and environmental significance. Palaeogeogr Palaeoclimatol Palaeoecol 23:199-229
Gardiner JS (1931) Coral reefs and atolls. MacMillan, London
Graf G, Rosenberg R (1997) Bioresuspension and biodeposition: a review. J Mar Systems 11:269-278
Gregory MR, Ballance PF, Gibson GW, Ayling AM (1979) On how some rays (Elasmobranchia) excavate feeding depressions by jetting water. J Sed Petrol 49:1125-1130
Hails J, Carr A (eds) (1975) Nearshore sediment dynamics and sedimentation. John Wiley, London, New York
Hall SJ, Basford DJ, Robertson MR, Rafaelli DG, Tuck I (1991) Patterns of recolonisation and the importance of pit-digging by the crab *Cancer pagurus* in a subtidal sand habitat. Mar Ecol Prog Ser 72:93-102
Hertweck G (1970) Die Bewohner des Wattenmeeres in ihren Auswirkungen auf das Sediment. In: Reineck H-E (ed) Das Watt, Ablagerungs- und Lebensraum. Kramer, Frankfurt/Main, pp 106-130
Hertweck G (1994) Zonation of benthos and lebensspuren in the tidal flats of the Jade Bay, southern North Sea. Senckenbergiana Marit 24:157-170
Hines AH, Loughlin TR (1980) Observations of sea otters digging for clams at Monterey Harbor, California. Fish Bull 78:159-163
Howard JD (1975) The sedimentological significance of trace fossils. In: Frey RW (ed) The study of trace fossils. Springer-Verlag, Berlin Heidelberg New York, pp 131-146
Howard JD, Reineck H-E (1972) Georgia coastal region, Sapelo Island, USA: Sedimentology and biology IV. Physical and biogenic sedimentary structures of the nearshore shelf. Senckenbergiana Marit 4:81-123
Howard JD, Reineck H-E, Rietschel S (1974) Biogenic sedimentary structures formed by heart urchins. Senckenbergiana Marit 6:185-201
Howard JD, Mayou TV, Heard RW (1977) Biogenic sedimentary structures formed by rays. J Sed Petrol 47:339-346

Hylleberg J (1975) Selective feeding by *Abarenicola pacifica* with notes on *Abarenicola vagabunda*, and a concept of gardening in lugworms. Ophelia 14:113–137

Jenkin PM (1957) The filter-feeding and food of flamingoes (Phoenicopteri). Philos Trans R Soc Lond B 240:401–493

Johnson KR, Nelson CH (1984) Side-scan sonar assessment of Gray whale feeding in the Bering Sea. Science 225:1150–1152

Kastelein RA, Mosterd P (1989) The excavation techniques for molluscs of Pacific walrusses (*Odobenus rosmarus divergens*) under controlled conditions. Aquat Mammals 15:3–5

Kastelein RA, Gerrits NM, Dubbeldam JL (1991) The anatomy of the walrus head (*Odobenus rosmarus*), part 2: description of the muscles and of their role in feeding and haul-out behaviour. Aquat Mammals 17:156–180

Kastelein RA, Muller M, Terlouw A (1994) Oral suction of a Pacific walrus (*Odobenus rosmarus divergens*) in air and under water. Z Säugtierkd 59:105–115

Katz LC (1980) Effects of burrowing by the fiddler crab *Uca pugnax* (Smith). Est Coast Mar Sc 11:233–237

Keith A (1942) A postscript to Darwin's "Formation of vegetable mould through the action of worms". Nature 147:716–720

Kornman BA, De Deckere EMGT (1998). Temporal variation in sediment erodability and suspended sediment dynamics in the Dollard estuary. In: Black KS, Paterson DM, Cramp A (eds) Sedimentary processes in the intertidal zone. Spec Publ Geol Soc Lond 139:231–241

Krejci-Graf K (1935) Beobachtungen am Tropenstrand. I. Bauten und Fährten von Krabben. Senckenbergiana 17:21–32

Krüger F (1959) Zur Ernährungsphysiologie von *Arenicola marina*. Zool Anz 22:115–120

Lee H, Swartz RC (1980) Biological processes affecting the distribution of pollutants in marine sediments, part II. In: Baker RA (ed) Contaminants sediments, vol 2. Ann Arbor Science Pub., Ann Arbor, MI, USA, pp 555–606

Leopold MF (1999) Na schelpdiervisserij in de Waddenzee en Noordzee-kustzone, nu ook schelpdiervisserij verder op zee. Nieuwsbrief Ned Zeevogelgroep 1(2):1–2

Levinton J (1995) Bioturbators as ecosystem engineers: control of the sediment fabric, inter-individual interactions, and material fluxes. In: Jones CG, Lawton JH (eds) Linking species and ecosytems. Chapman and Hall, New York, pp 29–36

Lindeboom HJ, de Groot SJ (1998) The effects of different types of fisheries on the North Sea and Irish Sea benthic ecosystems. NIOZ Rapport 1:1–404

Linke O (1939) Die Biota des Jadebusenwattes. Helgol Wiss Meeresunters 1:201–348

McGinitie GE (1934) The natural history of *Callianassa californiensis*. Am Midl Nat 15:166–177

McGinitie GE, McGinitie N (1968) Natural history of marine animals, 2nd edn. McGraw-Hill, New York

McManus J (1998) Temporal and spatial variations in estuarine sedimentation. Estuaries 21:622–634

Meadows PS, Meadows A (1991) The environmental impact of burrowing animals and animal burrows. Symp Zool Soc Lond 63:1–349

Myrick JL, Flessa KW (1996) Bioturbation rates in Bahía la Choya, Mexico. Cienc Mar 22:23–46

Nakaoka M, Aioa K (1999) Growth of seagrass *Halophila ovalis* at dugong trails compared to existing within-patch variation in a Thailand intertidal flat. Mar Ecol Prog Ser 184:97–103

Nelson CH, Johnson KR, Barber JH (1987) Gray whale and walrus feeding excavations on the Bering Shelf, Alaska. J Sed Petrol 57:419–430

Oliver JS, Slattery PN, O'Connor EF, Lowry LF (1983) Walrus, *Odobenus rosmarus*, feeding in the Bering Sea: a benthic perspective. Fish Bull 81:501-512

Paterson DM, Black KS (1999) Water flow, sediment dynamics and benthic biology. In: Nedwell DB, Raffaelli DG (eds) Estuaries. Adv Ecol Res 29:155-194

Pemberton SG, Frey RW (1990) Darwin on worms: the advent of experimental neoichnology. Ichnos 1:65-71

Piersma T, Koolhaas A (1997) Shorebirds, shellfish(eries) and sediments around Griend, western Wadden Sea, 1988-1996. NIOZ Rapport 7:1-118

Polderman PJG, Den Hartog C (1975) De zeegrassen in de Waddenzee. Wetensch Meded. Kon Ned Natuurh Ver 107:1-32

Preen A (1995) Impacts of dugong foraging on seagrass habitats: observational and experimental evidence for cultivation grazing. Mar Ecol Prog Ser 124:201-213

Reineck H-E (1963) Sedimentgefüge im Bereich der südlichen Nordsee. Abh Senckenb Naturf Ges 505 1-13.

Reineck H-E (1967) Parameter von Schichtung und Bioturbation. Geol Rundsch 56: 420-438

Reineck H-E (1976) Primärgefüge, Bioturbation und Makrofauna als Indikatoren des Sandversatzes im Seegebiet vor Norderney (Nordsee). I. Zonierung von Primärgefügen und Bioturbation. Senckenbergiana Marit 8:155-169

Reineck H-E (1977) Natural indicators of energy level in recent sediments: the application of ichnology to a coastal engineering problem. In: Crimes TP, Harper JC (eds) Trace fossils 2. Geol J Spec Issue 9:265-272

Reineck H-E, Cheng YM (1978) Sedimentologische und faunistische Untersuchungen an Watten in Taiwan. I. Aktuogeologische Untersuchungen. Senckenbergiana Marit 10: 85-115

Reineck H-E, Singh IB (1971) Der Golf von Gaeta (Thyrrhenisches Meer). III. Die Gefüge von Vorstrand- und Schelfsedimenten. Senckenbergiana Marit 3:185-201

Reineck H-E, Singh IB (1975) Depositional sedimentary environments. Springer-Verlag, Berlin Heidelberg New York

Reise K (1982) Long-term changes in the macrobenthic invertebrates of the Wadden Sea: are polychaetes about to take over? Neth J Sea Res 16:29-36

Reise K (1985) Tidal flat ecology. Springer-Verlag, Berlin Heidelberg New York

Rhoads DC (1963) Rates of sediment reworking by *Yoldia limatula* in Buzzards Bay, Massachusetts, and Long Island Sound. J Sed Petrol 33:723-727

Rhoads DC (1967) Biogenic reworking of intertidal and subtidal sediments in Barnstable Harbor and Buzzards Bay, Massachusetts. J Geol 75:461-476

Rhoads DC (1974) Organism-sediment relations on the muddy sea floor. Oceanogr Mar Biol Annu Rev 12:263-300

Rhoads DC, Boyer LF (1982) The effects of marine benthos on physical properties of sediments, a successional perspective. In: McCall PL, Tevesz MJS (eds) Animal-sediment relations. The biogenic alteration of sediments. Plenum, New York, pp 3-52

Rhoads DC, Young DK (1970) The influence of deposit-feeding organisms on sediment stability and community trophic structure. J Mar Res 28:150-178

Rhoads DC, Young DK (1971) Animal-sediment relations in Cape Cod Bay, Massachusetts. II. Reworking by *Molpadia oolithica* (Holothuroidea). Mar Biol 11:255-261

Richter R (1924) Flachseebeobachtungen zur Paläontologie und Geologie. VII. *Arenicola* von heute und "Arenicoloides", eine Rhizocorallide des Buntsandsteins, als Vertreter verschiedener Lebensweise. Senckenbergiana 6:119-140

Richter R (1928) Aktuopaläontologie und Paläobiologie, eine Abgrenzung. Senckenbergiana 10:285-292

Richter R (1952) Fluidal-Textur in Sediment-Gesteinen und über Sedifluktion überhaupt. Notizbl Hess L-Amt Bodenforsch 6:67–81

Riisgard HU, Banta GT (1998) Irrigation and deposit feeding by the lugworm *Arenicola marina*, characteristics and secondary effects on the environment. A review of recent knowledge. Vie Milieu 48:243–257

Rijken M (1979) Food and food uptake in *Arenicola marina*. Neth J Sea Res 13:406–421

Rowden AA, Jones MB, Morris AW (1998) The role of *Callianassa subterranea* (Montagu) (Thalassinidea) in sediment resuspension in the North Sea. Cont Shelf Res 18: 1365–1380

Schäfer W (1956) Wirkungen der Benthos-Organismen auf den jungen Schichtverband. Senckenbergiana Lethaea 37:183–263

Schäfer W (1962) Aktuo-paläontologie nach Studien in der Nordsee. Kramer, Frankfurt/Main. (English translation, 1972: Ecology and paleoecology of marine environments. Univ Chicago Press, Chicago)

Schäfer W (1963) Biozönöse und Biofazies im marinen Bereich. Kramer, Frankfurt/Main

Seilacher A (1964) Biogenic sedimentary structures. In: Imbrie J, Newell ND (eds) Approaches to paleontology. John Wiley, New York, pp 296–316

Seilacher A (1967) Fossil behavior. Sci Am 217:72–80

Suchanek TH (1983) Control of seagrass communities and sediment distribution by *Callianassa* (Crustacea, Thalassinidea) bioturbation. J Mar Res 41:281–289

Tamaki A (1988) Effects of the bioturbating activity of the ghost shrimp *Callianassa japonica* Ortmann on migration of a mobile polychaete. J Exp Mar Biol Ecol 120: 81–95

Thorson G (1957) Bottom communities (sublittoral or shallow shelf). In: Hedgpeth JW (ed) Treatise on marine ecology and paleoecology. Geol Soc Am Mem 67(1):461–534

Underwood GJC, Paterson DM (1993) Seasonal changes in diatom biomass, sediment stability and biogenic stabilization in the Severn estuary. J Mar Biol Assoc UK 73: 871–887

VanBlaricom G (1982) Experimental analysis of structural regulation in a marine sand community exposed to oceanic swell. Ecol Monogr 52:283–305

Van Deinse AB, Junge GCA (1937) Recent and older finds of the California Gray whale in the Atlantic. Temminckia 2:161–188

Verwey J (1930) Einiges über Biologie ost-indischer Mangrovenkrabben. Treubia 12: 169–261

Walker PA, Heessen HJL (1996) Long-term changes in ray populations in the North Sea. ICES J Mar Sc 53:1085–1093

Weitkamp LA, Wissmar RC, Simenstad CA, Fresh KL, Odell JG (1992) Gray whale foraging on ghost shrimp (*Callianassa californiensis*) in littoral sand flats of Puget Sound, USA. Can J Zool 70:2275–2280

Wells GP (1966) The lugworm (*Arenicola*) – a study in adaptation. Neth J Sea Res 3: 294–313

Wesenberg-Lund C (1905) Umformungen des Erdbodens. Beziehungen zwischen Dammerde, Marsch, Wiesenland und Schlamm. Prometheus 16:561–566; 577–582. (Translated from Danish)

Witbaard R, Duineveld GCA (1989) Some aspects of the biology and ecology of the burrowing shrimp *Callianassa subterranea* (Montagu) (Thalassinidea) from the southern North Sea. Sarsia 74:209–219

Wietfield F (1980) Trampelwannen von Brandgänsen und Eiderenten. Natur Museum 110:170–174

Wohlenberg E (1937) Die Wattenmeer-Lebensgemeinschaften im Königshafen von Sylt. Helgol Wiss Meeresunters 1:1–92

Woodin SA (1978) Refuges, disturbance, and community structure: a marine soft-bottom example. Ecology 59:274–284

7 Competitive Bioturbators on Intertidal Sand Flats in the European Wadden Sea and Ariake Sound in Japan

E. FLACH and A. TAMAKI

7.1 Introduction

Distribution patterns and changes in abundance of macrozoobenthic species on intertidal sand flats have often been interpreted in terms of the physical environment, such as tidal level, sediment composition and salinity. More recently, however, biotic interactions have been recognised to be a significant structuring force for intertidal sand flat communities (Reise 1985; Posey 1987). There are several kinds of biotic interactions (Arthur and Mitchell 1989). The most frequently studied is competition (mainly for food), a − − interaction where there are effects on both species. The reverse of competition is mutualism, in which both species profit from the interaction (+ +); this is, however, not often observed on intertidal sand flats. The other often observed interaction is predation, which is a + − interaction, termed contramensalism by Arthur and Mitchell (1989). In this case, only one species benefits. In addition to these interactions in which both species are influenced, there are also interactions in which only one species is influenced. This could be either commensalism (+ 0 interaction) or amensalism (− 0 interaction). For this type of interaction the size of the animals often determines the outcome of the interaction (Wilson 1981). An example of this is the mobility-mode hypothesis of Posey (1987), which states that dense aggregations of large organisms or those with strong sedimentary effects may exclude smaller species through modifications related to their mobility type. In this chapter we will concentrate on one type of these interactions, the either positive and/or negative effects of two large bioturbating species on the benthic community at two intertidal sand flats on different sides of the world.

7.2 Large Bioturbators

Macrobenthic animals affect both rates and spatial distribution of sediment processes by their feeding, burrowing and ventilation activities (Aller and Yingst 1985; Andersen and Kristensen 1991). These sediment-reworking activities are generally called bioturbation (see Chap. 6).

On tidal flats along the European coasts the lugworm *Arenicola marina* L. (Polychaeta) is one of the dominating species. Lugworms are large deposit feeders and live in 20- to 40-cm deep L-shaped burrows in the upper sediment layer (Fig. 7.1). The feeding and defecating activities of the lugworm (Cadée 1976) result in a landscape of mountains and depressions on an otherwise more or less smooth tidal flat surface. The faecal mounds (3–5 cm diameter, 1–2 cm height) may become flush with the ambient surface during each period of submergence due to tidal currents. An individual worm generates new casts every 30–60 min (Reise 1985). The sediment-reworking activity of the lugworm is highest during the summer and lowest during the winter with an annual average of ~4.35 dm^3 per worm. At a density of 17 lugworms per m^2 (a mean density for the tidal flats of the Dutch Wadden Sea according to Beukema 1976), this gives an annual

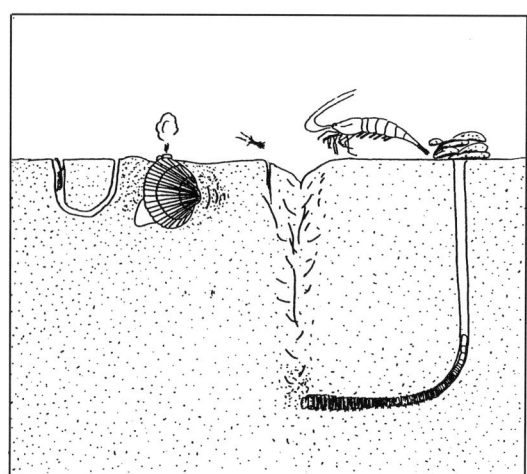

Fig. 7.1. The lugworm, *Arenicola marina* (*right*), lives in L-shaped burrows; while feeding, sand will sink down forming a funnel. For defecation it moves upward in the tail-shaft and produces the casting on the sediment surface. The amphipod *Corophium* (*left* and *middle*) lives in U-shaped burrows in the upper centimeters of the sediment. While feeding, the lugworm destroys the *Corophium* tube forcing the amphipod to migrate, making it an easy prey for epibenthic predators like the brown shrimp *Crangon crangon* (*upper right*)

reworking corresponding to a sediment layer of ~6.2 cm (Cadée 1976). At high lugworm densities, however, this annual sediment reworking can amount to a layer up to 33 cm. Lugworms also cause sediment instability at the surface, change the particle composition and stratification, irrigate the anoxic subsurface sediment with overlying water, release their excretions into the sediment, and produce large amounts of mucus (Reise 1985).

On the other side of the world, on the tidal flats of Japan, another large deposit-feeding species is found. The ghost shrimp, *Nihonotrypaea harmandi* (Bouvier) (Crustacea: Decapoda: Callianassidae), formerly misidentified as *Callianassa japonica* Ortmann, is an important member of the benthic community on shallow estuarine sand flats (Figs. 7.5, 7.6). It lives in a Y-shaped burrow occupying the entire sediment column (30–60 cm deep), under which lies an accumulation of large shell remains. Each adult burrow is inhabited by a single shrimp only. The ghost shrimp extrudes sediment through one burrow opening, which is deposited as a mound on the sand flat surface; on average, about 5–10 cm^3 of sediment (3–5-cm diameter mound) per shrimp is discarded per day.

7.3 Lugworms in the Wadden Sea

The lugworm *Arenicola marina* is found almost everywhere on the tidal flats in the Dutch Wadden Sea (Dankers and Beukema 1983) and accounts for about 20% of the benthic biomass (Beukema 1976). Densities of adult lugworms are usually of few tens per m^2; although locally they may reach densities up to about 100 per m^2, they generally do not surpass 50 per m^2 (Dankers and Beukema 1983). Adult lugworms show their highest densities at 1–4 km from the shore and are significantly less abundant both at shorter and longer distances from the coasts. These high densities fall within a zone from about MTL to about 6 dm below MTL and a silt content of the sediment between ~2 to 10% (Beukema and De Vlas 1979). The distribution of lugworms over the tidal flats is remarkably even and shows very little year-to-year variation (Beukema et al. 1983). Also, in the German part of the Wadden Sea, Reise (1985) reports a stable adult population at the tidal flats in Königshafen. This stable population size raises the question: how is this maintained? Reise (1985) reports two mechanisms; firstly a density-dependent mobility of the adults and, secondly, a migratory pattern of the juveniles in four steps avoiding physical stress, predation and competition with adults to guarantee a high survival rate. During the third phase, juvenile lugworms occupy, in very high densities, the upper tidal zone, the so-called nursery bed or Brutwatten (Reise 1985). Beukema and De Vlas (1979) observed high densities of young lugworms only on relatively high tidal flats where adults occurred only in low

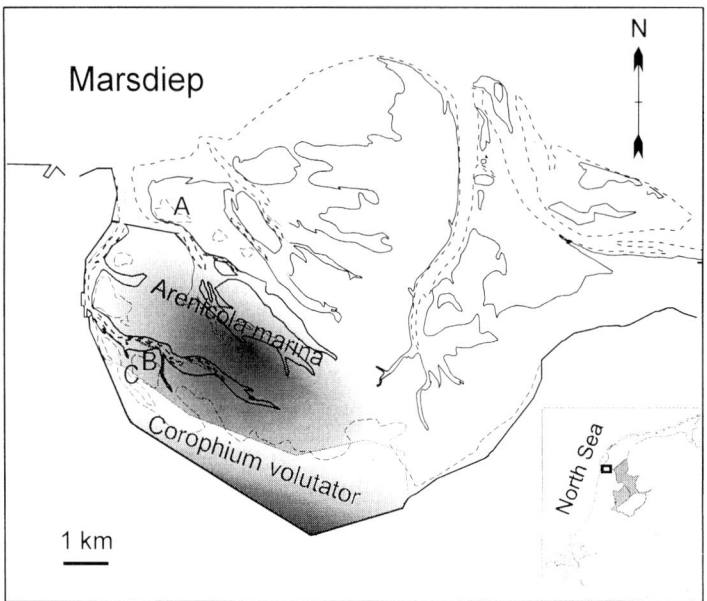

Fig. 7.2. The study area Balgzand in the westernmost part of the Wadden Sea, showing the zones with high densities of *Arenicola marina* and *Corophium volutator* and the experimental sites A, B and C

numbers or were even absent. Thus, some density-governing mechanisms seem to occur in the zone occupied by the adult population.

To study this phenomenon, experiments were carried out at Balgzand in the westernmost part of the Wadden Sea (Fig. 7.2) and lugworm densities were estimated in detail along a transect from the shore towards the tidal channel over 3 years (1990–1992). At two different sites (an exposed sandy site and a sheltered site with an intermediate sediment) lugworm densities were manipulated in small plots of 2.25 m^2 to obtain densities of 0, 25, 50, 75 and 100 adults per m^2 in April 1990. The enhanced densities showed a rapid decline to the level of the densities of the surrounding area within 1 year (i.e. ~10 per m^2 at the exposed site and ~20 per m^2 at the sheltered site; Flach 1992a; Flach and Beukema 1994). In the surrounding areas adult lugworm densities remained stable during the period of observation. The rates of decline were higher when starting from high than from low initial densities. Monthly declines amounted to ~25% of densities that were several times higher than the natural densities, but were only ~10% per month if densities were doubled (Flach and Beukema 1994). Thus intraspecific competition between the adults seems to occur, which maintains the population at the carrying capacity at a particular site.

In similar experimental plots the settlement of juvenile lugworms was studied. A significant ($r=-0.84$, $n=12$, $p<0.01$) negative relationship between the numbers of juveniles and the numbers of adults was observed. In addition, a significant ($r=-0.79$, $n=9$, $p<0.05$) negative relationship between the mean individual weight of the juveniles and the densities of the juveniles was also found (Flach and Beukema 1994). Thus, although the numbers of juveniles in plots with high adult densities were lower, they were larger individuals. This could be obtained either by different growth rates, or by migration of small individuals in the presence of adults, and/or by successful settlement of large juveniles between adult individuals.

Detailed observations of the lugworm population along the transect perpendicular to the coast showed the usual pattern of low densities close to the coast and high densities (~30–40 per m^2) in a broad zone nearly towards the tidal channel in May 1990. In October 1990, adult densities had declined dramatically (to ~10 per m^2) in the central part of the formerly high-density zone. During the summer of 1991 very high numbers of juveniles (up to ~400 per m^2) were found just in this area of low adult densities (Fig. 7.3), whereas they were nearly absent in the areas with high adult densities. Fishing in a nearby tidal channel during the subsequent winter showed that small juveniles were migrating, whereas the larger juveniles stayed on the tidal flats (Flach and Beukema 1994). This resulted in lugworm densities in this central part of the transect of ~30–40 per m^2 in the summer of 1992, thus restoring the situation as found in May 1990.

7.4 Effects of Lugworms on the Benthic Community

Field experiments were carried out at Balgzand (Fig. 7.2) and near the island of Schiermonnikoog in the Dutch Wadden Sea. Large experimental squares of 144 m^2 were defaunated during winter by placing a mat on top of the sediment, which was removed in early spring. Within these squares small plots of 2.25 m^2 were either restocked with different lugworm densities or left empty. Initial densities within the small plots were 0, 25, 50, 75 and 100 adult lugworms per m^2, all densities at least in duplicate per experimental plot. The repopulation of these plots was followed during the summer by taking three 86.6-cm^2 sediment samples to a depth of 15 cm per plot once a month. These samples were sieved through a 0.5-mm mesh sieve and all macrofaunal species counted. At the same time the numbers of lugworm casts were counted. In addition to these defaunated plots, lugworm densities were manipulated in 'natural plots' by either removing or adding lugworms in small 2.5-m^2 plots at Balgzand and near Schiermonnikoog. Moreover, the natural distribution patterns of lugworms and all other macrobenthic spe-

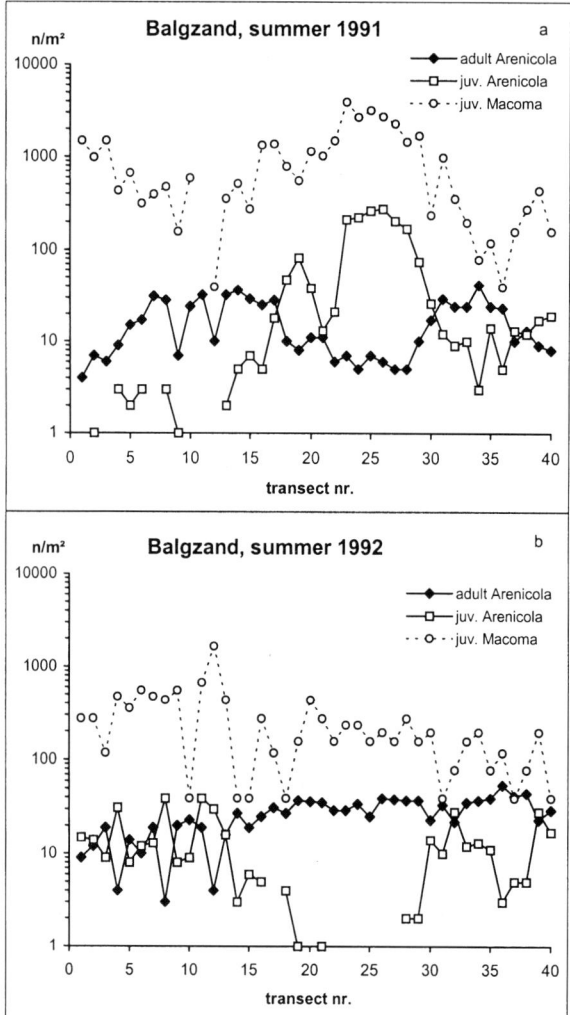

Fig. 7.3. Densities (n/m^2) on a log-scale of adult (*black symbols, solid line*) and juvenile (*open squares, solid line*) lugworms and juvenile *Macoma balthica* (*open circles, broken line*) along a transect from the shore (no. 1) to the tidal channel (no. 40) at Balgzand, situated close the experimental sites B and C, during the summer of **a** 1991 and **b** 1992

cies were studied by sampling transects from the coast towards the tidal channel.

Significant ($p<0.05$) negative relationships with lugworm densities were found in the numbers of recruits of nearly all common and abundant macrofauna species in the experimental plots (Flach 1992a). Mean densities of juveniles of all species were lower in all plots with lugworms than in the control plots without lugworms (Table 7.1). Although significant reductions in numbers of some species were already found at the lowest lugworm densities (~15 per m^2), higher lugworms densities (>30 per m^2) caused significant reductions in nearly all species. In the plots with the highest lugworm

Table 7.1. Absolute (n/m^2) and relative (% of numbers observed at 0 *Arenicola marina* densities) numbers of various species at different densities of *A. marina*. All data are averages of observations in two plots A and B in 1990

	Absolute numbers					Relative numbers (%)				
Density n/m^2 of A. marina ⇒	0	15	29	37	44	15	29	37	44
Cerastoderma edule	502	361	199*	254*	232*	72	40	51	46
Macoma balthica	454	314	260*	254*	180*	69	57	56	40
Angulis tenuis	376	232*	137**	141**	128**	62	36	38	34
Mya arenaria	318	172*	188*	163*	137*	54	59	51	43
Ensis spec.	111	81	38*	76	46	73	34	68	41
Nereis diversicolor	1524	1349	1067	1252	966*	89	70	82	63
Nephtys hombergii	817	536*	306*	203**	320**	66	37	25	39
Capitella capitata	719	523	439*	405*	333**	73	61	56	46
Heteromastus filiformis	4497	2976*	2536*	2618*	1575**	66	56	58	35
Scoloplos armiger	432	323	220**	250*	250*	75	51	58	58
Pygospio elegans	9169	5630*	4251*	4993*	4494*	61	46	54	46
Tharyx marioni	1039	595*	523*	392**	268**	57	50	38	26

* Different from control $p<0.05$; ** different from control $p<0.01$

densities (~45 per m^2) an average reduction of ~50 % in the number of recruits was found. As densities of >30 adult lugworms per m^2 are commonly observed in the Wadden Sea (Cadée 1976; Beukema and De Vlas 1979; Reise 1985), recruitment can therefore be strongly affected. In the middle part of the transect at Balgzand (Fig. 7.3), high numbers of recruits of different species (e.g. *Macoma, Nereis, Heteromastus*) were found during the summer of 1991 (when few adult lugworms were present there), but not in 1990 and 1992 (when adult lugworms were present in their normal density). Significant negative correlations ($p<0.05$) were found between the densities of adult lugworms and juvenile lugworms, as well as *Macoma, Mya, Nereis, Eteone, Capitella* and *Pygospio*, along various transects sampled during the summer of 1991 in the Dutch Wadden Sea (E. Flach, unpublished data; position of transects given in Flach 1993). The negative effects on the recruits were strongest later in summer, which implies that it was not the initial settlement of the juveniles that was prohibited by the presence of lugworms, but that, in the presence of high lugworm densities, either the survival rate of the juveniles was lower or the migration rate higher (Flach 1992a).

A strong negative impact of lugworms on the whole population of two *Corophium* species (*C. volutator* and *C. arenarium*) was also found (Flach 1992a,b, 1993). Within the experimental plots lugworm densities of only 18 per m^2 already caused a reduction in numbers in both *Corophium* species of ~50 %, whereas higher lugworm densities (~40–55 per m^2) caused reductions of between 80 and 95 % (Flach 1992a,b, 1993). Removal of adult lugworms in otherwise undisturbed 'natural' plots within the lugworm zone resulted in a significant increase in *Corophium*, whereas addition of lugworms in 'natural' plots within the *Corophium* zone caused a significant decrease in *Corophium* numbers (Flach 1992b, 1993). It was concluded that the commonly observed zonation pattern of *Corophium* dominating the upper tidal zone, and *Arenicola* the middle and lower zone, resulted from the strong negative impact of *Arenicola* on *Corophium*, restricting *Corophium* to the upper tidal zone, which for *Arenicola* is physically unfavourable (Beukema and Flach 1995).

This strong negative impact of *Arenicola* raises the question as to the mechanism by which *Arenicola* influences other species. To answer this question aquarium experiments have been carried out with *Corophium*. In the first experiment, settlement of *Corophium* was studied in the presence and absence of lugworms (Flach 1992a). In the absence of lugworms, *Corophium* settled about equally on both sides, but when lugworms were present on one side the numbers of *Corophium* on the lugworm-side were significantly lower (Fig. 7.4a). In a second experiment, migration of *Corophium* was studied in the presence and absence of lugworms (Flach 1993). About twice as many *Corophiums* had migrated to the empty side within 2 weeks when lugworms were present compared to the control treatment without lugworms (Flach 1993).

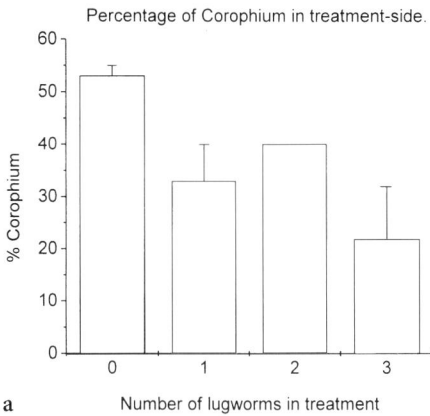

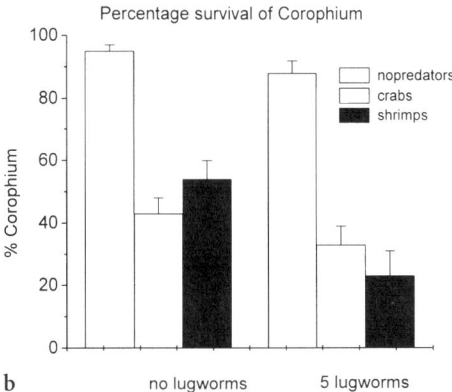

Fig. 7.4. Aquarium experiments with *Corophium volutator* and *Arenicola marina*. **a** Proportions (%) of total numbers of *Corophium* that settled in one half of the aquarium with varying numbers of lugworms (0 to 3), the other half contained no lugworms. **b** Proportions (%) of *Corophium* that survived with or without lugworms in the presence or absence of the shrimp *Crangon crangon* or the crab *Carcinus maenus*

Further experiments were carried out to study the survival of *Corophium*. The survival was always high (~90%) after 6 days and no significant difference was found between treatments with five and zero lugworms (Flach and De Bruin 1994). But when two crabs (*Carcinus*) or shrimps (*Crangon*) as epibenthic predators were added survival was lower and significantly lower in the presence of lugworms compared to the treatment without lugworms (Fig. 7.4b).

While feeding a lugworm can undermine a *Corophium* tube inducing the *Corophium* to swim away. During the time spent out of the sediment (between 13 s and 13 min, mean ~3 min) *Corophium* will be more vulnerable to epibenthic predators like crabs and shrimps, explaining the lower survival rate of *Corophium* in the presence of lugworms (Fig. 7.1). When a *Corophium* became buried under a lugworm cast, it simply extended its burrow till it reached the sediment surface again. The strong negative effect of *Arenicola* on the numbers of *Corophium* found in the field is thus caused by the feeding behaviour of *Arenicola*, which resulted in a high migration rate, which in the presence of epibenthic predators also lowered the survival rate of *Corophium* (Flach and De Bruin 1994).

7.5 Ghost Shrimps in the Ariake Sound Estuarine System in Japan

The study area is an estuarine system located in the middle part of western Kyushu, Japan (around 130°E and 32.5°N), spanning from Ariake Sound (1700 km^2 estuary) through Tachibana Bay (700 km^2) to the coastal waters of the East China Sea (open sea) (Fig. 7.5). There are two major water masses in Tachibana Bay, the northern lower-salinity one and the southern higher-salinity one, which are originally derived from the Ariake Sound waters and the East China Sea waters, respectively, with their time-averaged current directions being opposite (westward and eastward) (Matsuno et al. 1999). For 20 years, since 1979, Tamaki and colleagues have studied macrobenthic community dynamics on the intertidal sand flat in Tomioka Bay on the northwestern corner of Amakusa-Shimoshima Island; Tomioka Bay is a small part of Tachibana Bay, situated in the southwestern portion of the estuarine system. As the sand flat faces Ariake Sound, it is washed by much weaker waves than the exposed sandy beaches facing the East China Sea.

Until recently, it was believed that two species of *Callianassa* commonly occurred in Japan: *C. japonica* Ortmann, inhabiting bare intertidal sand flats, and *C. petalura* Stimpson, inhabiting boulder beaches (see Sakai 1969). However, in their taxonomic revision of the Japanese species, Manning and Tamaki (1998) revealed that Sakai's '*C. japonica*' was in fact a mixture of *C. japonica* and *C. harmandi* Bouvier. They also proposed a new genus, *Nihonotrypaea*, to include the three Japanese species. All the material studied by Tamaki and colleagues, so far described as *C. japonica*, has now proven to be *N. harmandi*, exclusively occurring on the Tomioka Bay sand flat; the distributions of *N. harmandi* and *N. japonica* are well separated around the boundary between the outer one-third and the inner two-thirds of Ariake Sound (Tamaki et al. 1999; Fig. 7.5). Of the local populations of *N. harmandi*, that on the Tomioka Bay sand flat is the largest in terms of both the density and the area occupied (Tamaki et al. 1997, 1999; Tamaki et al., in preparation).

The study site for the benthic community is situated in a moderately sheltered section of the Tomioka Bay sand flat (the black-coloured area in the inset in Fig. 7.5, about 300 m wide along shore and exposed for 310–325 m from the shoreline). The surface sediment consisted of well to moderately sorted, fine sand with 0.4%–2.6% silt-clay content. The first survey conducted in July 1979 revealed that the macrobenthic community in the study site

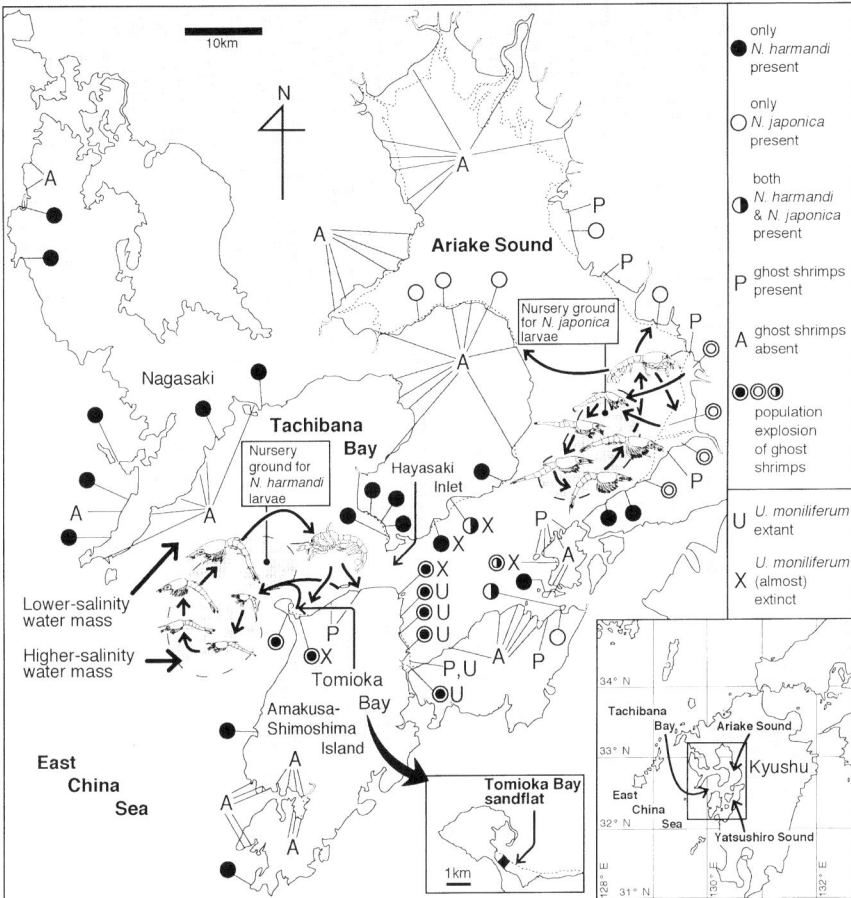

Fig. 7.5. Distributions of the ghost shrimps, *Nihonotrypaea harmandi/N. japonica*, at 103 locations and the trochid gastropod, *Umbonium moniliferum*, at 10 locations along the coastline from Ariake Sound through Tachibana Bay to the East China Sea in western Kyushu, Japan (adapted from Tamaki et al. 1999; Tamaki et al., in preparation). The *dotted line* just offshore of the coastline indicates the maximum extent of the extensive intertidal flats in Ariake Sound and of the other relatively large tidal flats in the sound and Tachibana Bay. The tidal flats with '*A*' marks in the innermost Ariake Sound are mudflats, where ghost shrimps cannot make their burrows. The Tomioka Bay sand flat is situated on the northwestern corner of Amakusa-Shimoshima Island, where the long-term monitoring of the benthic community has been conducted in the *black-coloured* area in the *inset*. The main nursery grounds for *N. harmandi* and *N. japonica* larvae are located in the southern Tachibana Bay and the central Ariake Sound, respectively. (Adapted from Tamaki and Miyabe 2000)

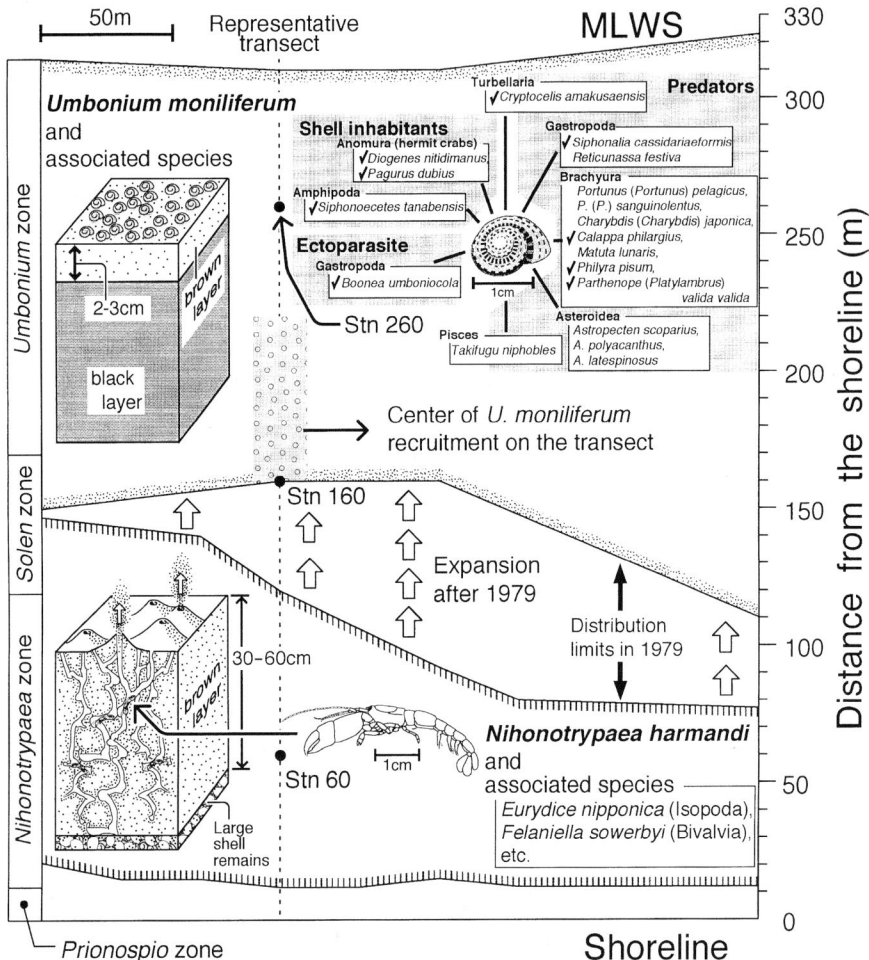

Fig. 7.6. Schematic representation of the macrobenthic community in the study site on the Tomioka Bay sand flat (the *black-coloured* area in the *inset* in Fig. 7.5) in summer, 1979 (adapted from Tamaki and Kikuchi 1983; Tamaki 1994). Four macrobenthic assemblage zones were recognised, with these positions on one representative transect designated by the generic name of each species on the left-hand side. *MLWS* Mean low water spring tide level. The subsequent regular sampling was conducted along the transect. The distribution range of *N. harmandi* expanded extensively after 1979, having occupied the entire sand flat in 1983. Following this, *Umbonium moniliferum* and nine of its associated species (*check marks*) that used to be on the lower sand flat became extinct

could be subdivided into four zonal assemblages, parallel to the shoreline (Tamaki and Kikuchi 1983; Fig. 7.6). They were identified according to their most characteristic species, including *N. harmandi* and the epibenthic filter-feeding trochid gastropod, *Umbonium (Suchium) moniliferum*. On one representative transect, the *Nihonotrypaea* and *Umbonium* zones occupied the upper one-third and lower half parts, respectively. Subsequent studies revealed that biological interactions either associated with or incompatible with *N. harmandi* and/or *U. moniliferum* were responsible for the determination of the distribution of several dominant species (e.g. Tamaki 1985a,b, 1987, 1988, 1994; Tamaki and Suzukawa 1991; Tamaki et al. 1992).

From 1979, the distribution range of *N. harmandi* expanded seaward, and, by summer 1983, the entire sand flat was densely populated by the ghost shrimp (Tamaki 1994), which subsequently caused considerable bioturbation effects on both the sediment and the benthic community. A population explosion of both *N. harmandi* and *N. japonica* occurred also on several other sand flats in the present estuarine system in these 20 years (indicated by double circles in Fig. 7.5). From 1984, on the Tomioka Bay sand flat, although no reduction in the ghost shrimp distribution range was observed, its population density varied.

Based on Tamaki and Ingole (1993) and Tamaki et al. (1997; in preparation), the change in the population density of *N. harmandi* on the Tomioka Bay sand flat over the course of 20 years (mostly in July or August each year) is summarised in Fig. 7.7a. Three stations were located along the transect, with each station number identical to the distance from the shoreline (Stns 60, 160, and 260 in Fig. 7.6). For the ghost shrimp collection, a corer of a 100-cm^2 unit-area was used, with 16–20 (in most cases) sediment columns to the base layer taken per station. Only adult members of the population are considered, and juveniles recruited in the same year are excluded from the present analysis. Accompanying the completion of the distribution expansion from 1980 to 1984, the mean density at Stn 60 increased by a factor of 3.6. After 1984, for a period of 10 years, the population density at each station was fairly stable, with the mean numbers per 100 cm^2 of 10–14 at Stn 260, 4–10 at Stn 160 (except in 1989), and 4–6 at Stn 60. Since 1995, the densities at Stns 260 and 160 have declined to 1.6–3.2/100 cm^2 in 1998, while that at Stn 60 has remained more or less constant.

For the three phases of the *N. harmandi* population dynamics on the Tomioka Bay sand flat (explosion, stability, and decline), several explanations have been proposed concerning their driving forces (Tamaki et al. 1992, 1997; Tamaki et al., in preparation; Tamaki and Ingole 1993). The *population explosion phase* involved: (1) the colonisation of a new zone seaward of the original habitat by emigrant adults prior to larval settlement period, and (2) the highest settlement of post-larvae in that zone, followed by far better survival of those newly recruited juveniles than those that settled in the zone

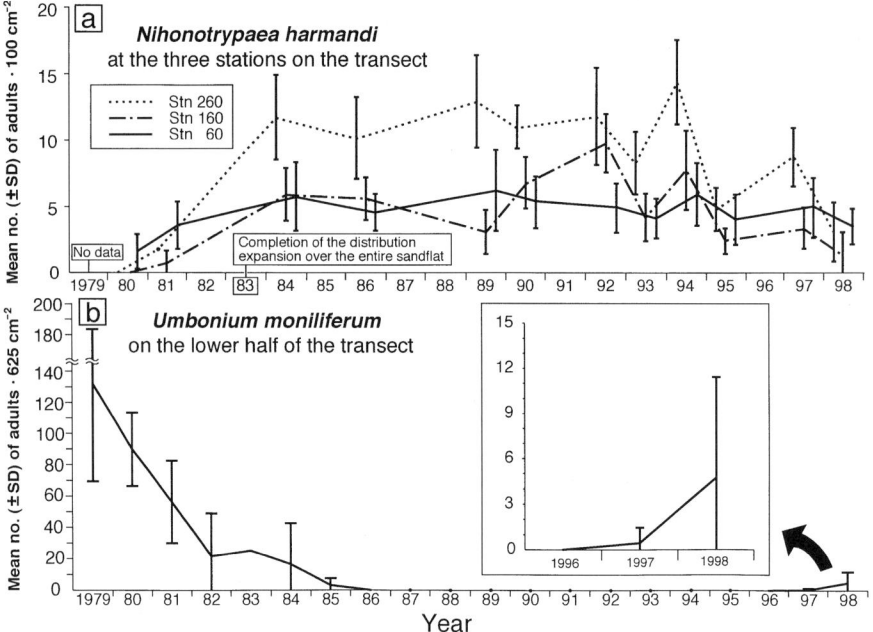

Fig. 7.7. a Long-term change in the densities of adults of *Nihonotrypaea harmandi* at the three stations along the transect on the Tomioka Bay sand flat (Fig. 7.6) (adapted from Tamaki and Ingole 1993; Tamaki et al., in preparation). For the definition of adults, see text. The shrimp collection was made with a 100-cm² corer to the substrate base, with 16–20 samples per station; **b** Long-term change in the density of adults of *Umbonium moniliferum* at all the stations on the lower half of the transect (adapted from Tamaki 1994; Tamaki et al., in preparation). Due to the annual sampling months (July or August) being prior to the gastropod recruitment, only adults were collected. The collection was made with a 25 × 25-cm quadrat to a depth of 10 cm, with one sample per station

without adults. The latter points to facilitation of larval settlement and survival by conspecific adults. Adults can indirectly assist in settlement by softening the substrate through bioturbation, with the presence of adult burrows also acting as a conduit helping the larvae reach the deeper portion of the sediment column. In addition to the above local-scale factors, it is supposed that much larger-scale factors would also have been involved in the population explosion, considering its widespread occurrence in western Kyushu in recent years (Fig. 7.5). One large-scale hypothesis is that recent changes in the water conditions of the estuarine system have increased the survival of pelagic larvae, whose main nursery grounds are located in the southern part of Tachibana Bay (the higher-salinity water mass for *N. harmandi*) and in the central part of Ariake Sound (for *N. japonica*) (Tamaki and Miyabe 2000; Fig. 7.5).

During the *stable period* for the population on the Tomioka Bay sand flat, the higher densities of newly recruited shrimps at the seaward stations (Tamaki and Ingole 1993; Tamaki et al. 1997) and its reflection in adult shrimp densities (Fig. 7.7a) could primarily be due to the above-mentioned positive intraspecific relationship. Intraspecific competition for space could also regulate the population density. In fact, the burrow space exclusively occupied by each shrimp is a good predictor of the carrying capacity for the population on the sand flat (Tamaki et al. 1997; Tamaki and Ueno 1998). For the *population decline* at the two seaward stations in recent years, one possibility is the sudden increase in the abundance of the stingray, *Dasyatis akajei*, from 1995, the fish voraciously feeding on ghost shrimps and disturbing the substrate to a considerable depth (up to 20 cm), especially in the lower two-thirds of the sand flat (A. Tamaki, unpublished data).

7.6 Effects of the Ghost Shrimp Expansion and Decline

7.6.1 Effects on Sediment Properties

Following the expansion of *N. harmandi* over the Tomioka Bay sand flat, sediment properties were remarkably altered by the shrimp bioturbation on the landscape scale (Tamaki and Suzukawa 1991; Tamaki 1994; Fig. 7.6). Destabilisation occurred because mounds deposited on the sand flat surface were easily dispersed by tidal currents and waves causing increased erodability. An estimated deposition rate of 2.6–12.7 mm/day would result in a mixing of the entire sediment column in 47–235 days. The silt-clay content at the surface sediment has been reduced due to the ghost shrimps 'blowing off' the fine particles resuspended into the water column, thereby reducing the cohesiveness of the sediment. Increased oxygenation of the entire sediment column has occurred due to the overlying water introduced into the burrows, such that the brown layer that was originally close to the surface is now extended to the substrate base. Finally, larger shell material, such as empty shells of *U. moniliferum*, was buried deeper by continual sediment deposition by ghost shrimps. Thus, the substrate properties of the entire dissipative sand flat were altered by the ghost shrimp to approximate those of exposed sandy beaches. Since 1995, the substrate properties have not been examined in detail, but are likely to have shifted due to the lowered ghost shrimp densities (Fig. 7.7a) and increased disturbance by stingrays.

7.6.2 Effects on Invertebrates

The expansion of *N. harmandi* over the Tomioka Bay sand flat also caused a change in the distribution and abundance of most macrofauna. Regular sampling of macrobenthos was carried out annually along the transect (Fig. 7.6) during the end of July to mid-August. Every 20 m, a 25 × 25 cm quadrat was sampled to a depth of 10 cm, and sieved through a 0.5-mm mesh sieve. One of the most dramatic effects from 1979 onwards was the steady decline of *U. moniliferum* from 2000 m^{-2} to 0 in 1986 (Tamaki 1994; Fig. 7.7b). It is likely that newly settled *U. moniliferum* larvae were buried and smothered under the sediment deposited by ghost shrimps (Fig. 7.6). Interestingly, following the decline of the *N. harmandi* population from 1995, the *U. moniliferum* population showed signs of recovery. Subsequent to the extinction of *U. moniliferum*, nine of its associated species (those with check marks in Fig. 7.6) also disappeared. In particular, those species using empty shells suffered from the deep burial of shells by ghost shrimps.

The effects of the ghost shrimp bioturbation on the smaller-sized macrobenthos were both positive and negative. The cirolanid isopod, *Eurydice nipponica*, which was confined to the *Nihonotrypaea* zone in 1979 (Fig. 7.6), increased its distribution range and density over the entire sand flat in 1984 (Tamaki and Suzukawa 1991), but the population has declined since 1995 (A. Tamaki, unpublished data). As the primary habitat of *E. nipponica* is exposed sandy beaches facing the East China Sea, the isopod on the Tomioka Bay sand flat appeared to have benefited from the destabilised substrate conditions created by ghost shrimps. In contrast, sediment extrusion by ghost shrimps caused active flight behaviour in the mobile, opheliid polychaete, *Armandia amakusaensis*, by disturbing its surface deposit-feeding behaviour (Tamaki 1988), resulting in its much lower densities in the *Nihonotrypaea* zone in 1979 and 1980 (Tamaki 1985a,b). Since 1995, this polychaete has declined, probably partly due to stingray disturbance (A. Tamaki, unpublished data). The abundance of the tube-building, spionid polychaete, *Pseudopolydora paucibranchiata*, which was initially the most dominant macrofaunal species, occurring exclusively in the *Solen* to uppermost *Umbonium* zones (Tamaki and Kikuchi 1983; Tamaki 1985a), has been quite low since 1984 (A. Tamaki, unpublished data). Most probably, the destruction of the substrate surface would have inhibited the establishment of the polychaete's tubes. The recent slight recovery of the *U. moniliferum* population on the Tomioka Bay sand flat (Fig. 7.7b) apparently owed its larval sources to the five extant local populations scattered along the east coast of Amakusa-Shimoshima Island (the sand flats with 'U' marks in Fig. 7.5; Tamaki et al., in preparation). The four local populations of *U. moniliferum* in Ariake Sound became extinct (the sand flats with 'X' marks in Fig. 7.5), and two of these cases were probably caused by the population explosion of *N. harmandi*, which had colonised a considerable part of each smaller sand flat (Tamaki et al., in preparation). Densities of 160 or more ghost shrimps per m^2 seemed to be crucial to inhibit the recruitment success of *U. moniliferum*. This density might well explain the recent recovery of the gastropod population on the Tomioka Bay sand flat, especially at Stn 160 (Figs. 7.6, 7.7). On four of the five relatively large sand flats where the *U. moniliferum* populations remained on the east coast of Amakusa-Shimoshima Island, the expansion of *N. harmandi* was not complete and a fairly large part of each sand flat was left as a refuge for the *U. moniliferum* recruitment (Tamaki et al., in preparation). *Umbonium moniliferum* larvae have a short planktonic period (within 2 days? cf. Berry 1986), so that larvae released from the east coast of the island could swiftly come to the Tomioka Bay sand flat by riding on a strong ebb tide current which can attain speeds as high as 3.6 m/s at spring tides (Fig. 7.5). Given the larger population size of *U. moniliferum* on the Tomioka Bay sand flat recorded in 1979, the above source-sink relationship would be reciprocal. This relationship is in stark contrast to the situation for *N. harmandi*. Tamaki and Miyabe (2000) have demonstrated that, in Tachibana Bay, larvae were retained only in the southern part of the bay, being flushed out of the northern part of the bay westward to the East China Sea (Fig. 7.5). Because of the longer planktonic period for *N. harmandi* (3–4 weeks), those originating from the local populations other than the Tomioka Bay sand flat and its neighbours would eventually be entrained by the lower-salinity water mass in the northern Tachibana Bay and hence would not reach the Tomioka Bay sand flat. Thus the *N. harmandi* population on the Tomioka Bay sand flat would be a source rather than a sink in the metapopulation of the Ariake Sound estuarine system.

7.7 Comparisons Between Biogeographic Regions

We have shown here that the bioturbative activities of two species, even belonging to two different major taxa (Polychaeta and Crustacea), can cause similar effects on benthic communities composed of different species on different sides of the world. Both the lugworm and the ghost shrimp provide good examples of phenomena denoted by Posey's *mobility-mode hypothesis*. Both species are relatively large and have a strong impact on sediment structure by their bioturbating activities producing a similar mountain-like landscape on the sediment surface, which in turn has a strong influence on the densities of other species. These affected species are often smaller and/or live in more or less permanent tubes and the effect of the large biotubators are in this case always negative and thus good examples of amensalism. For the lugworm, only negative influences were found in the present study. However, Reise (1985) also found positive effects of lugworms on meiofauna species.

Lackschewitz and Reise (1998) report a positive relationship between lugworms and the amphipod *Urothoe poseidonis,* and Beukema et al. (1999) between lugworms and the polychaete *Harmothoe sarsi.* On the intertidal flats of Tomioka Bay a positive relationship was found between the ghost shrimp and the isopod *Eurydice,* expansion of the ghost shrimp being followed by expansion of the isopod to the seaward zone. Thus, on both sides of the world, examples of both amensalism and of commensalism were found. In both cases the large bioturbators themselves did not seem to be influenced by the interactions. The reduction in numbers of the ghost shrimp in the years after the increase of the stingray population in Tomioka Bay appears to be an example of predator-prey interaction that is of importance in structuring the benthic community.

Predation seems to be important in the Wadden Sea as well, but as an additional factor reducing further densities of species that are already negatively influenced by the bioturbating activities of the lugworms. This is an example of interaction modification (Wootton 1993), where the relationship between a predator and its prey is changed by the effect of a third species (the lugworm in our case) that alters the behaviour of the prey (flight reaction in *Corophium*) therefore increasing its predation risk. In shallow water bays in the Gullmarsfjord on the Swedish west coast, this combined interaction was shown to strongly influence the distribution pattern of *Corophium* (Flach and De Bruin 1994). In the Finnsbovik a similar zonation pattern was observed to that in the Wadden Sea, *Corophium* dominating the most shallow zone and *Arenicola* strongly dominating the intermediate zone (Flach and De Bruin 1993). Removal of the lugworms and excluding predators by placing cages at two different levels in the lugworm zone resulted in roughly similar high *Corophium* densities as those found in the shallowest zone. The exclusion of predators and lugworms had a much stronger effect in the deeper part, where predation pressure was higher, than in the shallower part, thus demonstrating the strong impact of both types of interactions working together (Flach and De Bruin 1994).

The strong impact that expansion of the large bioturbator has on the benthic community, as was observed for the ghost shrimp in Tomioka Bay, was also found for expansion of the lugworm in the Finnsbovik in Sweden. In 1992 the lugworm was restricted in high densities to the intermediate part of a transect from the shore to the low water line. During summer 1992, high numbers of juvenile lugworms settled both in the shallow and the deep part, where adult numbers were low (Flach and Beukema 1994). The juvenile lugworms stayed in the shallow zone and survived the winter resulting in shoreward expansion of the lugworm zone (Fig. 8 in Flach and Beukema 1994), which had strong effects on the species that were previously living in this shallow zone. The numbers of *Corophium* were strongly reduced (Flach and De Bruin 1994) as were the numbers of the tube-building polychaete,

Pygospio elegans (E. Flach, unpublished data), and both species were in 1993 restricted to a very small zone close to the shore. It can thus be expected that an expansion of the lugworm in the Wadden Sea could have a similarly dramatic effect as that observed for the ghost shrimp in Ariake Sound. The American lugworm species *Abarenicola pacifica* also has a negative influence on *Pygospio* (Wilson 1981) and Woodin (1985) reports an active avoidance of the spionid *Pseudopolydora* of sites inhabited by *Abarenicola*. Interspecific interactions can thus be regarded as strongly influencing the benthic community on tidal flats.

Most of the reports on the effect of the bioturbation by callianassid shrimps on the abundance pattern of other macrobenthos have pointed out asymmetrical effects, not to mention shrimps harbouring obligate commensals within or close vicinity to their burrows (e.g. Ricketts et al. 1985), for example: (1) amensalism – against filter-feeding bivalves (Peterson 1977; Murphy 1985), corals (Aller and Dodge 1974), seagrass (Suchanek 1983), cultured penaeid shrimps (Nates and Felder 1998), and smaller macrobenthos with limited mobility such as tanaids and spionid polychaetes (Dayton and Oliver 1980; Posey 1986a); and (2) commensalism – for highly mobile species such as ostracods (Dayton and Oliver 1980; Riddle 1988), amphipods (Posey 1986a; Riddle 1988; Wynberg and Branch 1994), bivalves (Aller and Dodge 1974; Tudhope and Scoffin 1984), and polychaetes (Riddle 1988; Wynberg and Branch 1994). The effects on meifauna also seem to be dual (Branch and Pringle 1987; Dobbs and Guckert 1988; Wynberg and Branch 1994; Dittmann 1996). Although the abundance patterns of these affected species were quite distinct, some of the relationships with callianassid shrimps seem to be facultative. As regards the biological agents that limit the distribution of callianassid shrimps, only a few papers have been published (by predation, Posey 1986b; through inhibition of burrowing by seagrass, Brenchley 1982; Harrison 1987).

Intraspecific competition in key-species, such as the large bioturbating lugworm and ghost shrimp, can be important as well. For the lugworm both adult-adult, juvenile-juvenile and adult-juvenile interactions have been observed (Flach and Beukema 1994). In the case of the adult-adult and juvenile-juvenile interactions there seems to be competition for food (thus real competition, – – interaction Arthur and Mitchell 1989), keeping the adult population at the carrying capacity of a particular site and resulting in low mean individual weights for juveniles when they occur in high densities (Flach and Beukema 1994). Adult-juvenile interactions can be regarded as a form of amensalism; the adults do not seem to be influenced by the juveniles, but the juveniles are strongly negatively influenced by the adults probably again due to disturbance by the bioturbating activity of the adults. These intraspecific interactions seem to be responsible for the stable lugworm population in the Wadden Sea within the boundaries set by the physical environment (Beukema and De Vlas 1979). However, during a period of eutrophication, resulting in a higher food supply to the benthos (Beukema and Cadée 1986), the overall numbers and biomass of the lugworm in the Wadden Sea increased (Beukema 1991, 1992).

In Tomioka Bay, intraspecific competition for space seems to regulate the adult ghost shrimp population at the carrying capacity during the stable phase (Tamaki and Ueno 1998). The adult-juvenile interaction, however, appears to be positive, at least for juvenile ghost shrimps. The presence of adult ghost shrimps facilitates the larval settlement, just because of their bioturbating activities, which makes the sediment more easily penetrated by the larvae (Tamaki and Ingole 1993). This positive adult-juvenile interaction is one of the main differences between the ghost shrimp and the lugworm, and at least partly responsible for the population expansion of the ghost shrimp in Ariake Sound. The higher survival rate of the pelagic larvae, however, seems to be the result of a change in the physical environment.

It can thus be concluded that the physical environment is important in setting the boundaries for the populations of key-species, such as the large bioturbating ghost shrimp and lugworm, on a large scale. On a smaller scale, intraspecific interactions are important in regulating densities and distribution patterns of these key-species. Besides intraspecific interactions within the key-species, interspecific interactions, either positively or negatively related to the populations of the key-species, have a strong impact on the whole benthic community on intertidal sand flats all over the world.

References

Aller RC, Dodge RE (1974) Animal-sediment relations in a tropical lagoon Discovery Bay, Jamaica. J Mar Res 32:209-232

Aller RC, Yingst JY (1985) Effects of marine deposit-feeders *Heteromastus filiformis* (Polychaeta), *Macoma balthica* (Bivalvia), and *Tellina texana* (Bivalvia) on averaged sedimentary solute transport, reaction rates, and microbial distribution. J Mar Res 43:615-645

Andersen FØ, Kristensen E (1991) Effects of burrowing macrofauna on organic matter decomposition in coastal marine sediments. Symp Zool Soc Lond 63:69-88

Arthur W, Mitchell P (1989) A revised scheme for the classification of population interactions. Oikos 56:141-143

Berry AJ (1986) Daily, tidal and two-weekly spawning periodicity and brief pelagic dispersal in the tropical intertidal gastropod *Umbonium vestiarium* (L.). J Exp Mar Biol Ecol 95:211-223

Beukema JJ (1976) Biomass and species richness of the macrobenthic animals living on tidal flats in the western part of the Wadden Sea. Mar Biol 99:425-433

Beukema JJ (1991) Changes in composition of bottom fauna of a tidal-flat area during a period of eutrophication. Mar Biol 111:293-301

Beukema JJ (1992) Long-term and recent changes in the benthic macrofauna living on tidal flats in the western part of the Wadden Sea. Neth Inst Sea Res Publ Ser 20: 135–141

Beukema JJ, Cadée GC (1986) Zoobenthos responses to eutrophication of the Dutch Wadden Sea. Ophelia 26:55–64

Beukema JJ, De Vlas J (1979) Population parameters of the lugworm *Arenicola marina* living on tidal flats in the Dutch Wadden Sea. Neth J Sea Res 13:331–353

Beukema JJ, Flach EC (1995) Factors controlling the upper and lower limits of the intertidal distribution of two *Corophium* species in the Wadden Sea. Mar Ecol Prog Ser 125:117–126

Beukema JJ, Cadée GC, Hummel H (1983) Differential variability in time and space of numbers in suspension and deposit feeding benthic species in a tidal flat area. Oceanol Acta No Sp 21–25

Beukema JJ, Flach EC, Dekker R, Starink M (1999) A long-term study of the recovery of the macrozoobenthos on large defaunated plots on a tidal flat in the Wadden Sea. J Sea Res 42:235–254

Branch GM, Pringle A (1987) The impact of the sand prawn *Callianassa kraussi* Stebbing on sediment turnover and on bacteria, meiofauna, and benthic microflora. J Exp Mar Biol Ecol 107:219–235

Brenchley GA (1982) Mechanisms of spatial competition in marine soft-bottom communities. J Exp Mar Biol Ecol 60:17–33

Cadée GC (1976) Sediment reworking by *Arenicola marina* on tidal flats in the Dutch Wadden Sea. Neth J Sea Res 10:440–460

Dankers N, Beukema JJ (1983) Distributional patterns of macrobenthic species in relation to some environmental factors. In: Wolff WJ (ed) Ecology of the Wadden Sea, vol 1(4). Balkema, Rotterdam, pp 69–103

Dayton PK, Oliver JS (1980) An evaluation of experimental analyses of population and community patterns in benthic marine environments. In: Tenore KR, Coull BC (eds) Marine benthic dynamics. Univ South Carolina Press, Columbia, South Carolina, pp 93–120

Dittmann S (1996) Effects of macrobenthic burrows on infaunal communities in tropical tidal flats. Mar Ecol Prog Ser 134:119–130

Dobbs FC, Guckert JB (1988) *Callianassa trilobata* (Crustacea: Thalassinidea) influences abundance of meiofauna and biomass, composition, and physiologic state of microbial communities within its burrow. Mar Ecol Prog Ser 45:69–79

Flach EC (1992a) Disturbance of benthic infauna by sediment-reworking activities of the lugworm *Arenicola marina*. Neth J Sea Res 30:81–89

Flach EC (1992b) The influence of four macrozoobenthic species on the abundance of the amphipod *Corophium volutator* on tidal flats of the Wadden Sea. Neth J Sea Res 29:379–394

Flach EC (1993) The distribution of the amphipod *Corophium arenarium* in the Dutch Wadden Sea: relationships with sediment composition and the presence of cockles and lugworms. Neth J Sea Res 31:281–290

Flach EC, Beukema JJ (1994) Density-governing mechanisms in populations of the lugworm *Arenicola marina* on tidal flats. Mar Ecol Prog Ser 115:139–149

Flach EC, De Bruin W (1993) Effects of *Arenicola marina* and *Cerastoderma edule* on distribution, abundance and population structure of *Corophium volutator* in Gullmarsfjorden, western Sweden. Sarsia 78:105–118

Flach EC, De Bruin W (1994) The activity of cockles, *Cerastoderma edule* (L.), and lugworms, *Arenicola marina* (L.), make *Corophium volutator* (Pallas) more vulnerable to epibenthic predators: a case of interaction modification. J Exp Mar Biol Ecol 182:265-285

Harrison PG (1987) Natural expansion and experimental manipulation of seagrass (*Zostera* spp.) abundance and the response of infaunal invertebrates. Estuar Coast Shelf Sci 24:799-812

Lackschewitz D, Reise K (1998) Macrofauna on flood delta shoals in the Wadden Sea with an underground association between the lugworm *Arenicola marina* and the amphipod *Urothoe poseidonis*. Helgoländer Meeresunters 52:147-158.

Manning RB, Tamaki A (1998) A new genus of ghost shrimp from Japan (Crustacea: Decapoda: Callianassidae). Proc Biol Soc Wash 111:889-892

Matsuno T, Shigeoka M, Tamaki A, Nagata T, Nishimura K (1999) Distributions of water masses and currents in Tachibana Bay, west of Ari-ake Sound, Kyushu, Japan. J Oceanogr 55:515-529

Murphy RC (1985) Factors affecting the distribution of the introduced bivalve, *Mercenaria mercenaria*, in a California lagoon – the importance of bioturbation. J Mar Res 43:673-692

Nates SF, Felder DL (1998) Impacts of burrowing ghost shrimp, Genus *Lepidophthalmus* Crustacea: Decapoda: Thalassinidea, on penaeid shrimp culture. J World Aquacult Soc 29:188-210

Peterson CH (1977) Competitive organization of the soft-bottom macrobenthic communities of southern California lagoons. Mar Biol 43:343-359

Posey MH (1986a) Changes in a benthic community associated with dense beds of a burrowing deposit feeder, *Callianassa californiensis*. Mar Ecol Prog Ser 31:15-22

Posey MH (1986b) Predation on a burrowing shrimp: distribution and community consequences. J Exp Mar Biol Ecol 103:143-161

Posey MH (1987) Influence of relative mobilities on the composition of benthic communities. Mar Ecol Prog Ser 39:99-104

Reise K (1985) Tidal flat ecology. An experimental approach to species interactions. Ecological studies, vol 54. Springer, Berlin Heidelberg New York

Ricketts EF, Calvin J, Hedgpeth JW, revised by Phillips DW (1985) Between pacific tides, 5th edn. Stanford Univ Press, Stanford

Riddle MJ (1988) Cyclone and bioturbation effects on sediments from coral reef lagoons. Estuar Coast Shelf Sci 27:687-695

Sakai K (1969) Revision of Japanese callianassids based on the variations of larger cheliped in *Callianassa petalura* Stimpson and *C. japonica* Ortmann (Decapoda: Anomura). Publ Seto Mar Biol Lab 17:209-252, pls. 9-15

Suchanek TH (1983) Control of seagrass communities and sediment distribution by *Callianassa* (Crustacea, Thalassinidea) bioturbation. J Mar Res 41:281-298

Tamaki A (1985a) Inhibition of larval recruitment of *Armandia* sp. (Polychaeta: Opheliidae) by established adults of *Pseudopolydora paucibranchiata* (Okuda) (Polychaeta: Spionidae) on an intertidal sand flat. J Exp Mar Biol Ecol 87:67-82

Tamaki A (1985b) Zonation by size in the *Armandia* sp. (Polychaeta: Opheliidae) population on an intertidal sand flat. Mar Ecol Prog Ser 27:123-133

Tamaki A (1987) Comparison of resistivity to transport by wave action in several polychaete species on an intertidal sand flat. Mar Ecol Prog Ser 37:181-189

Tamaki A (1988) Effects of the bioturbating activity of the ghost shrimp *Callianassa japonica* Ortmann on migration of a mobile polychaete. J Exp Mar Biol Ecol 120:81-95

Tamaki A (1994) Extinction of the trochid gastropod, *Umbonium* (*Suchium*) *moniliferum* (Lamarck), and associated species on an intertidal sand flat. Res Pop Ecol 36:225-236

Tamaki A, Ingole B (1993) Distribution of juvenile and adult ghost shrimps, *Callianassa japonica* Ortmann (Thalassinidea), on an intertidal sand flat: intraspecific facilitation as a possible pattern-generating factor. J Crustacean Biol 13:175–183

Tamaki A, Kikuchi T (1983) Spatial arrangement of macrobenthic assemblages on an intertidal sand flat, Tomioka Bay, west Kyushu. Publ Amakusa Mar Biol Lab, Kyushu Univ 7:41–60

Tamaki A, Miyabe S (2000) Larval abundance patterns for three species of *Nihonotrypaea* (Decapoda: Thalassinidea: Callianassidae) along an estuary-to-open-sea gradient in western Kyushu, Japan. J Crustacean Biol 20 [Special number 2]:182–191

Tamaki A, Suzukawa K (1991) Co-occurrence of the cirolanid isopod *Eurydice nipponica* Bruce and Jones and the ghost shrimp *Callianassa japonica* Ortmann on an intertidal sand flat. Ecol Res 6:87–100

Tamaki A, Ueno H (1998) Burrow morphology of two callianassid shrimps, *Callianassa japonica* Ortmann, 1891 and *Callianassa* sp. (=*C. japonica*: de Man, 1928) (Decapoda: Thalassinidea). Crustacean Res 27:28–39

Tamaki A, Ikebe K, Muramatsu K, Ingole B (1992) Utilization of adult burrows by juveniles of the ghost shrimp, *Callianassa japonica* Ortmann: evidence from resin casts of burrows. Res Crustacea 21:113–120

Tamaki A, Ingole B, Ikebe K, Muramatsu K, Taka M, Tanaka M (1997) Life history of the ghost shrimp, *Callianassa japonica* Ortmann (Decapoda: Thalassinidea), on an intertidal sand flat in western Kyushu, Japan. J Exp Mar Biol Ecol 210:223–250

Tamaki A, Itoh J, Kubo K (1999) Distributions of three species of *Nihonotrypaea* (Decapoda: Thalassinidea: Callianassidae) in intertidal habitats along an estuary to open-sea gradient in western Kyushu, Japan. Crustacean Res 28:37–51

Tudhope AW, Scoffin TP (1984) The effects of *Callianassa* bioturbation on the preservation of carbonate grains in Davies Reef lagoon, Great Barrier Reef, Australia. J Sed Petrol 54:1091–1096

Wilson WH (1981) Sediment-mediated interactions in a densely populated infaunal assemblage: the effects of the polychaete *Abarenicola pacifica*. J Mar Res 39:735–748

Woodin SA (1985) Effects of defeacation by Arenicolid polychaete adults on spionid polychaete juveniles in field experiments: selective settlement or differential mortality. J Exp Mar Biol Ecol 87:119–132

Wootton JT (1993) Indirect effects and habitat use in an intertidal community: interaction chains and interaction modifications. Am Nat 141:71–89

Wynberg RP, Branch GM (1994) Disturbance associated with bait-collection for sandprawns (*Callianassa kraussi*) and mudprawns (*Upogebia africana*): long-term effects on the biota of intertidal sand flats. J Mar Res 52:523–558

8 Biological and Physical Processes That Affect Saltmarsh Erosion and Saltmarsh Restoration: Development of Hypotheses

R.G. Hughes

8.1 Introduction

Saltmarsh habitats are under increasing threat, particularly from sea-level rise (SLR) associated with global warming and increased wave action associated with climate change. The saltmarshes of southeastern England, which developed for several centuries under conditions of local SLR, are now disappearing rapidly. These marshes offer the potential for study of the processes that determine loss of vegetation, and of the processes necessary for the managed amelioration of these losses, as some experimental managed realignment schemes are in progress. Burd (1992) reported that, in the 15 years prior to 1988, the total losses of saltmarsh vegetation in the estuaries of southeastern England, from the River Orwell in Suffolk to the Swale in Kent, varied from 23 % in the Blackwater to 44 % in the Stour (Fig. 8.1). Most of the losses were from the pioneer zone where up to 74 % was lost during the same period. Loss of these marshes is causing concern because of the reduction in protection offered by the vegetation to the sea walls that surround most of this coastline, and because of the conservation importance of these habitats. These estuaries contain about 28,000 ha of mudflats and 8500 ha of vegetated saltmarshes, of which about 85 % is internationally important, largely because of their use by migrating and overwintering birds.

Enclosure of saltmarshes behind sea walls, primarily for agricultural use, has occurred in these estuaries for several centuries (e.g. Burd 1995). While these losses are attributable, the recent erosion is less easy to explain. Some factors that may contribute to saltmarsh loss include; removal of sediment by dredging, which may increase exposure to wave action; distant sea walls and groynes, which may limit erosion and transport of recharging sediment; and adjacent sea walls, which may accelerate local erosion by altering tidal flow and by reflecting wave energy back across the fronting saltmarsh. The loss of most of the intertidal eelgrass *Zostera marina* because of the wasting phenomenon from the 1930s, may have increased sediment erosion in front of

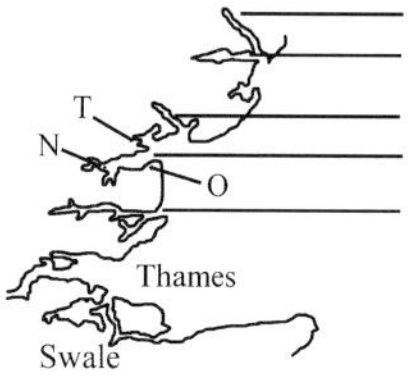

	Total	Pioneer Zone
Orwell	40	74
Stour	44	60
Colne	12	53
Blackwater	23	74
Crouch	26	25

Fig. 8.1. The proportion of the total saltmarsh area and pioneer zone area lost between 1973 and 1988 from some of the estuaries of SE England. (Data from Burd 1992)

the marshes and left them more exposed to wave action. Beardall et al. (1988) reported that the loss of *Zostera* from the River Stour led to the loss of 15 million m^3 of sediment and increased its tidal volume by 30 %.

Notwithstanding the possible impacts of the factors listed above, the generally accepted explanation for the erosion of these saltmarshes is coastal squeeze (Davidson et al. 1991; Burd 1992). It is generally considered that the upper limits of the distribution of saltmarsh plant species are determined by interspecific competition with plants of the next successional stage, and the lower limits by their tolerance to varied factors associated with increased periods of immersion by seawater. The natural response to SLR is for a progressive landward encroachment of the saltmarsh as the upper and lower distribution limits of the plant species move upward. Coastal squeeze is caused by the sea walls that prevent this landward encroachment of saltmarshes. There is little doubt that coastal squeeze is a problem of increasing importance, because of SLR associated with global warming. However, the hypothesis that coastal squeeze is already directly responsible for the losses of saltmarsh in SE England in this manner is unconvincing, and there is evidence to the contrary.

The hypothesis is unconvincing because isostatic SLR, currently about 3 mm year^{-1}, has been occurring for several thousands of years, and eustatic SLR for about 200 years. Until relatively recently, saltmarshes developed, often to seaward of claimed marshes, and their erosion began only 40–50 years ago (Boorman et al. 1989). Coastal squeeze should lead to the loss of the upper marsh species first, followed by mid marsh species leaving the pioneer zone species to disappear last. The main evidence for coastal squeeze in these marshes is from Burd (1992) who identified an increased dominance by lower zone communities from 1973 and 1998. However, Reed (1988) found that the rate of vertical accretion on the Dengie marsh (between the Blackwater and Crouch estuaries) was at least keeping pace with SLR, and the vegetation was not stressed by increased submergence. While the Dengie marsh may be unusual, in that it is experiencing severe wave erosion that may provide sediment for vertical accretion, recent evidence from a more sheltered marsh at Tollesbury also indicates that the rates of vertical accretion are greater than local SLR (Cahoon et al., 2000). Moreover, most of the loss of vegetation in

SE England was from the pioneer zone (Fig. 8.1), a result that is contrary to the effects of coastal squeeze. Houwing et al. (1999) concluded that the loss of pioneer zone vegetation in the Netherlands from 1976–1983 was not primarily due to increased tidal inundation. Although Burd (1992) attributed loss of pioneer zone communities to increased wave action, much of the loss of this (and other) vegetation has been from sheltered sites, particularly by lateral erosion of creeks within the marsh, without any apparent connection to wave action, nor indeed to coastal squeeze. For example, Fig. 8.2 shows that most of the erosion of marshes at Tollesbury identified by Burd (1992) was from within saltmarsh creeks. The vegetation is lost as undercut blocks of the saltmarsh slump into the creeks, as described by Gabet (1998). Creek erosion has resulted in the loss of large areas of vegetation within the marsh, especially in the back marsh, to leave extensive areas of mud with some residual mud mounds capped with filamentous algae (Fig. 8.3) (see also Allen and Pye 1992). This erosion may be more responsible for the loss of higher zone communities identified by Burd (1992), rather than the presumed upward encroachment of lower zone vegetation under the coastal squeeze hypothesis, for which there is no evidence. The evidence points to processes other than coastal squeeze being responsible for loss of vegetation.

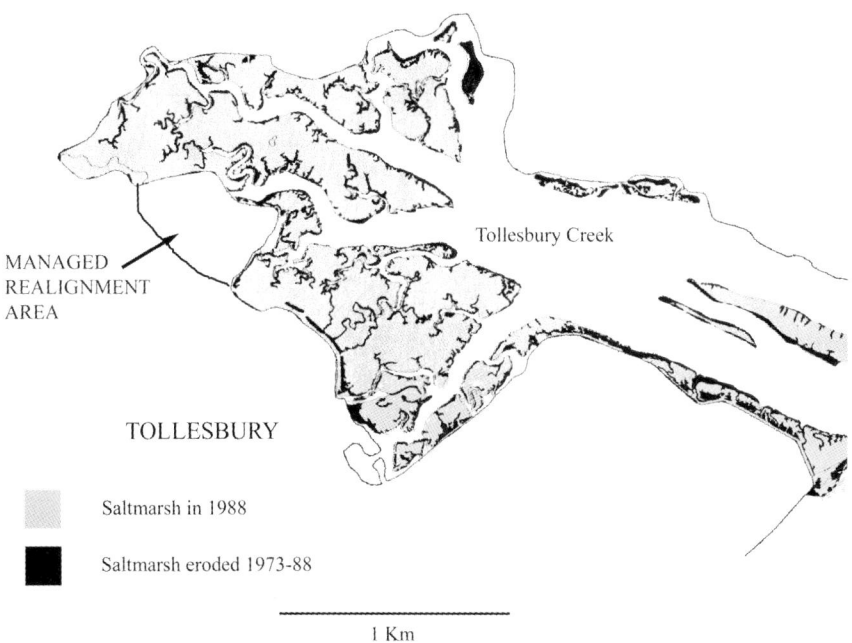

Fig. 8.2. Map of the Tollesbury marshes, modified from Burd (1992), showing the extent of creek erosion and the location of the managed realignment site established in 1995

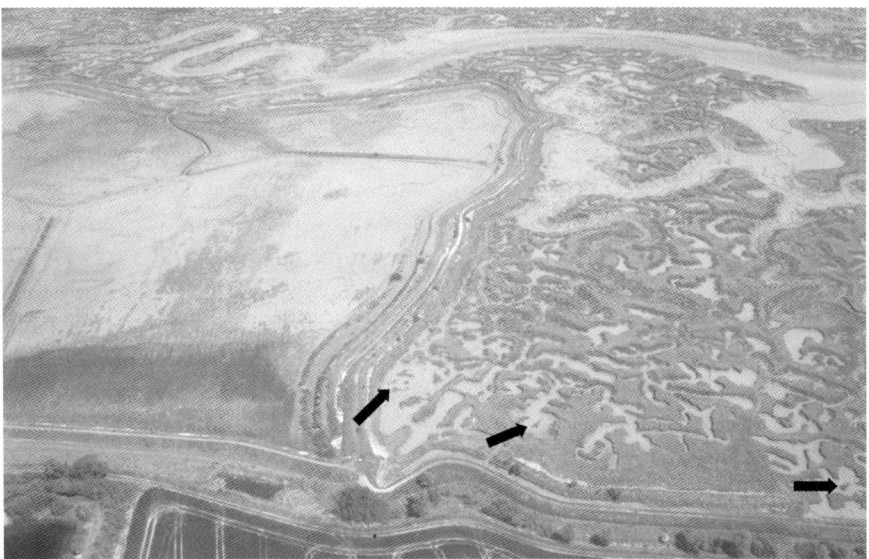

Fig. 8.3. Aerial photograph of part of the Tollesbury marshes, showing the creek systems and residual mud mounds in the back marsh (*arrows*), and the southeastern part of the managed realignment area (*to the left*)

The hypothesis that the saltmarsh losses are caused primarily by coastal squeeze has led to the adoption of managed realignment (also called managed retreat and set-back) as a possible policy for sustainable, and relatively inexpensive, coastal management. In a managed realignment scheme a sea wall will be breached, lowered, removed, or left to fail naturally, and the land behind the wall will become tidal up to a smaller, cheaper, sea wall built on higher ground. The newly tidal area will accumulate sediment and saltmarsh should develop and protect the new sea wall. The success of managed realignment depends on the development of saltmarsh vegetation, otherwise more expensive upgrading of the new sea wall may be required. Three experimental schemes have been established in the Blackwater estuary (see Fig. 8.1), the first was a small area on Northey Island (N) (Turner 1995; Underwood 1997), the second was at Tollesbury (T), and the third at Orplands (O).

Hughes (1999) questioned whether managed realignment will lead to saltmarsh development after examining the fate of two similar but older sites where sea walls were breached naturally. At Clementsgreen Creek the wall was breached in 1897 and the saltmarsh that developed has eroded recently (Burd 1992). At Wallasea Island the wall was breached in 1953 but saltmarsh did not develop, apart from a small patch opposite the breach. French (1996) and French et al. (1999) examined the fate of similar old sites in Suffolk, as analogues for realignment sites, and found that they remain largely as poorly drained unvegetated mud. These authors also concluded that the natural development of saltmarsh in realignment sites can not be taken for granted. In both Clementsgreen Creek and Wallasea Island, where pioneer zone plants could colonise the mud and begin the facilitated successional development of saltmarsh vegetation, the dominant processes were bioturbation and herbivory by the invertebrate infauna. The lower limits of the plant distributions were not determined by sea level (or water movement) directly, but by the activity of the invertebrates. The lower limits of their distributions were not those of their potential niches but had been raised to those of their realised niches.

Hughes (1999) proposed a hypothesis relating the losses of saltmarshes to the activities of the infauna. The hypothesis is that the pioneer zone, including that within saltmarsh creeks formed initially by drainage of water, will be colonised by invertebrates. In SE England these will be particularly the polychaete *Nereis diversicolor* (hereafter *Nereis*), which some evidence indicates has increased in range and abundance over the past few decades, and to a lesser extent the amphipod *Corophium volutator* (hereafter *Corophium*), and the gastropod *Hydrobia ulvae* (hereafter *Hydrobia*). Herbivory and bioturbation by these animals prevent colonisation by pioneer zone vegetation directly and render the sediment more erodable, by waves at the face of the marshes and by tidal currents in the creeks. Thus, if SLR is affecting the loss of vegetation it may be doing so indirectly, by allowing an upward displacement of the distributions of the infaunal species, and by increasing the duration and intensity of their activities in any given location. In addition to SLR, Burd (1992) reported an increase in the tidal range on this coast of 45 cm since 1870. A higher tidal range could increase erosion because of higher tidal currents that flow through the creeks.

Further evidence of the effects of *Nereis* on plant colonisation was provided by Hughes et al. (2000) who demonstrated that *Zostera noltii* transplants protected from the polychaetes had a higher survivorship, lower index of root damage, and greater biomass than unprotected ones. Emmerson (2000) reported that *Nereis* could also affect negatively transplanted *Spartina anglica*. *Z. noltii* and *Nereis* have almost non-overlapping distributions which led Hughes et al. (2000) to the hypothesis that there are two broadly alternative stable states on the upper mudflats. In one state plants are dominant and prevent colonisation by burrowing infauna. In this case the plants were *Z. noltii*, but the roots of a variety of plants, or filamentous algae, could also deter burrowing. In the other state the infauna dominate and prevent colonisation by plants, through herbivory and bioturbation. Managing these two states could be the key to establishing pioneer zone plants and the subsequent successional development of saltmarsh.

In this chapter, the interactions of the important biological and physical processes relevant to sediment stability and saltmarsh development are reviewed. The hypotheses of Hughes (1999) and Hughes et al. (2000) are developed and refined, and new hypotheses proposed, with particular reference to managed realignment schemes, and to the internal erosion of established saltmarshes. The biological factors, particularly the effects of the invertebrates, are emphasised as these have tended to be ignored in this context.

8.2 Managed Realignment

Much of the land behind the sea walls in SE England was previously saltmarsh that has been enclosed over the past few centuries (e.g. Burd 1995). Since the land was claimed its level would have fallen with respect to local sea level, through compaction, shrinkage and because of the subsequent SLR. Consequently, in this area, as in others, managed realignment land usually will be lower than the existing saltmarsh (see Reed et al. 1999). Therefore, factors that affect sediment accretion, which has to be sufficient to reach an elevation high enough to allow plants to colonise, are crucial to the successful development of new saltmarsh.

The following hypothesis is proposed. In low-lying managed realignment sites sediment will accumulate but as the surface of the sediment becomes higher the rate of accretion will decline and the rate of erosion will increase until an equilibrium elevation is reached determined by the interaction of physical and biological factors, as described below. If the sediment is suitable invertebrates will colonise the accreted sediment first and prevent establishment of pioneer zone plants, directly by herbivory and bioturbation, and indirectly by making the sediment more erodable and decreasing the equilibrium height of the sediment.

8.2.1 Physical Factors

After the sea wall is breached deposition of sediment will occur from a layer of water immediately above the sediment, the height of which will depend on the settling velocity of the particles and the length of time when water movements are sufficiently slow to allow sedimentation to occur. In general, water movements will be slowest at slack water on either side of high tide and when wave-generated water movement does not reach down to this boundary layer. As the height of the sediment rises the mean rate of sedimentation will decline and the mean rate of erosion will increase because of the increasing effect of internal wind-generated waves, the main source of water movement in such semi-enclosed sites (French et al. 1999). As the height of the sediment surface gets closer to the peak of the tidal curve the sediment remains close to the

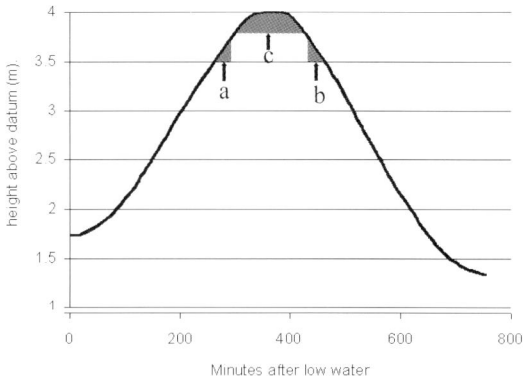

Fig. 8.4. Diagram of the predicted tidal curve for West Mersea on 1 January, 2000. See text for explanation of the *shaded areas a*, *b*, and *c*

water surface, and under the erosive influence of waves, for longer periods. This is illustrated by reference to Fig. 8.4, which shows a tidal curve for a neap tide at West Mersea (close to the Tollesbury realignment site) on 1 January 2000, when high water was at 4.0 m above CD. [Mean high water neap tide level (MHWNTL) is at 3.8 m]. Resuspension of sediment occurs when wave and/or tidal currents impart a shear stress (τ_0) to the sediment that exceeds the critical bed shear stress ($\tau_{0\,crit}$). As a first-order approximation, assume the wind generates waves under which $\tau_0 > \tau_{0\,crit}$ to a water depth of 20 cm. If the height of the sediment surface is 3.5 m, resuspension could occur for 48 min, 24 min each on the flood ("a" in Fig. 8.4) and ebb tide ("b" in Fig. 8.4). Moreover, between these periods, when the depth of water is more than 20 cm, deposition could occur during slack water around high tide. If the sediment surface is raised to 3.8 m, under similar tide and wave conditions, sediment resuspension could occur for 129 min over the period of high tide ("c" in Fig. 8.4) and no deposition would occur around high tide either. Therefore, it is expected that in low-lying realignment sites sediment will generally accumulate until a dynamic equilibrium is reached where the height of the sediment will fluctuate above and below a mean; sedimentation may occur, for example, on days with low wind speeds and a high sediment load in the incoming tide following a storm, whereas sediment may be exported from the site on windy days.

8.2.2 Biological Factors

8.2.2.1 Effects of the Flora

If the sediment surface is sufficiently high it may become colonised by microphytobenthos in high densities (see Underwood 1997), filamentous algae (particularly *Enteromorpha* sp.) or pioneer zone vascular plants (e.g. *Salicornia* spp. and *Spartina* spp.). The secretion of extracellular polymeric substances (EPS) by the microphytobenthos leads to the development of biofilms that increase sediment cohesion and stability by increasing $\tau_{0\,crit}$ (see reviews by Paterson 1997; Paterson and Black 1999). Sediment that accretes in this way may induce a positive feedback, for the biomass of microphytobenthos generally increases with increased elevation (Austen et al. 1999) so, as the biomass increases further, so will the erosion threshold and accretion is further enhanced. While it is generally understood that biofilms help stabilise sediments, the erosional characteristics of cohesive sediments are less well understood than those of non-cohesive sediments (e.g. sands), principally because the unit of erosion is a floc, a variable heterogeneous mix of inorganic particles in an organic matrix. Consequently, a large number of physical, chemical and biological variables can affect the stability of cohesive sediments (Paterson and Black 1999), to the extent that sedimentological techniques alone cannot be used to predict their behaviour.

In nutrient-enriched estuaries, like those of SE England, colonisation of mud by mats of cyanobacteria (e.g. *Oscillatoria* spp.) and algae, such as *Enteromorpha* spp., is typical. These filamentous forms grow into the sediment and stabilise it. They also protect the surface from erosive water movement by slowing the currents immediately above the surface, and may enhance deposition by trapping suspended sediment. Vascular plants may further aid accretion by increasing deposition, by slowing water movement at the sediment surface, and reducing erosion by binding sediment with their roots (van Eerdt 1985). The effect of the plants may be dependent on their morphology and on their density. For example, *Spartina* plants, which are relatively large and complex, may have a greater effect than the smaller, relatively simple, smoother *Salicornia*. Individual plants may have little effect as the currents may simply flow around them increasing velocity and causing local scouring of the sediment. A thicker cover of vegetation will increase the depth of the boundary layer of slow moving water above the sediment surface.

8.2.2.2 Effects of Invertebrates

As the sediment accumulates it may become colonised by invertebrates that may have a major effect on sediment stability, and ultimately the development of saltmarsh vegetation. Infaunal invertebrates may help sediment accretion, by compacting the sediment as they burrow, and by binding sediment grains together with the mucous secretions used in the construction of their burrows (e.g. Meadows and Tait 1989). However, the weight of evidence is that the overall effect of infaunal invertebrates is to reduce sediment stability. In the unmanaged and managed realignment sites in SE England four species are particularly abundant, the polychaete *Nereis diversicolor* (hereafter *Nereis*), the amphipod *Corophium volutator* (hereafter *Corophium*), the gastropod *Hydrobia ulvae* (hereafter *Hydrobia*) and the bivalve *Macoma balthica* (hereafter *Macoma*). *Nereis* may be predators of other invertebrates, they may also filter-feed, but they are predominantly surface deposit feeders (see Smith et al. 1996 for review). They excavate burrows to a depth of 10–40 cm (Davey 1994) from which they partially emerge to engulf sediment with epipelic diatoms (Smith et al. 1996), filamentous algae, seeds, seedlings, and *Zostera* leaves (Olivier et al. 1996; Hughes et al. 2000; Paramor and Hughes, in preparation). Exclusion of *Nereis* in experiments at Clementsgreen Creek allowed colonisation by microphytobenthos and filamentous algae, and promoted sediment accretion (Smith et al. 1996; Hughes 1999).

Corophium form U-shaped burrows to a depth of 5 cm and are predominantly surface deposit feeders too (Meadows and Reid 1966). They partially emerge to scrape surface material into their burrows including microphytobenthos, organic matter and detritus (Gerdol and Hughes 1994a). *Corophium* reduce the shear strength of the sediment (Gerdol and Hughes 1994b) and may prevent the establishment of larger plants because their bioturbation buries seeds and disturbs seedlings (Gerdol and Hughes 1993).

Hydrobia, which crawl on, or burrow just under, the sediment surface, are deposit feeders consuming bacteria, small organic particles, and epipelic and epipsammic diatoms (Fenchel et al. 1975; Austen et al. 1999). Austen et al. (1999) concluded that *Hydrobia* increased the erodability of mud in the Danish Wadden Sea by consuming microphytobenthos, and by the production of faecal pellets, which lack cohesive properties. Laboratory experiments have demonstrated that *Hydrobia* prevent establishment of small seedlings, by physical disturbance as the gastropods move over and through the surface sediments (Hughes et al., in preparation). In field experiments, *Salicornia* seedlings that were protected from the effects of *Hydrobia*, by surrounding them with pellets containing molluscicide, had a higher survivorship rate than unprotected seedlings (Hughes et al., in preparation).

Widdows et al. (1998) concluded that the erodability of the sediment in the Humber Estuary was related to the activities of *Macoma* and the cockle *Cerastoderma edule*. *Macoma* are facultative deposit feeders that feed by directing their inhalent siphon onto the sediment surface, where, by sucking up surface deposits, they may have a negative effect on sediment accretion (Paterson and Black 1999). Sometimes *Macoma* are suspension feeders and may aid deposition by incorporating inorganic particles, previously in suspension, into pseudofaeces. The pseudofaecal pellets and faecal pellets may become incorporated into the sediment although they are easily transported by currents and may contribute to the removal of surface sediments (Davey and Partridge 1998). Kang et al. (1999) reported that suspension feeding *Cerastoderma* consume microphytobenthos taken into suspension by wave action. The same may be true of *Macoma*, and both species may reduce the potential for these microphytobenthos to be re-deposited and contribute to sediment stability.

Cerastoderma edule and *Arenicola marina* occur elsewhere in these estuaries but are not common either in the managed and unmanaged realignment sites, or in the saltmarsh creeks. The reasons are not known, but, were these species to colonise sediment in realignment areas, they might reduce further the potential for saltmarsh establishment. *Cerastoderma* often move over the sediment, where they disturb other invertebrates (Flach 1996), and probably also the microphytobenthos and the integrity of the sediment surface, increasing erosion. *Arenicola* could deter development of vegetation, as Philippart (1994) concluded that the worms restricted colonisation by *Zostera*, because of their bioturbatory activities.

8.3 Managing Sediment Accretion and Development of Saltmarsh Vegetation in Managed Realignment Sites

In managed realignment sites, particularly low-lying ones, further management of the biological and physical processes which affect sediment stability may be necessary to raise the equilibrium height of the sediment to a level where saltmarsh vegetation can develop. Even then further management action may be required to shift the community from an animal-dominated state to a plant-dominated state.

8.3.1 Physical Factors

French et al. (1999) identified exposure to wave action as a major factor affecting sediment elevation in unmanaged realignment sites. In managed realignment sites reducing wave action, or the effects of wave action, would increase the probability of saltmarsh development, indirectly by promoting sedimentation to a level where plants could survive, and directly by reducing disturbance of seeds and seedlings. Wiehe (1935) concluded that the lower vertical limit of *Salicornia* was above MHWNTL because a threshold time of 2 or 3 days undisturbed by tides was necessary for the establishment of the seedlings. This significance of tidal level has been extrapolated to other species (Davidson et al. 1991); however, Gerdol and Hughes (1993) challenged its ubiquity when they found that transplanted *Salicornia* seedlings could survive to maturity below MHWNTL. Although in their sheltered site bioturbation by *Corophium* prevented establishment of *Salicornia*, rather than water movement, in more open areas tidal or wave disturbance may wash the seeds away, bury them, or prevent seedlings establishing (see Houwing et al. 1999).

Reducing the fetch over which the wind can generate internal waves would reduce erosion and disturbance to seeds and seedlings. These measures could take a variety of forms, including the use of brushwood groynes, such as those used on the coast of the Wadden Sea (e.g. Houwing et al. 1999), or by excavating earth banks prior to the initial flooding. Obviously, the position of any wave breaks needs to be perpendicular to the longest fetch and/or the direction of the prevailing wind, but not in positions where they would create bottlenecks that generate rapid, and erosive, tidal water movement. Possible designs include single rows of detached short banks, or double rows of the "open chevron" discussed below.

In some unmanaged realignment sites the mud remains flat, unconsolidated and saturated throughout periods of low tide with no apparent natural creek formation (French 1996; R.G. Hughes, personal observation). The tide floods as a "sheet" and ebbs in the same manner. The accreted sediment in managed realignment sites may develop the same characteristics and become too wet for plant colonisation. It has been suggested that the digging of creeks, a practice in some man-made marshes in the Wadden Sea (e.g. Dijkema 1997), would aid development of vegetation by draining the sediment and by aiding transport of sediment into the developing marsh (French 1996). However, channelling tidal water into creeks will increase the frictional erosive forces of the tides, which may lead to the lateral creek erosion prevalent on some established marshes (see below). Whether artificial creeks are effective or not may be site-related and specific research would be required to estimate their benefits.

8.3.2 Biological Factors

If invertebrates can reduce sediment accretion and plant colonisation then the factors that affect their abundance may be managed to reduce their effects. Figure 8.5 shows an interaction web that summarises some of the biological interactions that may affect the abundance of animals and plants on these mudflats. Some interactions are not different to a food web; for example, birds eating invertebrates that would otherwise eat microphytobenthos. A similar trophic cascade was described by Daborn et al. (1993) where sandpipers contributed to sediment accretion by eating *Corophium*. An example of a non-trophic interaction is the possible negative effect of birds on sediment stability through physical disturbance of the sediment with their feet. Reducing disturbance by humans could increase bird predation on invertebrates. At Tollesbury the old and the new sea walls have footpaths on the top and wading

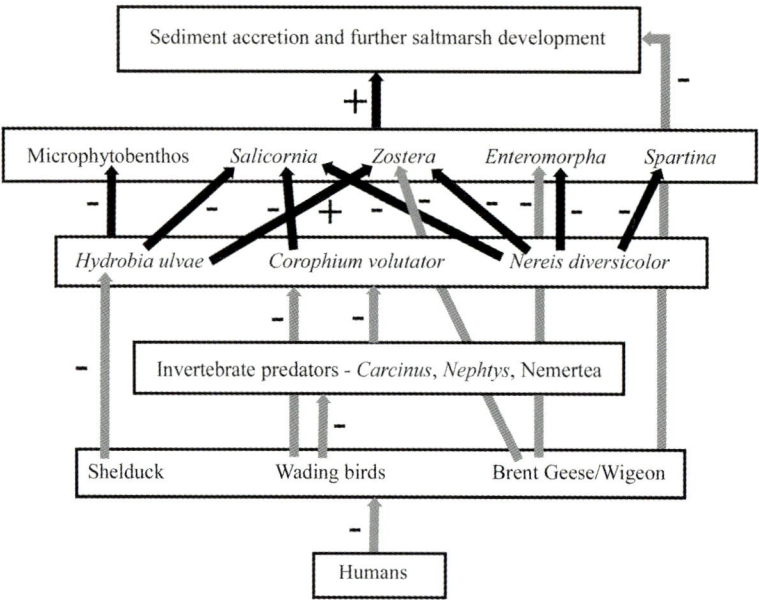

Fig. 8.5. Diagrammatic web showing some of the important positive (+) and negative (−) interactions between the organisms on the mudflats and saltmarshes of SE England. Not all the possible interactions are shown, particularly those between organisms at the same level (e.g. between *Corophium* and *Nereis*). Interactions shown by *black arrows* are referred to specifically in the text, those in *grey* are not

birds are disturbed by people walking on them (R.G. Hughes, personal observation). Simply diverting the footpaths to behind the sea walls may allow the birds to feed undisturbed for longer. Some interactions are more complex; for example, if birds reduce the abundance of invertebrate predators, this could have a positive effect on the abundance of the herbivorous invertebrates. More research is required on factors affecting bird use of realignment sites, particularly in quantifying their effects on the invertebrates.

Colonisation by invertebrates may be reduced or prevented by other means. The type of sediment and the depth of sediment suitable for burrowing may deter some invertebrates or affect their abundance indirectly. Managed realignment sites in which the land is already at around MHWNTL might be more successful at developing saltmarsh because burrowing invertebrates may not colonise the soil. While *Hydrobia* inhabit the surface layers of a variety of sediment types from soft muds to hard sands, *Corophium* prefer fine sediments (Meadows 1964) to a depth of 5 cm, and *Nereis* and *Macoma* are often found even deeper. The selectivity of *Nereis* and *Macoma* for different types of sediment is not known in detail but may be important in determining the fate of these sites. If research shows that these deep-burrowing invertebrates are deterred by coarse or shallow sediment, raising the surface of the land appropriately prior to flooding may prevent them colonising it. Burd (1995) discussed the direct advantage of using dredged material to raise the level of the sediment prior to flooding and, if this sediment deterred burrowing invertebrates, then a double advantage may accrue. Even if a shallow layer of finer sediment were to develop, any invertebrates that burrowed into it could be more prone to predation by birds.

Artificial barriers to burrowing invertebrates may offer another management solution. In experiments on small spatial and temporal scales removing *Corophium*, or preventing *Nereis* reaching the surface, has enabled plants to colonise sediments lower than they occur naturally (see above). Expanding the size and duration of these experiments, for example, by use of matting of the sort used to stabilise the banks of new road cuttings, may favour longer term sediment accretion and colonisation by vascular plants. This may affect a switch from an invertebrate-dominated community to a plant-dominated community proposed by the alternative state hypothesis of Hughes et al. (2000).

8.4 Loss of Saltmarsh Vegetation by Lateral Erosion of Creeks

Adam (1990) reviewed several possible reasons for the formation of creeks in marshes but concluded that in specific cases the causes were unknown. Creek density and shape may be related to sediment type and to tidal range. In SE England, where fine sediments predominate, and the tidal range is comparatively moderate, the marshes have a high creek density. In some other regions saltmarsh creeks are notably absent. Adam (1990) noted that although creeks give the impression of instability, with cliffs and eroding banks, many are remarkably stable over periods of many decades, a view supported by Gabet (1998). However, this is not the case in many locations in SE England (see above) (Burd 1992). At Tollesbury creek erosion is noticeable as severely undercut creek banks followed by the slumping of sections of the saltmarsh surface into the creeks. This erosion is apparent even in the head of the creeks, where an explanation based on purely physical processes is untenable because of the virtual absence of waves and very slow tidal water movement, even during spring tides. The slumped blocks are characterised by the downward displacement of the vegetation, which may change subsequently because of the greater degree of inundation by seawater. A possible successional sequence of vegetation is apparent; on recent slumps the vegetation is similar to that of the saltmarsh surface, while on some older slumps the original flora has been replaced by pioneer zone species. The older slumps have also been eroded, as they are relatively low lying and smoothly contoured.

At Tollesbury the saltmarsh creeks are inhabited predominantly by *Nereis* and *Hydrobia* (Paramor and Hughes, in preparation). *Corophium* is not found here but does occur in creeks elsewhere (Hughes and Gerdol 1997). Hughes (1999) suggested that the creeks, formed initially by water draining from the marsh, were colonised by invertebrates that contributed to further erosion. The channels increase in width progressively until they merge to leave large areas of mud with residual mud mounds, within the marsh, as at Tollesbury (Fig. 8.3), or ultimately with no original marsh surface remaining, as at Clementsgreen Creek (Hughes 1999). This hypothesis is extended here. In extensive creek systems such as those at Tollesbury, where large volumes of tidal water flood into, and ebb from, several hundred meters of creek, erosion of the creeks initiates a positive feedback in which erosion leads to more erosion by increasing tidal flows, as described below. SLR may increase saltmarsh loss in this manner indirectly as the invertebrates may colonise the sediment higher up the sides of the creeks, contributing even more to their erosion.

Consider a natural creek system, as in Fig. 8.6. The creeks are generally 2–4 m wide and 1–1.5 m deep, but are deeper and wider nearer to the front of the marsh. Generally, there are no barriers, or ridges, at the mouths of the creeks that usually drain completely leaving little standing water. Within the creek systems waves are rare, but in some creeks the tidal currents are rapid. Assume the flood tide first reaches the outer part of this creek system at mid-tide and rises for a further 3 h and reaches the surface of the marsh but does not flood over it – a "bank-full" tide. During the flood and ebb tides the volumes of water that flow past point X will have to fill the creeks at Y and Z. The mean and maximum current speeds at X will be much higher than at Y and Z. The slumping and subsequent erosion of soil at Y will, on a future similar tide, increase the volume of water that flows past X, and the mean and maximum current speed. Thus the erodability of the sediment at X will be increased by an event that has occurred elsewhere on the marsh. The slumping of a section of the saltmarsh into the creek at X will reduce the cross-sectional area of the creek and increase the velocities of the tidal currents, further eroding the sediment in the vicinity, usually on the bank opposite to the slump.

One potential benefit of creeks is that they facilitate the transport of suspended sediment, brought into the marsh by the tide and eroded from the creeks, deep into the marsh where it would be available for deposition on the surface, a process enhanced by the vegetation. Reed (1988) found that some of the sediment that accreted on the marsh surface was transported into the marsh via the creeks, but some was derived directly from the eroding creeks. With rising sea level saltmarshes must accrete vertically at a rate at least equal

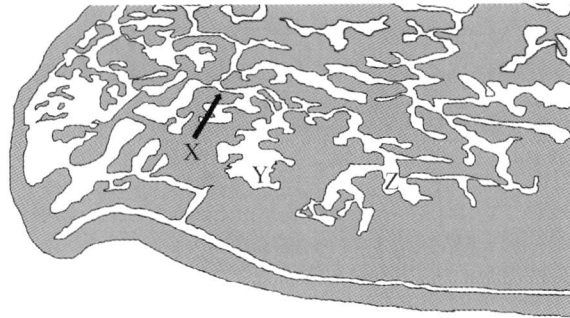

Fig. 8.6. Diagram of part of the creeks system of the Tollesbury saltmarsh (drawn from Fig. 8.3). See text for explanation of X, Y and Z

to that of mean SLR. However, if such a rate of vertical accretion is maintained by redistribution of sediment from within the creeks to the marsh surface, it is of little benefit. This seems to be the case at Tollesbury where the surface is accreting at a rate greater than SLR (Cahoon et al., 2000) but the creeks are getting wider.

Pethick (1992) discussed the morphology and mechanisms of formation of saltmarsh creeks. He suggested that mud mounds were a "morphological device that forced the wind-generated waves into hydrodynamically inefficient channels, thereby dissipating wave energy". This view contrasts with that which considers that mud mounds are erosional remnants of a previously more substantial marsh (Allen and Pye 1992; Hughes 1999). Pethick (1992) further argued that saltmarsh creeks were also devices to dissipate energy. On open coast marshes the creeks were a response to wave action and had a relatively simple morphology, linear and elongate, to dissipate wave energy. In sheltered marshes the creeks develop a series of long intricate meanders that allow more retardation of the tidal flow. Pethick's (1992) view, that the longer the creeks the more energy is dissipated, contrasts with the hypothesis described above, that the longer the creeks are more water will flow to fill them and *more* (erosive) frictional energy is produced. If the saltmarsh face were simply a cliff undissected by creeks, the tide would simply rise against it and ebb away from it. Creeks transform the potential energy of the tides into erosive energy of flowing water.

8.5 Managing Reduction of Lateral Creek Erosion

Creek erosion has been apparent for many years but the extent of the problem may have been underestimated, because of the general view that the rates were low, and no concerted measures to slow or reverse the processes have been attempted. Sometimes woven willow or brushwood baffles have been placed across some creeks, but these will be largely ineffective as the amounts of water moved by the tides past the baffles will not be reduced by their presence, nor will its erosive capacity. Baffles reduce the effects of wave action not those of tidal currents that erode the sediment in the creeks. Reducing creek erosion requires one or both of two strategies, reducing the volumes of water flowing through the creeks, and/or reducing their erosive effects. Reducing the erodability of the sediment could be achieved by reducing the effects of the invertebrates, as discussed above. There may be one difference, however, as birds and their footprints are seen rarely in the creeks at Tollesbury indicating that the invertebrates in the creeks suffer less predation than those on more open mudflats. Paramor and Hughes (in preparation) have stimulated sediment accretion in some eroding creeks by using matting to exclude invertebrates in experiments similar to those of Hughes (1999).

Reducing the water flow in creeks could be achieved simply by infilling, perhaps with dredged sediment if it were available. If this sediment were relatively coarse (sand, gravel, shingle) the tidal volumes would be reduced but the drainage of the marsh by the creeks, which enhances plant development, could be maintained to some extent. A partial infilling with coarse sediment, as a lining to the creek bottoms, could reduce erosion in three ways; the flows would be reduced, the $\tau_{0\,crit}$ will be reduced because of the larger particle sizes, and the invertebrates may be deterred. If this measure leads to a natural accumulation of fine sediment, which invertebrates could then colonise and destabilise, it could be repeated to create an increasing depth of alternate layers of coarse and fine sediments.

The factors that determine creek erosion and stability have relevance to current management schemes designed to increase sedimentation. One example is on the Hindenburg Dam that joins Sylt (northern Wadden Sea, Germany) to the mainland. Periodically the saltmarsh is excavated in varied ways but often in parallel channels, sometimes separated from the sea by a perpendicular wall with small openings (Figs 8.7, 8.8A). These designs contain bottlenecks, at the gaps in the wave breaks and at the mouths of the creeks, where tidal movement will produce relatively high current speeds and deter sedimentation, if not cause erosion. A better design to minimise current velocities is shown in Fig. 8.8B. The excavated creeks are triangular in plan, where the width increases progressively to seaward, so that the ratio of the cross-sectional area of the creek to the volume above (inland) remains constant. Any wave breaks should be a single row of short walls set at an angle appropriate for the particular location, or a double row of overlapping relatively short walls, to create an "open chevron" pattern (Fig. 8.8C) which will prevent ingress of waves from any offshore direction. These arrangements of short wave breaks are advantageous as there are many gaps that allow the tidal currents to flow relatively unimpeded with slow velocities.

Fig. 8.7. Photograph of one of a series of recently dug parallel channels in the marsh on the Hindenburg Dam on the Island of Sylt. In the background a low ridge of sediment reinforced with wooden stakes protects the excavations from waves

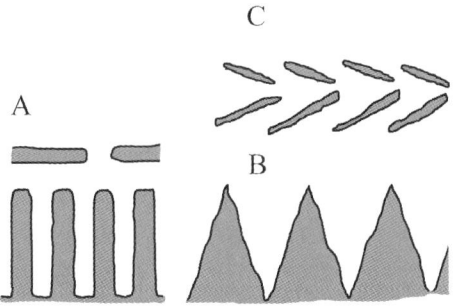

Fig. 8.8. **A** Diagram of the excavated channels shown in Fig. 8.7. **B** Diagram of suggested improved design with triangular channels (see text for details). **C** Diagram of a double row of wave breaks to create an "open chevron" design

8.6 Conclusions

The generally accepted hypothesis that SLR and coastal squeeze are responsible for the losses of vegetation in saltmarshes of SE England is not supported by data, which indicate that vertical accretion of marsh surfaces is at least sufficient to keep pace with SLR, or by the pattern of erosion, which is largely by the lateral extension of saltmarsh creeks. The hypothesis that the rarity of developing pioneer zone vegetation and saltmarsh erosion is, at least in part, due to the infaunal invertebrates, which exacerbate sediment erosion and loss of vegetation by bioturbation and herbivory, is developed in the context of increasing the success of managed realignment schemes and in reducing erosion of extensive saltmarsh creek systems.

A hypothesis is proposed that in low-lying managed realignment sites sediment will accumulate initially, but, as the surface of the sediment becomes elevated, the rate of accretion will decrease and the rate of erosion increase. The sediment elevation will reach a dynamic equilibrium, the height of which will be affected by the interaction of physical and biological factors. Invertebrates may colonise the accreted sediment and prevent establishment of vegetation, indirectly by making the sediment more erodable, and directly by herbivory and bioturbation. The invertebrates that colonise the sediments of extensive creek systems may exacerbate erosion that initiates a positive feedback, in which erosion leads to more erosion because of increased tidal flows. Consequently, it is predicted that further management will be necessary to promote sediment accretion and ensure successional development of saltmarsh vegetation, in managed realignment sites and in eroding saltmarsh creeks.

References

Adam P (1990) Saltmarsh ecology. Cambridge Univ Press, Cambridge
Allen JRL, Pye K (1992) Coastal saltmarshes: their nature and importance. In: Allen JRL, Pye K (eds) Saltmarshes. Morphodynamics, conservation and engineering significance. Cambridge Univ Press, Cambridge, pp 1–18
Austen I, Anderson TJ, Edelvang K (1999) The influence of benthic diatoms and invertebrates on the erodability of an intertidal mudflat, the Danish Wadden Sea. Estuar Coast Shelf Sci 49:99–111
Beardall CH, Dryden RC, Holzer TJ (1988) The Suffolk Estuaries. The Suffolk Wildlife Trust, Saxmundham, Suffolk
Boorman L, Goss-Custard JD, McGrorty S (1989) Climatic change, rising sea level and the British coast. ITE research publication 1. HMSO, London, pp 1–24
Burd F (1992) Erosion and vegetation change on the saltmarshes of Essex and north Kent between 1973 and 1988. Research and survey in nature conservation, vol 42. Nature Conservancy Council, Peterborough, 116 pp
Burd F (1995) Managed retreat: a practical guide. English Nature, Peterborough
Cahoon DR, French JR, Spencer T, Reed DJ, Moller I (2000) Vertical accretion versus elevational adjustment in UK saltmarshes: an evaluation of alternative methodologies. In: Pye K, Allen JRL (eds) Coastal and estuarine environments. Geological Society special publication. Geological Society, London, 175:223–238
Daborn GR, Amos CL, Berlinsky M, Christian H, Drapeau G, Faas RW, Grant J, Long B, Paterson DM, Perillo GME, Piccolo MC (1993) An ecological "cascade" effect. Migratory birds affect stability of intertidal sediments. Limnol Oceanogr 38:225–231
Davey JT (1994) The architecture of the burrow of *Nereis diversicolor* and its quantification in relation to sediment-water exchange. J Exp Mar Biol Ecol 179:115–129
Davey JT, Partridge VA (1998) The macrofaunal communities of the Skeffling muds (Humber Estuary), with special reference to bioturbation. In: Black KS, Peterson DM,

Kramp A (eds) Sedimentary processes in the intertidal zone, Special publication 139. Geological Society, London, pp 115–124

Davidson NC, Loffoley Dd'A, Doody JP, Way LS, Gordon J, Key R, Drake CM, Pienkowski MW, Mitchell RL, Duff KL (1991) Nature conservation and estuaries in Great Britain. Nature Conservancy Council, Peterborough

Dijkema KS (1997) Impact prognosis for salt marshes from subsidence by gas extraction in the Wadden Sea. J Coastal Res 13:1294–1304

Emmerson M (2000) Remedial habitat creation: does *Nereis diversicolor* play a confounding role in the colonisation and establishment of the pioneering saltmarsh plant *Spartina anglica*? Helg Mar Res 54:110–116

Fenchel T, Kofoed LH, Lappapainen A (1975) Particle size-selection of two deposit feeders; the amphipod *Corophium volutator* and the Prosobranch *Hydrobia ulvae*. Mar Biol 30:119–128

Flach EC (1996) The influence of the cockle, *Cerastoderma edule*, on the macro-zoobenthic community of tidal flats in the Wadden Sea. Mar Ecol Pubbl Staz Zool Napoli 17:87–98

French JR (1996) Function and design of tidal channel networks in restored saltmarshes. In: Gardiner P (ed) Proceedings tidal 96 – interactive symposium for practising engineers. Univ Brighton, Department of Civil Engineering, pp 128–137

French CE, French JR, Clifford NJ, Watson CJ (1999) Abandoned reclamations as analogues for sea defence re-alignment. In: Kraus NC, McDougal WG (eds) Coastal sediments 99: proceedings of the 4th international symposium on coastal engineering and science of coastal sediment processes, vol 3. American Society of Civil Engineers, New York, pp 1912–1926

Gabet EJ (1998) Lateral migration and bank erosion in a saltmarsh tidal channel in San Francisco Bay, California. Estuaries 21:745–753

Gerdol V, Hughes RG (1993) Effect of the amphipod *Corophium volutator* on the colonisation of mud by the halophyte *Salicornia europaea*. Mar Ecol Prog Ser 97: 61–69

Gerdol V, Hughes RG (1994a) Feeding behaviour and diet of *Corophium volutator* in an estuary in southeastern England. Mar Ecol Prog Ser 114:103–108

Gerdol V, Hughes RG (1994b) The effect of *Corophium volutator* on the abundance of benthic diatoms, bacteria and sediment stability in two estuaries in southeastern England. Mar Ecol Prog Ser 114:109–115

Houwing E-J, Willem E, Smit-van der Waaij Y, Dijkema KS, Terwindt JHJ (1999) Biological and abiotic factors influencing the settlement and survival of *Salicornia dolichostachya* in the intertidal pioneer zone. Mangroves Saltmarshes 3:197–206

Hughes RG (1999) Saltmarsh erosion and management of saltmarsh restoration; the effects of infaunal invertebrates. Aquatic Conserv Mar Freshw Ecosyst 9:83–95

Hughes RG, Gerdol V (1997) Factors affecting the distribution of the amphipod *Corophium volutator* (Pallas) in two estuaries in southeastern England. Estuar Coast Shelf Sci 44:621–627

Hughes RG, Lloyd D, Ball, L, Emson D (2000) The effects of the polychaete *Nereis diversicolor* on the distribution and transplanting success of *Zostera noltii*. Helg Mar Res 54:129–136

Kang CK, Sauriau P-G, Richard P, Blanchard GF (1999) Food sources of the infaunal suspension-feeding bivalve *Cerastoderma edule* in a muddy sandflat of Marennes-Oléron Bay, as determined by analyses of carbon and nitrogen stable isotopes. Mar Ecol Prog Ser 187:147–158

Meadows PS (1964) Substrate selection by *Corophium* species: the particle size of substrates. J Anim Ecol 33:387–394

Meadows PS, Reid A (1966) The behaviour of *Corophium volutator* (Crustacea; Amphipoda). J Zool Lond 150:387–399

Meadows PS, Tait J (1989) Modification of sediment permeability and shear strength by two burrowing invertebrates. Mar Biol 101:75–82

Olivier M, Desrosiers G, Caron A, Retiere C, Caillou A (1996) Juvenile growth of *Nereis diversicolor* (O.F. Müller) feeding on a range of marine vascular and macroalgal plant sources under experimental conditions. J Exp Mar Biol Ecol 208:1–12

Paterson DM (1997) Biological mediation of sediment erodability, ecology and physical dynamics. In: Burt N, Parker R, Watts J (eds) Cohesive sediments. Wiley Interscience, Chichester, pp 215–229

Paterson DM, Black KS (1999) Water flow, sediment dynamics and benthic biology. Adv Ecol Res 29:155–193

Pethick JS (1992) Saltmarsh geomorphology. In: Allen JRL, Pye K (eds) Saltmarshes. Morphodynamics, conservation and engineering significance. Cambridge Univ Press, Cambridge, pp 41–62

Philippart CMJ (1994) Interactions between *Arenicola marina* and *Zostera noltii* on a tidal flat in the Wadden Sea. Mar Ecol Prog Ser 111:251–257

Reed DJ (1988) Sediment dynamics and deposition in a retreating coastal saltmarsh. Estuar Coastal Shelf Sci 26(1):67–81

Reed DJ, Spencer T, Murray AL, French J, Lynn L (1999). Marsh surface sediment deposition and the role of tidal creeks: implications for created and managed coastal marshes. J Coastal Conserv 5:81–90

Smith D, Hughes RG, Cox E (1996) Predation of epipelic diatoms by the amphipod *Corophium volutator* and the polychaete *Nereis diversicolor*. Mar Ecol Prog Ser 145:53–61

Turner KM (1995) Managed re-alignment as a coastal management option. Biol J Linn Soc 56 [Suppl]:217–219

Underwood GJC (1997) Microalgal colonisation in a saltmarsh restoration scheme. Estuar Coast Shelf Sci 44:471–481

van Eerdt MM (1985) The influence of vegetation on erosion and accretion in salt marshes of the Oosterschelde, The Netherlands. Vegetatio 59/60:367–375

Widdows J, Brinsley M, Elliot M (1998) Use of an *in situ* flume to quantify particle flux (biodeposition rates and sediment erosion) for an intertidal mudflat in relation to current velocity and benthic macrofauna. In: Black KS, Peterson DM, Kramp A (eds) Sedimentary processes in the intertidal zone. Special publication 139. Geological Society, London, pp 85–97

Wiehe PO (1935) A quantitative study of the influence of the tide upon populations of *Salicornia europaea*. J Ecol 23:323–333

Part III

Seagrasses and Benthic Fauna of Sediment Shores

9 Common Structures and Properties of Seagrass Beds Fringing the Coasts of the World

C. den Hartog and R.C. Phillips

9.1 Introduction

Seagrasses are aquatic angiosperms which are completely confined to the marine environment. The name refers to a superficial resemblance to grasses, because of the linear leaves of most of the approximate 60 species (Den Hartog 1970). They are of paramount importance in the coastal environment as, when they occur, they generally form dense beds that may cover extensive areas. One of the most conspicuous ecological functions of the seagrass beds is their capacity to stabilize and to modify the unconsolidated substrates in which they are rooted. By their dense growth, they protect the substrate to a considerable extent from erosion, but they also trap floating coarse and suspended fine materials. Further, they produce litter. The accompanying organisms contribute to the alteration of the substrate by contributing fecal material, shells or other protective structures. Part of this material leaves the system, and may be washed up onto the beach, or may be deposited on the bottom of adjacent systems in deeper waters. There are even records of seagrass remains from depths of thousands of metres (Wolff 1976). Another function is that the seagrass beds offer a substrate to many algae and animals which otherwise could not establish themselves on the sandy and muddy bottoms. Furthermore, the seagrass beds serve as a nursery for many fish, crustaceans, and other invertebrates, providing food not only for these animals, but also for a considerable number of migrating birds, sirenians, turtles, and many organisms from neighbouring habitats.

9.2 Distribution and Zonation

Seagrasses have a wide geographical distribution along the coasts of most continents, extending from above the Arctic Circle southward through the temperate areas into the tropics. They are absent only in Antarctica. Their vertical distribution extends from the midlittoral zone down to considerable depths. The greatest depth recorded for seagrass is 90 m (Taylor 1928). This, however, may be exceptional and possibly indicates an allochthonous origin of the material. The occurrence of dense seagrass beds down to depths of 40–50 m has been repeatedly recorded. On a local scale, seagrass beds demonstrate great variation where the habitat is concerned. They can occur in the open sea, unprotected from the ocean swell, to deep into the sheltered environments of estuaries with reduced salinity. There are also some species that inhabit sand-pocketed rocky shores.

The 'ideal' pattern of zonation (Table 9.1) consists of communities of small and narrow-leaved seagrasses in the intertidal zone. Near low-water mark, these become replaced by larger species, which in deep waters can be followed by small species again. In regions where the larger species do not occur, small species may replace them, as in the Persian Gulf and along the coast of Brazil (Den Hartog 1977).

Table 9.1. Idealized zonation pattern of seagrasses along the coasts of the world seas

Temperate coasts	
North temperate intertidal	*Zostera* subgen. *Zosterella* + annual *Zostera marina*
North temperate subtidal	*Zostera* subgen. *Zostera* (replaced by *Zostera* subgen. *Zosterella*)
South temperate intertidal	*Zostera* subgen. *Zosterella* + *Heterozostera*
South temperate subtidal	*Posidonia, Amphibolis, Heterozostera* (replaced by *Zostera* subgen. *Zosterella* + *Halophila*)
Tropical coasts	
Tropical Atlantic intertidal	*Halodule*
Tropical Atlantic shallow subtidal	*Thalassia, Syringodium, Halophila, Halodule* (replaced by *Halodule, Halophila*)
Tropical Atlantic deep subtidal	*Halophila*
Tropical Indo-West Pacific intertidal	*Halodule, Halophila, (Cymodocea)*
Tropical Indo-West Pacific shallow subtidal	*Thalassia, Syringodium, Cymodocea, Enhalus, Thalassodendron, Halodule, Halophila* (replaced by *Halodule, Halophila*)
Tropical Indo-West Pacific deep subtidal	*Halophila*

In the intertidal zone of western Europe, the narrow-leaved *Zostera noltii* is generally dominant. This species is usually a biennial. It survives the winter in a green condition, but in less dense beds than in the summer. In many areas, this species is accompanied by a narrow-leaved morph of *Z. marina*. This latter species develops particularly in small depressions on sandy or muddy flats, where a thin film of water prevents the total desiccation of the plants during low tide. In contrast, *Z. noltii* inhabits mainly the inconspicuous elevations on these flats.

Around the low-water mark and deeper in the sublittoral zone, the beds of the perennial morph of *Z. marina* are the most conspicuous feature. In many areas, beds comprised of annual and perennial seagrasses are separated by miles depending on the slope of the tidal flats.

In the Mediterranean, where tidal movements are more or less restricted to the northern part of the Adriatic Sea, *Z. noltii* forms the upper belt where it is permanently submerged under very sheltered conditions. It borders on beds of *Cymodocea nodosa* that border in this turn on beds of *Posidonia oceanica*. However, it appears that *Posidonia* beds extend over large areas, down to at least 40 m, and that areas unsuitable for this species are often occupied by *C. nodosa*. A *Posidonia* bed bordered on its upper limit as well as on its lower limit by *C. nodosa* has been described by Bay (1984) from Corsica. *Z. marina* is very restricted in its occurrence in the Mediterranean, being mainly confined to the northern Adriatic and northern Aegean Sea and to the Black Sea.

In the Baltic Sea with no tides, and a salinity gradient to fresh-water, beds of *Z. marina* occur fully submerged, forming mosaics with brackish-water plants, such as *Ruppia cirrhosa* and *Potamogeton pectinatus*.

On the North American Atlantic coast, the situation is simpler, as only *Z. marina* occurs with annual and perennial morphs. In the Gulf of California (Mexico), where high water temperatures exceed the lethal point by June or early July each year, all *Z. marina* is annual (Phillips et al. 1983a). Along the North American Pacific coast, originally only *Z. marina* occurred, but during the last century, the mainly intertidal species *Z. japonica* has been introduced (Harrison and Bigley 1982). This species has more or less the same ecological niche as *Z. noltii* in Western Europe. In more recent times, Phillips and Wyllie-Echeverria (1990) have discovered the sublittoral *Z. asiatica* in North America, but still very little is known about the ecological demands of this species.

On the western side of the North Pacific, *Z. marina* is the common species in the sublittoral, while *Z. japonica* occupies the intertidal zone. In East Asia there may be large gaps between the communities of these two species, but *Z. japonica* extends much further to the south, viz. to Vietnam, where its stands may border on growths of tropical species. Of the other three *Zostera* species in this region, e.g. *Z. asiatica, Z. caulescens* and *Z. caespitosa*, very little is known.

The broad occurrence of seagrasses over a wide range of latitudes and in a diversity of habitats suggests genetic differentiation among the species populations (McMillan and Phillips 1979). Laboratory and field experimental results imply that local habitat conditions have had a selective influence.

In the temperate regions of the southern hemisphere, *Zostera* species are also prominent in some areas. In South Africa, *Z. capensis* forms dense beds in suitable locations in the intertidal zone, but it is also able to occupy the upper part of the sublittoral zone, possibly because there is no counterpart for *Z. marina*. Sublittoral morphs of *Z. capensis* are considerably coarser than those from the intertidal. The intertidal morph of this species extends along the east coast of Africa up to the equator. Its beds do not intermix with those of the tropical species.

Along the temperate coasts of Australia the situation is more complex. There are three species of *Zostera*, each with its own area of distribution. *Z. mucronata* is confined to the southwestern part of Australia. *Z. muellerii* is confined to the southeastern part, while *Z. capricorni* is confined to the eastern part of Australia. Only the latter species is able to establish beds in the sublittoral zone where its morphs are very robust. The other two species are mainly restricted to the intertidal zone. *Z. muellerii* and *Z. mucronata* can border on beds of *Heterozostera tasmanica*, and even intermingle with them around the low-water mark. However, the sublittoral seagrass beds in temperate Australia are not composed of *Zostera* species, but of species of *Posidonia* and *Amphibolis*. These communities show a considerably higher diversity in species than do those of *Zostera*, and they are structurally more complex. In disturbed places in the sublittoral, *H. tasmanica* and *Halophila ovalis* may dominate.

In New Zealand the seagrass vegetation is not as developed as in Australia. There is one species, *Z. novozelandica,* which occurs in the intertidal and upper sublittoral zones. *Z. capricorni* is present, but only in the extreme north.

The temperate west coast of South America is almost devoid of seagrasses. Three beds of *H. tasmanica* have been found near Coquimbo, Chile. No seagrass beds have been found from the temperate east coast of South America. There is a record of a *Zostera* leaf washed ashore near Montevideo, Uruguay (Setchell 1935).

In the tropical waters, the place of *Zostera* is taken by species of other genera. In the lower littoral zone, there are narrow-leaved morphs of species of *Halodule*, and here and there tiny species of *Halophila*. In the sublittoral, dense beds of species of *Cymodocea*, *Syringodium* and *Thalassia* occur.

In the Caribbean and the Gulf of Mexico, beds of *Halodule wrightii* or *H. beaudettei* occur near the low-water mark bordering on sublittoral beds of *Thalassia testudinum*. *Syringodium filiforme* may be interspersed in the *Thalassia* bed, but it may also form stands of its own. The same holds for the

Halophila species in the region. However, *Halophila* can extend to great depths. *Halodule* species also occur in the sublittoral, occasionally in unispecific stands in water deeper than *Thalassia*. Along the coast of Brazil, *Thalassia* and *Syringodium* are absent, while sublittoral beds of *H. wrightii* or *H. emarginata* are present (De Oliveira et al. 1983). *Halophila* is represented by two species, but these are too rare to be of significance.

Along the tropical Atlantic coast of Africa, the occurrence of *H. wrightii* has been recorded, but nothing is known about its habitat in that region.

The seagrass communities of the Indo-Pacific have many common traits. In the intertidal, narrow-leaved *Halodule* and *Halophila* species have been found over the entire area. *H. uninervis* is common in most of these communities, while *H. pinifolia* is restricted to the western Pacific and the eastern part of the Indian Ocean. *H. wrightii* is found along the coast of East Africa. Narrow-leaved morphs of *Halophila ovalis* often co-exist with *Halodule* species. Other *Halophila* species in the intertidal zone are generally restricted to specific habitats. *H. beccarii* is found on muddy flats and creeks in mangrove areas. In the sublittoral, larger seagrasses form dense beds, often consisting of a mixture of several species. The dominant species of these 'mixed lawns' are *Thalassia hemprichii*, *Cymodocea serrulata*, *C. rotundata*, the wide-leaved morph of *Halodule uninervis*, *Syringodium isoetifolium*, *Enhalus acoroides*, the robust morph of *Halophila ovalis*, and other species with a less wide area of distribution. In habitats exposed to ocean swell, *Thalassodendron ciliatum* beds may be found in unispecific stands. Where large seagrasses are absent, their place is taken by the small species, as in the Persian Gulf.

9.3 Structure of the Seagrass Community

The structure of seagrass beds is usually regarded as quite simple, as they consist of one or a few plant species of a rather similar morphology. However, they show differences when inspected more closely. At least six growth forms have been recognized (Den Hartog 1977). The frame element is the seagrass itself. As a consequence of its dominance and seasonal rhythm, it determines, to a considerable degree, which algae and aquatic animals can accompany it.

In Table 9.2, a survey is given of the various structural elements that can be distinguished. The term structure comprises at least the following components: (1) floristic and faunistic composition in a qualitative and a quantitative sense, (2) the arrangement of the organisms in space and time, and (3) the relationships between the organisms within the community and their relationships with the surrounding biotic and abiotic environment (Den Hartog and Van der Velde 1988).

Table 9.2. Structural elements of seagrass beds

1. The seagrass itself
2. Epiphytes
3. Other associated algae and fungi
4. Epifauna and mining fauna of seagrasses
5. Bottom fauna
6. Swimming and temporary visitors
7. Microbial communities

Epiphytic algae form an important structural element. Seagrass leaves are usually covered by diatoms, which form food for numerous invertebrates. Due to grazing, the leaves look generally bright green, but, if the growth of algae is not controlled, the leaves become covered by a brown slimy mess, consisting of detritus agglomerated by the algae, which can easily be wiped from the leaves. This material hampers the photosynthetic capacity of the seagrass. Uncontrolled diatom growth has been observed mostly in estuarine environments subjected to organic pollution.

In sublittoral beds of the perennial *Zostera marina*, particularly when they occur under oceanic conditions, small epiphytic Rhodophyta and Phaeophyta occur abundantly on the leaf-tips and -edges. Many of these algae are specific to the leaves. There is a balance between epiphyte growth and the seagrass growth. Only the oldest leaf parts, which are possibly not functioning optimally, are affected. In waters affected by eutrophication, epiphyte growth may become excessive, but the troublesome epiphytic algae are common fouling algae, such as *Ceramium rubrum* and *Ectocarpus confervoides*. The true *Zostera* epiphytes, in contrast, are very sensitive to eutrophication and are among the first organisms to disappear.

The rhizomes of the *Zostera* beds are generally devoid of algae. Only on dead rhizomes can some algal development take place. In this case, the algae concerned are species of *Enteromorpha*, *Ulva* and *Asperococcus*. In several other seagrass communities, viz. those dominated by species of *Posidonia*, *Thalassodendron* and *Amphibolis*, which all have hard persistent rhizomes, two epiphyte communities can be recognized, e.g. one on the leaves and another on the rhizomes and lignified upright shoots. The first community is related to the epiphytic community as described for *Zostera* leaves. The other community seems comparable to the community occupying shaded sublittoral rocky substrates in sheltered habitats (Van der Ben 1969, 1971). Among other associates are the rhizophytic algae. In open tropical and subtropical seagrass beds, mosaics are often found of seagrass and rhizophytic algae of *Caulerpa*, *Avrainvillea*, *Penicillus*, *Udotea*, and other members of the Codiaceae. In the Mediterranean, *Caulerpa prolifera* is able to form mixed beds with *Cymodocea nodosa*. In open *Zostera* communities, particularly on muddy sand, the bottom may be covered by a fine film of algae, mainly diatoms, and Cyanophyta. Under estuarine conditions, *Vaucheria* occasionally takes part in the algal cover of the substrate.

In the permanently submerged *Zostera marina* beds, many detached algae are caught in the leaf canopy and become deposited on the bottom. Along the Atlantic coast of France, these secondary algal growths may locally comprise 10%–20% of the total aboveground phytomass. When the environment is enriched with nutrients, large blankets of Chlorophyta may develop in the intertidal zone. These are not a part of the seagrass system, but may be a threat to them (Den Hartog 1994).

A few endophytes have occasionally been found in seagrass tissue. Of more importance are the fungi, particularly *Labyrinthula* and *Plasmodiophora*. Each seagrass genus seems to have its specific *Labyrinthula* taxon (Vergeer and Den Hartog 1994). The most well known is *L. zosterae*, which has been indicated by many authors to be the cause of the so-called wasting disease (Den Hartog 1987; Muehlstein et al. 1991). In our opinion, it is one of the first decomposers of senescent seagrass tissue (Den Hartog 1996). The species of *Plasmodiophora* form characteristic galls on some seagrass species, but are never abundant. An Ascomycete, *Lulworthia*, has also been ascribed to be a significant decomposer of *Zostera* leaf tissue.

A considerable number of sessile invertebrate species occur together with epihytic algae on seagrass leaves. There are also many vagile ones in the seagrass canopy, particularly grazing snails. Mining organisms have rarely been encountered in seagrass beds. However, in some *Posidonia* species from southwestern Australia, minute isopods of the genera *Lyseia* and *Limnoria* mine within the leaf parenchyma (Brearley and Walker 1995, 1996). In the rhizomes of the tropical seagrass *Thalassia hemprichii*, very small teredoid molluscs have been observed.

Some species come to graze the seagrass itself. In the tropics sirenia and green turtles graze the seagrasses and have a considerable impact on them. Geese and ducks have a very heavy impact on the seagrass beds in the temperate zone. The number of species that feed directly on seagrass is limited. Most animals depend indirectly on seagrass, feeding on the accompanying epiphytes and small faunal organisms, which, in their turn, graze the epiphytes, and the detritus produced by the joint community of seagrass, algae, and fauna. There are two trophic pathways, viz., grazing and detrital pathways. The latter is by far not only the most complex, but also the most important one.

Seagrass beds are spatially positioned between the inshore mangroves and marshes, and the offshore coral reefs and open sea. Thus, as well as a large number of animals that are permanent inhabitants, many others find temporary shelter and/or food within the leaf canopy at some stage of their life cycle or some portion of the day. In the tropics, a number of fish and sea urchins perform a diurnal migration from adjacent coral reefs to nocturnal feeding grounds of the seagrass beds, and vice versa. There are also seasonal visitors that shelter and feed during their migratory movements in the seagrass beds. Finally, there are a great number of juvenile fish species which migrate from the mangrove and marshes to the seagrass beds to find refuge and food there before migrating to offshore habitats (Thayer et al. 1984).

The number of structural elements in the intertidal beds is smaller than in the sublittoral counterparts. Some of these differences depend not only on the degree of tolerance to desiccation, but also depend on salinity and temperature fluctuations. A more important part, however, is often played by the dominant seagrass species itself, as its activity rhythm, and thus the temporal development of the bed, determines to a high degree whether a species can inhabit the seagrass bed permanently or not. In the case that species can only live temporarily in the seagrass bed it is of importance that they can shift to other communities where they meet favourable conditions. The richness of species is thus largely dependent on the ecological quality of other habitats in the immediate surroundings. It leads to the conclusion that large seagrass beds may not be richer in species than small beds that form mosaics or that border on other communities.

9.4 Seagrass Production

There are many data available on the functioning of seagrass beds, particularly of the performance of the dominant species. Most seagrass species form vast dense meadows in shallow coastal areas. They are documented as belonging to the richest and most productive ecosystems in the world, rivalling in productivity cultivated tropical agricultural crops (Zieman and Wetzel 1980). The physical stability and shelter supplied by the seagrass bed structure provide the basis for a highly productive ecosystem (Wood et al. 1969; Fonseca et al. 1998). The ability of seagrasses to exert a major influence on the marine seascape is due to a large extent on their rapid growth and high net productivity (Tables 9.3, 9.4).

The biomass of the leaves varies, depending on water depth, substrate, nutrient availability, and season. Zieman (1975) reported that leaves of *Thalassia testudinum* constituted 15%–22% of the total dry weight of the plants, but this proportion could vary between 10% and 45%. Most of the biomass is in the sediment and is often difficult to sample because of the penetration of the roots (Zieman 1972, 1975; Zieman and Wetzel 1980). Most of the roots are found in the upper sediment layers. Zieman (1972) found in south Florida that the roots of *T. testudinum* were able to penetrate to 4 m depth in the sediment to the underlying bedrock. In annual populations, the

Table 9.3. Phytomass of selected seagrass species

Species	g dry weight m^{-2}	Source
Zostera marina	120–1020	Phillips (1969, 1972); Mann (1972)
Halodule wrightii	200	Dillon (1971)
Posidonia oceanica	740	Drew (1971)
Thalassia testudinum	150–3100	Odum (1963); McRoy (1974); Phillips (1960); Bauersfield et al. (1969); Taylor et al. (1973)

Table 9.4. Annual net production of selected seagrass species

Species	g C m^{-2} year^{-1}	Source
Zostera marina	84–800	Sand-Jensen and Borum (1983); Phillips (1984); Thorne-Miller and Harlin (1984)
Halodule wrightii	280	Zieman and Wetzel (1980)
Thalassia testudinum	825	Zieman and Wetzel (1980)
Spp. in temp. Australia	120–690	Hillman et al. (1989)

leaves constitute the larger part of the biomass. The investment in the root system is often no more than 30%.

Rates of annual net production compare with the maximum values for terrestrial plants, but are somewhat lower than the highest rates of heavily subsidized crops such as sugar cane and sorghum (3000 g C m^{-2} year^{-1}) and mangroves (2000 g C m^{-2} year^{-1}; Hillman et al. 1989).

Data on seagrass epiphyte production are sparse, as are those on the benthic macrophyte and microphyte components. Seagrass epiphyte production can attain up to 50% of that of the seagrass plants (Penhale 1977; Morgan and Kitting 1984; Brouns and Heijs 1986). However, high productivity of epiphytes is generally linked to eutrophication of the water layer and/or absence of algae-consuming herbivores. Loose-lying and attached macro-algae within the meadow may account for 2%–39% of the total above-ground seagrass bed production at times and in various places, sometimes ranging to almost 75% (Dawes 1987).

Leaves typically grow at rates of 5 mm day^{-1} but, under favourable conditions, rates of over 10 mm day^{-1} are not uncommon (Zieman 1982; Brouns 1987). The leaf turnover rates of seagrasses have a profound effect on the entire ecosystem. Leaf life spans affect epiphyte diversity, primary productivity, abundance and biomass. More critically, they also determine the amount of dissolved and particulate organic matter available to fuel the detritus food chains. The rate of leaf replacement differs somewhat with habitat and varies seasonally with changes in growth rate, with life spans of leaves ranging from 11–24 days in *Halophila ovalis,* over about 50 days in *Zostera marina,* and up to 125 days in *Posidonia australis* (West and Larkum 1979; Hillman 1985; Borowitzka and Lethbridge 1989). These rates are also affected by grazing, and differences in susceptibility to grazing (Hootsmans and Vermaat 1985). Life spans of leaves range from 1–4 months. Ott (1980) demonstrates that a plant of *Posidonia oceanica* produces about 10 leaves per year, but the leaves produced in autumn may persist up to two to five times longer than those produced in spring. West and Larkum (1979) found that the turnover rate of leaves of *P. australis* is about 2.5 times greater in summer than in winter. There are also significant differences in the leaf turnover rates at different sites in the same area.

The percentage of flowering stalks and the number of eelgrass seeds produced are directly related to the degree of stress in the habitat, e.g. more seeds were produced per square metre in the Gulf of California and in the southerly ends of the area of distribution along the Pacific coast of North America and in Alaska than in the mid-range locations in Washington State (Phillips et al. 1983a). Also, more seeds were produced at any one site in the intertidal zone than in the sublittoral. The same relationships appear to be valid for the Atlantic coasts of North America (Phillips et al. 1983a) and Europe (Jacobs and Pierson 1981).

9.5 Seagrass Dynamics

Very little attention has been paid to successional concepts (Molinier and Picard 1951, 1952). It has become evident that seagrass beds do not function as a prelude to terrestrial ecosystems. Seagrass beds are almost always separated spatially from marshes, and, in the tropics, the same is true for seagrass beds and mangrove stands.

The seagrasses, however, show year-to-year patterns in their development. In the intertidal zone of the Wadden Sea, most of the *Zostera marina* is annual. Whether this annual behaviour is genetically or environmentally determined has not yet been established. Probably both options are possible. In normal winters, *Z. marina* does not survive, but, after mild winters, survivors are often found. The accompanying *Z. noltii* rhizomes survive; the above-ground biomass is considerably reduced in winter. These seagrass beds are heavily grazed by Brent geese in autumn. According to Jacobs et al. (1981), the geese and ducks (widgeon, mallard and pintail) consume almost the total above-ground biomass as well as part of the underground biomass, i.e. approximately half the annual production of the seagrass. As a consequence of ice scour in winter, sediment movement under the influence of gales, and grazing by birds, the sediment is levelled each year. As a result, the geomorphological character of the area does not change from year to year (Nacken and Reise 2000). Sedimentation and erosion are more or less in equilibrium. In areas where frost is less frequent in winter, the seagrass may also maintain a coverage of 20%–30%, in spite of grazing by geese and ducks (Tubbs and Tubbs 1983). In the most western reaches of the Channel where frost is an exception, and which are off the main migration routes of the birds, the seagrass beds remain green in winter and are not or only partly levelled by gales. As a consequence, the bottom becomes raised, and, as more coarse material settles along the margins of the beds than in the center, the beds become heterogeneous and show a considerable relief. In the highest situated beds, in particular, this leads to a decline of the seagrasses, because of increasing drainage during low tide and the impact of erosion.

Almost all available data on the dynamics of seagrass beds relate to their expansion and decline. The existence of a bed is taken for granted. In spite of that, there are still no answers to the question why in a certain area a seagrass bed occurs, and why the direct surroundings are without seagrass, even as there does not seem to be any obvious physical cause for the difference. In some areas, the beds seem to have more or less stable positions, but, in other areas, the beds show continuous changes in size and topographical position, and 'move' through the area. Establishment of new beds has not yet been studied. The opportunity to study this process exists along the North American Pacific coast, where the introduced *Zostera japonica* is still expanding in area, or in the Mediterranean where *Halophila stipulacea* is still expanding (Lipkin 1975; Den Hartog and Van der Velde 1993; Di Martino 1999). Between the actual first settlement and the formation of a well-established bed, many thresholds have to be overcome. Successful settlement is probably only possible in years or series of years with very special climatic conditions. Another factor of importance is whether a species is completely absent in an area or whether there are some small relict populations present.

9.6 Worldwide Decline of Seagrass Beds

Seagrass beds are declining all over the world. In many places this is gradual as a consequence of progressive pollution of the seawater. At first the epiphytic growth on the leaves increases to the point that the plants begin to suffer, and after that a gradual decrease of the bed sets in. This is characterized by a decrease in the density of the shoots in the bed, later by fragmentation of the bed. Often this process is not recognized in the first stages because it is so gradual.

A 'wasting disease' reduced the beds of *Zostera marina* in the North Atlantic region in the 1930s (Milne and Milne 1951; Martin 1954; Rasmussen

1977; Den Hartog 1987). The disease was ascribed to the activity of some fungi, particularly to a virulent pathogenic organism, *Labyrinthula macrocystis*, recently recognized as a separate species, *L. zosterae* (Muehlstein et al. 1991). In the last decade, the *Zostera* beds along the North American Atlantic coast suffered again from a wasting disease epidemic (Short et al. 1987, 1988). Along the European west coast, no decline attributed to *L. zosterae* could be ascertained, although this organism was isolated from almost every population that was investigated, even from geographically isolated populations of *Z. marina* in the Adriatic Sea (Den Hartog et al. 1996) and the White Sea (Maximova and Den Hartog, unpubl.). We have grave doubts that *L. zosterae* is the primary cause of the wasting disease. Damage patterns similar to those in *Labyrinthula*-infected plants have been found in herbarium specimens collected long before wasting disease was ever observed. Further, *Labyrinthula* has also been isolated from *Zostera noltii* and *Z. japonica*. Its very wide distribution in the northern Atlantic and in the northern Pacific, and also in populations isolated from the main distributional area over thousands of years, does not lend support for considering it an organism dangerous enough to bring *Zostera* to the brink of extinction. It is one of the first organisms to develop in senescent *Zostera* tissue. The main feature of the wasting disease is that it develops not only in the senescent oldest leaves, but also in the youngest leaves, confronting the plants with a serious metabolic problem.

Tutin (1938) correlated periods of eelgrass reduction with excessively cloudy years, implying that eelgrass became weakened due to light reduction in the North Atlantic in the early 1930s. In 1996, Den Hartog called this the light-deficit hypothesis. In Western Europe, the wasting disease started in spring 1932. This was the second of two consecutive years with a considerable light deficit during the main growing season of *Zostera marina*. Starved plants were again exposed to low light in combination with rising water temperatures (higher water temperatures also noted by Rasmussen 1977), stressing the plant's metabolism to the limit. Add to this the effects of pollution, turbidity, and other human activities, and circumstances were right for damage to the plants. In this scenario, the precocious development of *Labyrinthula* is a consequence of the environmentally induced poor metabolic condition of the plants and not the cause of the disease. For this reason, the term 'wasting disease' should be replaced by the more neutral term 'wasting phenomenon'.

Other large-scale losses of seagrass beds have been recorded from the eastern coast of northern America and Australia. Orth and Moore (1983) recorded a massive decline of all submerged macrophytes, including the seagrasses, from Chesapeake Bay, and regarded it to be the result of human impact by agricultural and industrial waste. Poiner et al. (1989) described the disastrous effects of the cyclone "Sandy" that struck the seagrass beds along the Gulf of Carpentaria in March, 1985. Of the 183 km² of seagrass beds, 151 km² were completely destroyed. In April 1986, it became clear that the beds that had survived the storm had disappeared too. As a result of scouring, the remaining shallow beds had been eroded away, and the remaining deeper beds had disappeared as a result of deposition of fine silt that smothered the seagrasses. This disaster had a purely natural cause. However, Preen et al. (1995) described the effects of another cyclone, "Fran", on the seagrass beds of Hervey Bay in 1992. In this case, the cyclone was followed by extremely large river discharges, which caused a considerable decrease in salinity. The sediment discharged by the floods in its turn caused a low-light environment and deposition of material on the bottom. In a period of 6 weeks, more than 1000 km² of seagrass beds were lost, along with the accompanying organisms. The authors mentioned that there has been a slight recovery of seagrass in the deep water but, in the shallow environments, no recovery has been observed. In this case, the damage by the floods can be ascribed without hesitation to a century of human mismanagement of the adjacent terrestrial environment: deforestation with the unavoidable result of erosion, degradation of the soil, and a highly decreased water-holding capacity of the bottom.

Small-scale disappearances of seagrass beds are repeatedly observed. Their cause is not always clear. Usually, they are the result of one or other type of pollution, dredging, or construction works such as dams, ports, etc. Due to the continually increasing pollution, the green alga *Enteromorpha radiata* has increased in the last decades. This alga grows loosely anchored in the mud and can form large blankets under sheltered conditions, which may float and become deposited in other places. If these blankets become deposited on a seagrass bed, this may lead to the complete destruction of the bed owing to suffocation, as has been observed in Langstone Harbour, Hampshire, England (Den Hartog 1994). Although the event took place in 1991, the seagrass has not returned (Den Hartog, personal observation, summer 1999).

Other impacts to seagrass beds are adventive algae, which accidentally arrive in other areas and upset the existing balance between the available native species of the system. In Western Europe, the brown alga *Sargassum muticum* gradually replaces *Zostera* beds in the lower part of the intertidal zone on gravelly or mixed substrata (Den Hartog 1997). In the Mediterranean, the green alga *Caulerpa taxifolia* is disastrous to the beds of *Posidonia oceanica* (Meinesz and Hesse 1991).

In the past decades, we have seen the disappearance of many large and small seagrass beds. Unfortunately, we have to admit that we do not know of any newly formed seagrass bed by native species in their own area of distribution. This is an alarming observation. For this reason, it is urgently necessary that the study of the dynamics of seagrass beds be promoted, from the very beginnings through maturity to senility. This knowledge is necessary to understand this very valuable ecosystem. Without this knowledge, it will not be possible to make any sensible interventions in order to reverse the gradual deterioration of the seagrass beds.

9.7 Conclusions

A general picture is sketched of the 'ideal' zonation pattern as found along the coasts of the world. This picture has been worked out in more detail for the intertidal seagrass beds that are composed of *Zostera* species along the temperate coasts, and of *Halodule* and *Halophila* in the tropics. These genera can dominate in the sublittoral where the larger sublittoral seagrasses are absent.

Intertidal seagrass beds usually have less species and have less structural elements developed than the sublittoral beds. An important part is played by the dominant seagrass, as its activity rhythm determines whether associated species can live in the seagrass bed as permanent inhabitants or not. Where species live only temporarily in the bed, the species richness of the bed is largely dependent on the quality of other habitats in the immediate vicinity. It means that large, uninterrupted beds are not necessarily richer in associated species than small beds that can form mosaics with other communities.

Since seagrasses display high primary productivity and have a relatively rapid leaf turnover rate, they produce a prodigious amount of dissolved and particulate organic matter which forms the basis for the detrital foodweb. This is by far the most important and complex trophic pathway in the seagrass beds. The other trophic pathway, direct consumption, is generally less important.

Most of the questions in relation to the dynamics of the seagrass beds are still unanswered, despite their relevance if lost beds are to be restored. Finally, it has been pointed out that seagrass beds are declining the world over. A number of cases of almost 'terminal' loss of seagrass are discussed. Even when 'purely natural' causes are involved, the conditions are often human-related, as in the case of the greatest seagrass loss ever recorded, e.g. more than 1000 km^2 in Hervey Bay, Queensland, Australia.

References

Bauersfield P, Kleer RR, Durrant NW, Sykes JE (1969) Nutrient content of turtle grass (*Thalassia testudinum*). Proc Int Seaweed Symp 6:637–645

Bay D (1984) A field study of the growth dynamics and productivity of *Posidonia oceanica* (L.) Delile in Calvi Bay, Corsica. Aquat Bot 20:43–64

Borowitzka MA, Lethbridge RC (1989) Seagrass epiphytes. In: Larkum AWD, McComb AJ, Shepherd SA (eds) Biology of seagrasses; A treatise on the biology of seagrasses with special reference to the Australian region. Elsevier, Amsterdam, pp 458-499

Brearley A, Walker DI (1995) Isopod miners in the leaves of two Western Australian *Posidonia* species. Aquat Bot 52:163-181

Brearley A, Walker DI (1996) Burrow structure and effects of burrowing Isopods (Limnoriidae) in southwestern Australian *Posidonia* meadows. In: Kuo J, Phillips RC, Walker DI, Kirkman H (eds) Seagrass biology: Proceedings of an international workshop, Rottnest Island, Western Australia, 25-29 Jan 1996, pp 261-268

Brouns JJWM (1987) Aspects of production and biomass of four seagrass species (Cymodoceoideae) from Papua New Guinea. Aquat Bot 27:333-362

Brouns JJWM, Heijs FML (1986) Production and biomass of the seagrass *Enhalus acoroides* (L.f.) Royle and its epiphytes. Aquat Bot 25:21-45

Dawes CJ (1987) The dynamic seagrasses of the Gulf of Mexico and Florida coasts. In: Durako MJ, Phillips RC, Lewis RR (eds) Proceedings of the symposium on subtropical-tropical seagrasses of the Southeastern United States. Fla Mar Res Publ 42:25-30

Den Hartog C (1970) The seagrasses of the world. Verh Kon Ned Akad Wetensch Afd Natuurk Ser II 59 (1):1-275

Den Hartog C (1977) Structure, function, and classification in seagrass communities. In: McRoy CP, Helfferich C (eds) Seagrass ecosystems, a scientific perspective. Dekker, New York, pp 90-121

Den Hartog C (1987) "Wasting disease" and other dynamic phenomena in *Zostera* beds. Aquat Bot 27:3-14

Den Hartog C (1994) Suffocation of a littoral *Zostera* bed by *Enteromorpha radiata*. Aquat Bot 47:21-28

Den Hartog C (1996) Sudden declines of seagrass beds: "wasting disease" and other disasters. In: Kuo J, Phillips RC, Walker DI, Kirkman H (eds) Seagrass biology: Proceedings of an international workshop, Rottnest Island, Western Australia, 25-29 Jan 1996, pp 307-314

Den Hartog C (1997) Is *Sargassum muticum* a threat to eelgrass beds? Aquat Bot 58:37-41

Den Hartog C, Van der Velde G (1988) Structural aspects of aquatic plant communities. In: Symoens JJ (ed) Vegetation of inland waters. Handbook of vegetation science, vol 15 (1). Kluwer Academic Publishers, Dordrecht, pp 113-153

Den Hartog C, Van der Velde G (1993) Occurrence of the seagrass *Halophila stipulacea* (Hydrocharitaceae) along the Mediterranean coast of Turkey. Posidonia Newsletter 4(2):5-6

Den Hartog C, Vergeer LHT, Rismondo AF (1996) Occurrence of *Labyrinthula zosterae* in *Zostera marina* from Venice Lagoon. Bot Mar 39:23-26

De Oliveira FEC, Pirani RJ, Giulietti AM (1983) The Brazilian seagrasses. Aquat Bot 16:251-267

Di Martino V (1999) Espansione di vegetali marini tropicali in Sicilia ed in Calabria. In: Gravez V, Boudouresque C-F, Meinesz A, Scabbia G (eds) 4th International workshop on *Caulerpa taxifolia*. Lerici, Italy, 1-2 Feb 1999. Abstract p 81

Dillon CR (1971) A comparative study of the primary productivity of estuarine phytoplankton and macrobenthic plants. PhD dissertation. Univ North Carolina, Chapel Hill, 112 pp

Drew EA (1971) Botany, chap 6. In: Woods JA, Lythgoe JN (eds) Underwater science. Oxford University Press, London

Fonseca MMS, Kenworthy WJ, Thayer GW (1998) Guidelines for the conservation and restoration of seagrasses in the United States and adjacent waters. NOAA Coastal

Ocean Program Decision Analysis Ser 12. NOAA Coastal Ocean Office, Silver Springs, Maryland, 222 pp

Harrison PG, Bigley RE (1982) The recent introduction of the seagrass *Zostera japonica* Aschers. & Graebn. to the Pacific coast of North America. Can J Fish Aquat Sci 39: 1642–1648

Hillman K (1985) The production ecology of the seagrass *Halophila ovalis* (R.Br.) Hook.f. in the Swan/Canning estuary, Western Australia. PhD Thesis, Univ Western Australia, Perth

Hillman K, Walker DI, Larkum AWD, McComb AJ (1989) Productivity and nutrient limitation. In: Larkum AWD, McComb AJ, Shepherd SA (eds) Biology of seagrasses. A treatise on the biology of seagrasses with special reference to the Australian region. Elsevier, Amsterdam, pp 635–685

Hootsmans MJM, Vermaat JE (1985) The effect of periphyton-grazing by three epifaunal species on the growth of *Zostera marina* L. under experimental conditions. Aquat Bot 22:83–88

Jacobs RPWM, Pierson ES (1981) Phenology of reproductive shoots of eelgrass, *Zostera marina* L., at Roscoff (France). Aquat Bot 10:45–60

Jacobs RPWM, Den Hartog C, Braster BF, Carrière FC (1981) Grazing of the seagrass *Zostera noltii* by birds at Terschelling (Dutch Wadden Sea). Aquat Bot 10:241–259

Lipkin Y (1975) *Halophila stipulacea*, a review of a successful immigration. Aquat Bot 1:203–215

Mann KH (1972) Macrophyte production and detritus food chains in coastal waters. Mem Ist Ital Idrobiol Dott Marco Marchi 29 [Suppl]:353–383

Martin AC (1954) A clue to the eelgrass mystery. Trans 19th North Am Wildlife Conf Washington, DC, pp 441–449

McMillan C, Phillips RC (1979) Differentiation in habitat response among populations of New World seagrasses. Aquat Bot 7:185–196

McRoy CP (1974) Seagrass productivity: carbon uptake experiments in eelgrass, *Zostera marina*. Aquaculture 4:131–137

Meinesz A, Hesse B (1991) Introduction et invasion de l' algue tropicale *Caulerpa taxifolia* en Méditerranée nord-occidentale. Oceanol Acta 14:415–426

Milne JL, Milne MJ (1951) The eelgrass catastrophe. Sci Am 184 (1):52–55

Molinier R, Picard J (1951) Biologie des herbiers de Zostéracées des côtes françaises de la Méditerranée. CR hebd Seances Acad Sci Paris 233:1212–1214

Molinier R, Picard J (1952) Recherches sur les herbiers de phanérogames marines du littoral méditerranéen français. Ann Inst Océanogr (Monaco) 27:157–234

Morgan MD, Kitting CL (1984) Productivity and utilization of the seagrass *Halodule wrigthii* and its attached epiphytes. Limnol Oceanogr 29:1066–1076

Muehlstein LK, Porter D, Short FT (1991) *Labyrinthula zosterae* sp. nov., the causative agent of wasting disease of eelgrass *Zostera marina*. Mycologia 83:180–191

Nacken M, Reise K (2000) Effects of herbivorous birds on intertidal seagrass beds in the northern Wadden Sea. Helgoland Mar Res 54:87–94

Odum HT (1963) Productivity measurements in Texas turtle grass and the effects of dredging an intracoastal channel. Publ Inst Mar Sci Texas 9:48–58

Orth RJ, Moore KA (1983) Chesapeake Bay: an unprecedented decline in submerged aquatic vegetation. Science 222:51–53

Ott JA (1980) Growth and production in *Posidonia oceanica* (L.) Delile. PSZNI Mar Ecol 1:47–64

Penhale PA (1977) Macrophyte-epiphyte biomass and productivity in an eelgrass (*Zostera marina* L.) community. J Exp Mar Biol Ecol 26:211–224

Phillips RC (1960) Observations on the ecology and distribution of the Florida seagrasses. Prof Pap Ser Fla Bd Conserv 2:1–72

Phillips RC (1969) Temperate grass flats. In: Odum HT, Copeland BJ, McMahon EA (eds) Coastal ecological systems of the United States. Water Pollution Control Admin. Contract RFP 68-128, 2:737-773

Phillips RC (1972) Ecological life history of *Zostera marina* L. (eelgrass) in Puget Sound, Washington. PhD dissertation, Univ of Washington, Seattle

Phillips RC (1984) The ecology of eelgrass meadows in the Pacific Northwest: a community profile. US Fish and Wildl Serv FWS/OBS-84/24, 85pp

Phillips RC, Wyllie-Echeverria S (1990) *Zostera asiatica* Miki on the Pacific coast of North America. Pacif Sci 44:130–134

Phillips RC, Grant WS, McRoy CP (1983a) Reproductive strategies of eelgrass (*Zostera marina* L.). Aquat Bot 16:1–20

Phillips RC, McMillan C, Bridges KW (1983b) Phenology of eelgrass, *Zostera marina* L., along latitudinal gradients in North America. Aquat Bot 15:145–156

Poiner IR, Walker DI, Coles RG (1989) Regional studies – seagrasses of tropical Australia. In: Larkum AWD, McComb AJ, Shepherd SA (eds) Biology of seagrasses. A treatise on the biology of seagrasses with special reference to the Australian region. Elsevier, Amsterdam, pp 279–303

Preen AR, Lee Long WJ, Coles RG (1995) Flood and cyclone related loss, and partial recovery, of more than 1000 km^2 of seagrass in Hervey Bay, Queensland, Australia. Aquat Bot 52:3–17

Rasmussen E (1977) The wasting disease of eelgrass (*Zostera marina*) and its effects on environmental factors and fauna. In: McRoy CP, Helfferich C (eds) Seagrass ecosystems, a scientific perspective. Dekker, New York, pp 1–51

Sand-Jensen K, Borum J (1983) Regulation of growth in eelgrass (*Zostera marina* L.) in Danish coastal waters. Mar Tech Soc J 17(2):15–21

Setchell WA (1935) An occurrence of *Zostera* on the east coast of South America. Rev Sudamer Bot 2:1–3

Short FT, Muehlstein LK, Porter D (1987) Eelgrass wasting disease: cause and recurrence of a marine epidemic. Biol Bull 173:557–562

Short FT, Ibelings BW, Den Hartog C (1988) Comparison of a current eelgrass disease to the wasting disease in the 1930s. Aquat Bot 30:295–304

Taylor JL, Saloman CH, Priest KW (1973) Harvest and regrowth of turtle grass (*Thalassia testudinum*) in Tampa Bay, Florida. US Natl Mar Fish Serv Fish Bull 71: 145–148

Taylor WR (1928) The marine algae of Florida, with special reference to the Dry Tortugas. Carnegie Inst Wash, Publ 379; papers from the Dry Tortugas Lab 25, 219 pp (see p 37)

Thayer GW, Kenworthy WJ, Fonseca MS (1984) The ecology of eelgrass meadows of the Atlantic coast: a community profile. US Fish Wildl Serv FWS/OBS-84/02, 147 pp

Thorne-Miller B, Harlin MM (1984) The production of *Zostera marina* L. and other submerged macrophytes in a coastal lagoon in Rhode Island, USA. Bot Mar 27: 539–546

Tubbs CR, Tubbs JM (1983) The distribution of *Zostera* and its exploitation by wildfowl in the Solent, southern England. Aquat Bot 15:223–239

Tutin TG (1938) The autecology of *Zostera marina* in relation to its wasting disease. New Phytologist 37:50–71

Van der Ben D (1969) Les épiphytes des feuilles de *Posidonia oceanica* sur les côtes françaises de la Méditerranée. Proc Int Seaweed Symp 6:79–84

Van der Ben D (1971) Les épiphytes des feuilles de *Posidonia oceanica* Delile sur les côtes françaises de la Méditerranée. Mem Inst R Sci Nat Belg 168:1-101

Vergeer LHT, Den Hartog C (1994) Omnipresence of Labyrinthulacae in seagrasses. Aquat Bot 48:1-20

West RJ, Larkum AWD (1979) Leaf productivity of the seagrass *Posidonia australis* in eastern Australian waters. Aquat Bot 7:57-65

Wolff T (1976) Utilization of seagrass in the deep sea. Aquat Bot 2:161-174

Wood EJF, Odum WE, Zieman JC (1969) Influence of seagrasses on the productivity of coastal lagoons. In: Coastal lagoons, a symposium, UNAM-UNESCO. Univ Nacional Autonomia Mexico, Mexico DF, pp 459-502

Zieman JC (1972) Origin of circular beds of *Thalassia* (Spermatophyta: Hydrocharitaceae) in South Biscayne Bay, Florida, and their relationship to mangrove hammocks. Bull Mar Sci 22:559-574

Zieman JC (1975) Seasonal variation of turtle grass, *Thalassia testudinum* König, with reference to temperature and salinity effects. Aquat Bot 1:107-123

Zieman JC (1982) The ecology of the seagrasses of south Florida: a community profile. US Fish Wildl Serv, Office of Biological Services, Washington, DC. FWS/OBS 82/85, 185 pp

Zieman JC, Wetzel RG (1980) Productivity in seagrasses: methods and rates. In: Phillips RC, McRoy CP (eds) Handbook of seagrass biology: an ecosystem perspective. Garland STPM Press, New York, pp 87-116

10 The Leaf Canopy of Seagrass Beds: Faunal Community Structure and Function in a Salinity Gradient Along the Swedish Coast

S.P. BADEN and C. BOSTRÖM

10.1 Introduction

10.1.1 *Zostera marina* L.

The distribution of eelgrass (*Zostera marina*, hereafter *Zostera*) is mainly concentrated along temperate coasts of the Northern Pacific and the Atlantic oceans, and it is the only seagrass extending into Arctic areas (71°N) (Den Hartog 1970). Eelgrass commonly inhabits muddy and sandy bottoms and forms continuous meadows or patchy beds in non-tidal, intertidal as well as subtidal areas.

Zostera is considered euryhaline and tolerates salinities from about 32 psu to 5 psu (Luther 1951; Mathiesen and Nielsen 1956). In contrast to salinity tolerance, eelgrass has relatively narrow temperature requirements as suggested many years ago by Setchell (1929). However, seagrasses may form temperature-adapted populations and seagrass flowering has been observed to occur below 10°C (Phillips and Menez 1988). Eelgrass meadows are generally more luxuriant in sheltered areas, the main habitats for *Zostera* on the Swedish west coast, while it mainly occurs at exposed sites along the Swedish east coast and in Finland (Baltic Sea). Leaf area, shoot length and leaf width increase with higher levels of sediment ammonium concentrations, while shoot density and abundance of flowering plants show an inverse relationship to interstitial ammonium content (Den Hartog 1970; Short 1983a,b). However, recent experimental work in the western Baltic Sea (Kiel area) implies that eelgrass at exposed, sandy sites is not nutrient-limited, and that the positive effects of shoot density (Worm and Reusch 2000) or fertilization by mussels (Reusch et al. 1994) are more important than sediment nutrient availability for plant growth and bed expansion. In regions with low environmental stress in terms of salinity, temperature, tidal range and wave exposure, vegetative propagation is important whereas seedling recruitment is more important in extreme environments where disturbances occur (Phillips and Menez 1988).

Zostera populations living near their temperature and salinity limits in the Baltic Sea, and thus under what normally could be considered as environmental stress, represent an interesting example of the opposite. Here, sexual reproduction at exposed sandy bottoms has minor importance for recruitment and population maintenance, whereas vegetative propagation (clonal growth) and dispersal by drifting plants are more common. At the Åland Islands (northern Baltic Sea) almost complete monoclonality has been found (Boström 1996; Boström and Bonsdorff 1997; Reusch et al. 1999a); however, similar results have been obtained in fully saline cold-temperate *Zostera* meadows (Nova Scotia, Canada: Reusch et al. 1999b).

Seagrasses are three-dimensional structures rich in microhabitats and niches hosting abundant and diverse animal assemblages. One focus within seagrass research has been the "bottom-up" (Sand-Jensen 1977; Borum 1985; Tomasko and Lapointe 1991) versus "top-down" (Nelson 1979, 1981; Robertson 1984; Summerson and Peterson 1984; Brönmark 1985; Heck et al. 2000) regulating processes, investigating the interrelationship and energy flow between nutrients, primary producers, herbivores, detritivores, filter feeders and finally primary and secondary consumers (predators). However, very few of these investigations have aimed at a holistic view on food webs in the seagrass meadows (Möller et al. 1985; Heck et al. 2000) as most studies involve only one or two trophic levels (see references above, Heck and Crowder 1991).

10.1.2 Aims of the Study

In this study we demonstrate features of the flora and fauna associated with the *Zostera* leaf canopy along a >1500 km salinity gradient reaching from the fully saline (30 psu) Swedish/Norwegian border in the Skagerrak, through the polyhaline Kattegat and the Danish straits and Kiel Bight (10–15 psu), to the Åland Islands (Finland) in the brackish (6 psu) northern Baltic Sea (Fig. 10.1). The investigated area encompasses five large marine and estuarine sea areas (Kattegat and Skagerrak, Gulf of Kiel, Baltic Proper, Åland Sea; Nielsen et al. 1995) thus covering both a latitudinal (60–54°N) and a longitudinal (12–20°E) gradient (Fig. 10.1). Localities 1–5 are hereafter referred to as the Swedish west coast, while the Swedish east coast refers to localities 7–9. Kiel is referred to as locality 6. Here we first describe the structural features of the leaf canopy i.e., shoot density, biomass, leaf area and co-occurrence with other phanerogams. Then we compare species composition, abundance, biomass, diversity and functional aspects (mobility and feeding groups) of the leaf faunal assemblage in different regions. Possible couplings between the leaf fauna and the sediment sub-systems are discussed. In many areas, predation is considered as an important structuring factor for seagrass leaf fauna (Nelson 1979, 1981; Heck and Orth 1980; Heck and Crowder 1991, review;

The Leaf Canopy of Seagrass Beds: Faunal Community in a Salinity Gradient 215

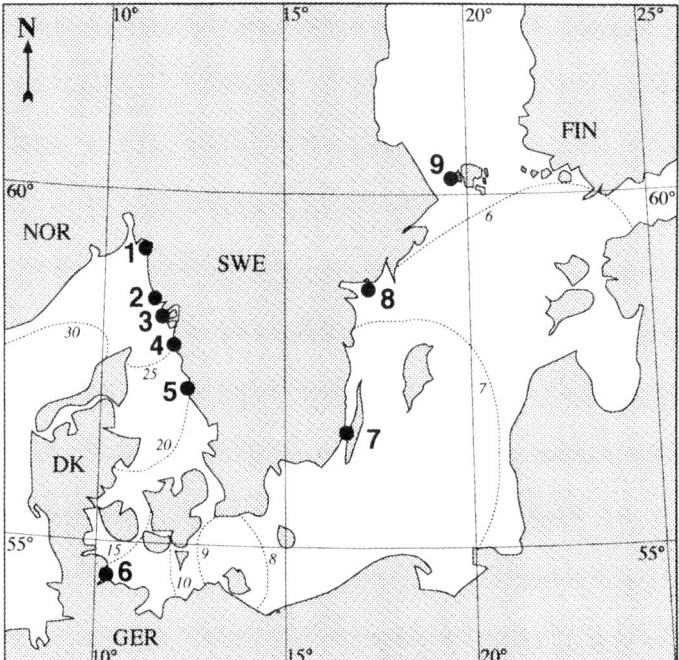

Fig. 10.1. Map of study area showing the *Zostera marina* localities studied. *1* Tjärnö; *2* Gullmarsfjord, *3* Skaftö, *4* Stenungssund, *5* Vendelsöfjord, *6* Kiel, *7* Öland, *8* Askö, *9* Åland. Salinity (psu) is indicated with *dashed lines*

Connolly 1994), therefore a brief description and comparison of the composition and importance of the larger mobile predatory fauna is given for two of the regions studied, i.e. localities 3 and 9 (Fig. 10.1).

10.2 Features of the Study Area

10.2.1 Effects of Salinity

Few areas in the world enable comparative studies of structure and function of the same seagrass species, in this case *Zostera*, and its associated biota along a continuous and geographically permanent salinity gradient. However, Orth (1973) compared the infauna of six *Zostera* beds along an estuarine salinity gradient (Chesapeake Bay, USA). The Baltic Sea has often been classified as a giant estuary Jansson 1978; Rumohr et al. 1996). However, its large transitional sea districts (see Elmgren 1984; Nielsen et al. 1995) do not undergo large temporal and spatial fluctuations in salinity (Kautsky and

Van der Maarel 1990; Wallentinus 1991) and are fairly stable both during the growing season and from year to year. The Swedish west coast has a stratified water mass consisting of Baltic water on top of the more saline water from the North Sea, and generally surface salinity varies between 15 and 25 psu. However, after heavy rainfalls the surface water (down to 2 m) can show reduced salinities for short periods and after upwelling situations high saline bottom water can reach the surface. Thus, *Zostera* can experience salinity extremes between 0 and 30 psu.

Changes in salinity along the gradient are expressed in both plants and animals at the ecological, morphological and physiological levels. For macroalgae of marine origin the number of species decreases from about 300 to about 100 going from the poly-saline environment in the Eastern Skagerrak and Kattegat to the brackish Archipelago Sea, northern Baltic Sea. Among these macroalgae the number of green algal species shows a less dramatic decline due to a complementary increase of freshwater species in the northern Baltic Sea (Nielsen et al. 1995). The macroalgal populations have, since their recruitment into the Baltic Sea in ca. 7500 BP (Russell 1988), diverged considerably from their marine origin, and adapted to brackish water conditions by changes in physiology and morphology, sometimes creating new ecotypes (Russell 1987). Salinity stress is expressed as altered cell volume, ion content or changed concentration of organic solutes (Russell 1987). However, salinity cannot be directly linked to such differences, since reduced plant sizes do not necessary show reduction in cell size, and plants with smaller cells are not always smaller in thallus size (Russell 1985). The aquatic angiosperms show an opposite trend to macroalgae. As all the angiosperms in the study area, except *Zostera*, *Zannichellia palustris* and *Ruppia* spp., are of limnetic origin, the number of flowering plants increases with decreasing salinity, being 5 in the Kattegat and around 20 in the Baltic Archipelago Sea area (Leppäkoski et al. 1999; Snoeijs 1999). In fully saline water, carbonate is the dominating form of carbon and both macroalgae and angiosperms rely on an uptake of carbon via carbonate instead of carbon dioxide. With decreasing salinity, the availability of carbon changes from carbonate to carbon dioxide, causing carbon limitation, size reduction and lowered productivity in *Zostera* (Pinnerup 1980; Stevenson 1988; Ohlsson and Andersson 1990; Hellblom and Björk 1999).

The fauna of the Baltic can be divided into three groups dependent on their evolutionary background: marine immigrants, indigenous brackish water species and limnetic immigrants. Most species have either marine or limnetic background and live close to their physiological tolerance limit with regard to salinity. The physiological strain of salinity affecting, for example, the osmoregulation can result in increased energy expenditure and reduced size as found for the marine bivalve *Mytilus edulis* (Remane 1934; Tedengren and Kautsky 1986; Tedengren et al. 1990), which recently has been divided into two species, *M. edulis* (Swedish west coast) and *M. trossulus* (Baltic Sea) (Väinölä and Hvilsom 1991; Seed 1992). As only a restricted number of marine and limnetic animals have adapted to the brackish environment the total faunal assemblage decreases from >1500 to ≈50 in the study area (Kautsky and van der Maarel 1990; Bonsdorff and Pearson 1999). This reduction in species diversity allows for development of generalistic behaviour and expanded niche breadth of marine species in brackish water (Dahl 1974). Such structural and functional changes in biodiveristy reduce ecosystem complexity and lead to fewer alternative energy flow pathways (Elmgren 1984; Tedengren 1990).

10.2.2 Physical Settings and Substrate Characteristics

On the Swedish west coast (localities 1–5), the main habitats for *Zostera* are sheltered muddy bottoms; however, it may also grow on more exposed localities. The vertical distribution of *Zostera* beds in this area is mainly between 0.7 and 3 m (Baden and Pihl 1984) but low-density patches can grow down to 8 m. In the Baltic Sea (localities 7–9) *Zostera* is found from 1 to 7 m depth and, in contrast, the habitat requirements are exclusively exposed, or moderately exposed shores with sandy sediments. The Kiel Bight (site 6) lies in a transition zone between marine (North Sea) and brackish (Baltic Proper) water and show fluctuations (10–18 psu) in salinity. In this region *Zostera* is found both at exposed sandy shores and in sheltered muddy embayments (Reusch 1994; Lotze 1998). Such enclosed bays north of this region are generally dominated by marsh plants (*Scirpus* spp.) in the upper littoral, succeeded by reed belts (*Phragmites australis*) and fresh water angiosperms (*Potamogeton* spp., *Myriophyllum* spp, *Ranunculus* spp.) or charophytes (*Chara* spp.). For all localities the openness of the bay, measured as effective fetch (Lf in km: Håkansson and Jansson 1983), shows that localities 1–3 are sheltered having Lf of 0.2–1.6 km, localities 4–5 are exposed having Lf close to 3 km and localities 6–9 are all very exposed with Lf more than 4 km. The effective fetch of localities 4–5 is similar to the exposure of Baltic localities. The temperature range at all localities is similar being between 0 and 25 °C (Pihl and Rosenberg 1982; Baden and Pihl 1984) with about 0–3 months of ice cover on the West coast and 1–4 months of ice cover in the Baltic (Leppäkoski and Bonsdorff 1989).

Sediment and particle trapping are characteristic features of seagrass beds (Ward et al. 1984). As a consequence of exposure, the localities on the Swedish west coast have higher sedimentation rates compared with the Baltic localities, building up a muddy substrate with an organic content of 2–25 % (Baden and Pihl 1984) compared with <2 % (Boström and Bonsdorff 1997) at the Baltic localities (Fig. 10.2). On the west coast, the C/N ratio of sedimenting material is lower (≈5, Baden et al. 2001) compared with older material (C/N ≈10, Pihl et al. 1999) incorporated in the detritus-based *Zostera* sediment. This indicates an input of pelagic allocthonous material with high nutritional value. In the more erosive, sandy substrates in the Baltic *Zostera* beds, where resuspension and transport processes are important (Boström and Bonsdorff 2000), no data on C/N ratios from *Zostera* beds exist. Due to transport (winds and waves), the importance of decaying *Zostera* leaves as a source of organic material might be small in both systems but needs to be investigated. The low nutrient value of the sandy substrates in the Baltic Sea may be partly compensated by increasing amounts of decaying drift algae (filamentous green and brown algae originating from the littoral rocky shores) caught between the eelgrass plants (Norkko and Bonsdorff 1996; Boström and Bonsdorff 2000). Also on the Swedish west coast, floating mats of ephemeral green algae

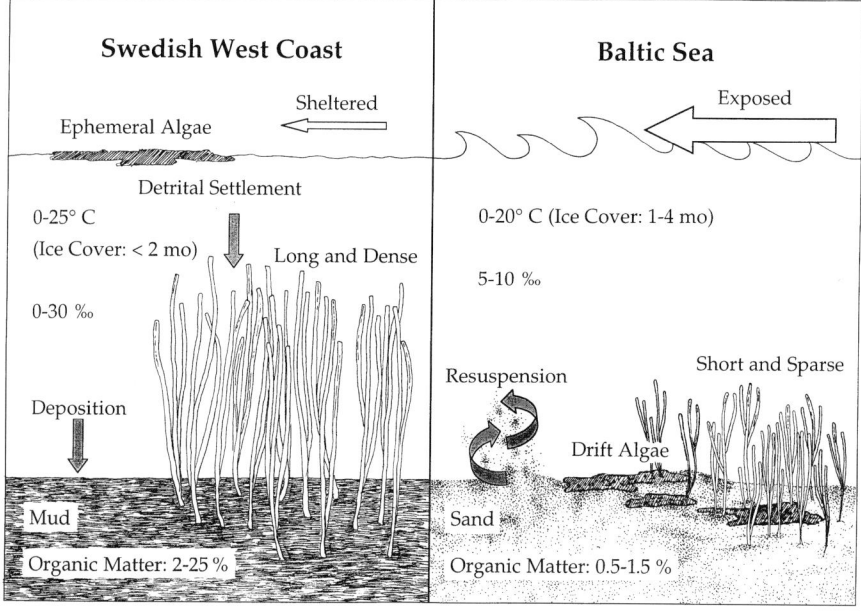

Fig. 10.2. Generalized drawing of environmental differences in *Zostera marina* habitats along the Swedish west coast, i.e. localities 1–5, Kattegat and Skagerrak (*left*) and east coast, i.e. localities 7–9, Baltic Sea (*right*). Abiotic and biotic factors summarized include wind exposure, temperature, duration of ice cover, salinity, substrate characteristics (percentage organic content), and growth forms of ephemeral algae in terms of fast growing floating mats (Swedish west coast) and decaying, benthic algal mats (Baltic Sea). Data from Pihl and Rosenberg (1982); Baden and Pihl (1984); Boström and Bonsdorff (1997). Illustration by Johanna Liljekvist

may contribute to the enrichment of shallow coastal sediments (Pihl et al. 1999).

The growing conditions of *Zostera* along the salinity gradient are also reflected in the morphology and bed structure of the plants (Fig. 10.2). On the Swedish west coast, *Zostera* forms dense and continuous meadows and has generally long, wide leaves as described for plants living under sheltered conditions (Den Hartog 1970). The local physical settings determine the coverage pattern of Baltic *Zostera* beds, which are far more patchy in character compared to the Swedish west coast, and support shorter shoots in lower densities. Similarly, strong water movements in exposed areas create small sized rocky shore algae with low biomass compared to algal belts in sheltered areas (Snoeijs 1999). Two growth forms of eelgrass are common both on the west coast and in the Baltic: a shorter dense shallow water form and a sparse taller form at the maximum depth of distribution. Such within-site size differences in morphology may be explained by compensation in vertical growth in response to reduced light levels, or by differing sediment characteristics, i.e. higher stability, organic content and increased nutrient levels at the lower distribution limit in contrast to the less-stable and organic-poor sediment at the upper limit of distribution (Den Hartog 1970).

10.3 Methods

10.3.1 Vegetation and Leaf Canopy Fauna

Along the Swedish west coast (1997) and in the Baltic (1998) the leaf fauna was sampled semi-randomly by a diver using a net bag on a frame, covering the depth range of the eelgrass bed. Similar bag-sampling techniques have also been used for collection of *Fucus* fauna in the Baltic (Haahtela 1969; Fagerholm 1978). The sampling (bag size and number of samples) was optimized according to the height and density of the leaf canopy. Thus, at localities 1–5 plants and animals were sampled with a bag covering an area of $0.123\,m^2$ (height: 75 cm, mesh size: 200 µm, $n=6$) and at localities 6–9 with a bag covering $0.031\,m^2$ (height: 50 cm, mesh size 250 µm, $n=10$). Samples were obtained by placing the net bag over the leaf canopy, and cutting the leaves above the sediment surface. At localities 1–5 the percentage coverage of *Zostera* was estimated in a randomly thrown frame of $0.25\,m^2$ ($n=15$), while a measure of coverage at localities 6–9 was performed by counting shoot densities within a randomly thrown frame covering $0.0625\,m^2$ ($n=40$). Visual observations of mobile epifauna (shrimps, crabs, benthivorous fish) were carried out during all sampling. In the laboratory the leaves were washed with fresh water after having been frozen and detritus, fauna and epiphytes were scraped off and sieved on a 250-μm mesh. *Zostera* biomass dry weight (12 h at 60 °C) and leaf area per m^{-2} were measured and the fauna classified to species (except for harpacticoids and insect larvae). The epiphytic fouling at localities 1–5 was sorted under a dissecting microscope into three fractions, namely detritus, fauna and filamentous epiphytes, before weighing and drying. For detailed information on the procedure, see Baden et al. (2000). At localities 6–9, the epiphytic covering was practically absent.

10.3.2 Predators

Sampling of predators does not allow for quantitative comparisons but gives an indication of the relative importance and composition of the mobile predatory faunal assemblages. Besides predators caught in the net bag and the visual observations made during sampling, the most reliable quantitative samples from fully marine *Zostera* meadows (locality 3) were taken with a drop-trap covering 0.5 m^2 (1.5 m high) released from a boat and emptied with a hand-net (method described in detail by Baden and Pihl 1984). On the Åland Islands, a survey comparing methods for catching seagrass associated fish (Sillanpää 2001) was used to evaluate the importance of fish in brackish water *Zostera* meadows. Data presented here are based on results from the site used for collection of leaf fauna (locality 9). Gear used were drop-trap (area 1 m^2, height 80 cm), survey nets (10–50 mm), haul seine, push net, passive non-baited traps and visual observation by SCUBA along transects.

10.4 Results

10.4.1 *Zostera marina*: Standing Stock and Leaf Area

The leaf canopy of *Zostera* meadows provide both substrate and hiding places for special faunal communities. Generally, this three-dimensional habitat supports higher species diversity and animal abundances than nearby on vegetated habitats (Heck and Thoman 1984; Orth et al. 1984; Möller et al. 1985; Mattila et al. 1999). Some type of linear or stepwise relationship between the standing stock and the leaf area of *Zostera* and different categories of fauna is found in many areas (Adams 1976; Stoner 1980; Orth et al. 1984; Baden 1990; Mattila et al. 1999).

On the Swedish west coast *Zostera* starts growing in May and reaches a maximum shoot (turion) density of 1000–2000 m^{-2} at 0.7 m depth in August.

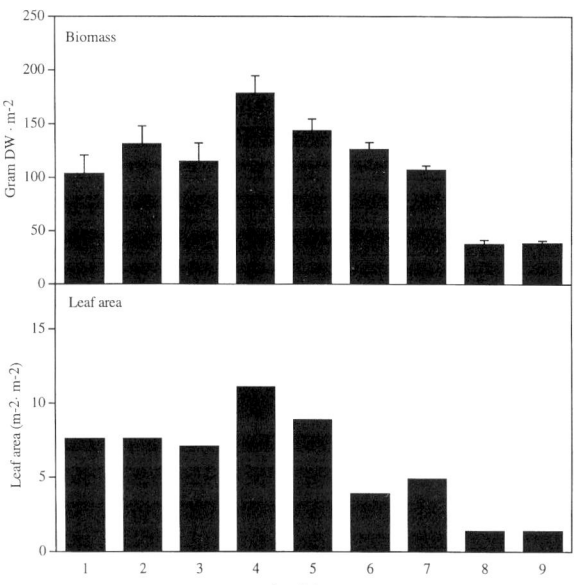

Fig. 10.3. *Zostera marina* standing stock (g DW · m^{-2}) and leaf area (m^{-2} · m^{-2}) at localities 1–9

Turion density decreases at about 2 m depth to 500–700 m^{-2}, depending on locality. The visible parts of *Zostera* are reduced in December. The leaf length is 20–40 cm with a maximum length of about 1 m (flowering shoots). The mean percentage of flowering shoots is low (1–4%), but increases with decreasing exposure of the locality (Baden and Pihl 1984). The mean (±1SE) biomass and leaf area of west coast *Zostera* (based on 19 localities, including localities 1–5) has been estimated to be 131±9.2 g DW and 8.45±0.52 m^{-2} bottom, respectively (Fig. 10.3). Within the meadows the mean coverage of *Zostera* is about 60%. The remaining area is usually dominated by drifting filamentous algae, *Fucus vesiculosus*, *Fucus serratus* and unvegetated sediment (Baden et al. 2001). The mean biomass of *Zostera* at the western localities in 1997 was about half the values found earlier in the Øresund (August 1978; Wium-Andersen and Borum 1980) and in the Gullmarsfjord area (July 1981, 1986; Baden and Pihl 1984; Baden 1990). However, the leaf area was about the same (7 m^2 · m^{-2}) in all investigations. The differences in biomass might depend on a mean coverage of *Zostera* in 1997 of only 60% combined with a high water temperature, giving earlier leaf senescence and thus lower biomass. During the last decade both naked and vegetated bottoms have temporarily been covered with ephemeral algae (Pihl et al. 1999), which affects both vegetation, fauna and human recreation (Isaksson and Pihl 1992).

When moving into the Baltic, the standing stock, shoot length and leaf area decrease (Figs. 10.3, 10.7). However, the biomass values of localities 6 and 7 are comparable to the west coast, mainly due to high individual shoot biomass in

Kiel and high shoot density at Öland. Biomasses of *Zostera* at the two northernmost localities (Askö, Åland) are low and similar, i.e. around 40 g DW · m^{-2}. The leaf area of the *Zostera* follows the biomass pattern (Fig. 10.3) but decreases by half in the transition zone (Kiel area) (3.9 and 4.9 m^2 · m^{-2} at Kiel and Öland, respectively) and is further reduced in the northern Baltic (1.4 m^2 · m^{-2} at Askö and Åland). The complexity of the leaf canopy in terms of shoot density is less than half at localities 6–9 compared to the west coast thus ranging between 200 and 800 shoots · m^2 in the Baltic Sea. As the contribution of limnetic phanerogams increases with decreasing salinity, co-occurrence of *Zostera* with other flowering plants is common at the northernmost localities (8–9). *Potamogeton pectinatus*, in particular, may form mixed stands with *Zostera* at shallower depths. Association with brown algae (*Fucus vesiculosus*, *Chorda filum*) is less common, but may increase the above-ground complexity of *Zostera* meadows at sandy bottoms with gravel or boulders, while mat-forming, loose-lying algae constitute a serious threat to *Zostera* sites in the northern Baltic Sea (Boström and Bonsdorff 1997) and elsewhere (Kruk-Dowgiallo 1991; Schramm 1996).

10.4.2 Leaf Fauna

The *Zostera* leaf fauna is here defined as the sessile or motile fauna living on the leaves, in contrast to the intermediate or large predators moving among the leaves (see Sect. 10.4.5). The fauna is often associated with epiphytic filamentous algae, detritus, microbial diatoms and bacteria, as described by Jacobs (1982) and Baden et al. (2001), and consists mainly of detritivores, grazers and suspension/filter feeders (Brönmark 1985; Baden 1990).

The abundance of leaf fauna decreases successively from being high (250–80 10^3·m^{-2}) at localities 1–3, intermediate (6–40 10^3·m^{-2}) at localities 4–7 and low (1–1.5 10^3 m^{-2}) at localities 8 and 9 (Figs. 10.4, 10.7). At all localities, the fauna is mainly composed of crustacean amphipods and isopods, molluscs and copepods. At localities 1–5 crustaceans are dominated by the detritivorous amphipods *Erichtonius difformis, Corophium insidiosum* and *Microdeutopus gryllotalpa*. At localities 6–9 the crustacean assemblage is dominated by *Idotea baltica*, while *E. difformis* (only locality 6), *Jaera albifrons* and *Gammarus* spp. occur in smaller numbers. However, the brackish northern Baltic *Zostera* meadows support only one species with an analogous function to the tube building amphipods at other localities, namely *Leptocheirus pilosus*. This amphipod is generally rare, but has been found on *Zostera* leaves in the Tvärminne area at densities around 300 individuals · m^{-2} (Lappalainen et al. 1977; Boström 1996). In the fully marine regions (localities 1–6) the grazing gastropod fauna is dominated by *Hydrobia* spp., *Rissoa* spp., *Littorina* spp. and *Gibbula cineraria*. In brackish water (localities 7–9),

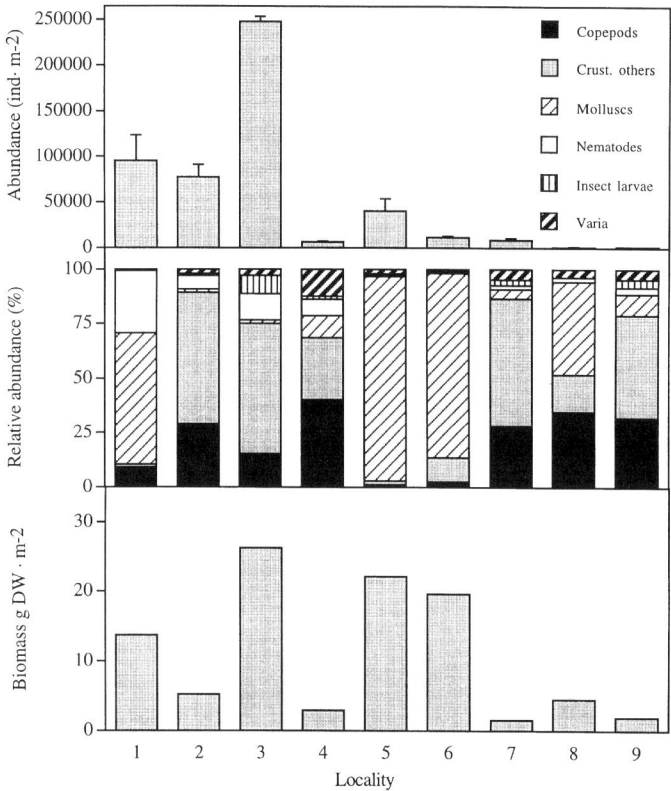

Fig. 10.4. Abundance (individuals · m^{-2}), composition [relative abundance (%) of dominating groups] and biomass (g DW · m^{-2}) of leaf fauna associated with *Zostera marina* meadows at the localities studied. The key to symbols refers to the relative abundance only

Hydrobia spp. still dominates, but other marine gastropods are replaced by the limnetic immigrants *Theodoxus fluviatilis* and *Lymnaea* spp. (Table 10.1). The blue mussel, *Mytilus edulis*, is a common component of the leaf fauna in all regions, but clearly peaks in terms of abundance and biomass in the Kiel Bight, where the mutualistic co-occurrence of blue mussel beds and *Zostera* patches is a key feature of sandy bottoms (Reusch et al. 1994; Reusch and Chapman 1995; Reusch 1998).

The total biomass of leaf fauna, with a mean of about 8 g DW · m^{-2} for localities 1–6 (maximum of 25 g DW · m^{-2} at locality 3) and 3 g DW · m^{-2} in localities 7–9 (Fig. 10.4), generally correlates with abundance. However, localities dominated by molluscs (e.g. Kiel) had overall higher biomass than localities dominated by crustaceans (Baden 1990; Baden et al. 2001). Orth et al. (1984) and Jernakoff et al. (1996) reviewed the role of leaf fauna in seagrass

Table 10.1. Abundance of dominating taxa of *Zostera marina* leaf fauna along the Swedish west coast (localities 1–5) and east coast (localities 7–9). X= sparse, XX=common, XXX=abundant

Taxonomic Group/Species	West	East
Crustaceans		
Erichtonius difformis	XXX	
Corophium insidiosum	XX	
Microdeutopus gryllotalpa	XX	
Copepoda	XXX	XXX
Cladocera		XXX
Ostracoda		XX
Idotea spp.		XX
Gammarus spp.		X
Molluscs		
Rissoa spp.	XX	
Littorina spp.	XX	
Cardium edule	X	
Mytilus spp.	XXX	XX
Hydrobia spp.	XX	XXX
Cerastodenna glaucum		X
Lymneae spp.		X
Theodoxus fluviatilis		XX
Nematodes	XXX	XX
Oligochaetes	X	XX
Chironomids	X	XX
Polychaetes	X	X

beds. Abundance and species composition found along the Swedish west coast in this study are generally higher than in most of the earlier studies reviewed. This can partly be explained by the small size of the mesh (250 µm) used in this study (500-µm or 1-mm sieves used in most other studies), which collects copepodes and many of the newly recruited amphipods. A similar fauna to the Swedish west coast localities has been described for an *Amphibolis antarctica* bed off Western Australia (Edgar 1990). In this subtropical bed the leaf fauna is dominated by the detritivorous crustaceans *Ericthonius pugnax* and *Leprochelia ignota* reaching maximum abundances of 100×10^3 individuals $\cdot$ m^{-2} and biomass of 21 g DW $\cdot$ m^{-2}, correlating to densities at locality 3 in this study. However, *Amphibolis* plants differ considerably in morphology from *Zostera* and may support independent faunal assemblages associated with the terminal leaf cluster, epiphytes or plant stems (Edgar and Robertson 1992). The quantitatively much poorer epifauna of the Baltic *Zostera* meadows (Fig. 10.7) has been described by Göthberg and Röndell (1973); Lappalainen et al. (1977); Asmus et al. (1980); Boström (1996) and Boström and Bonsdorff (1997). Lappalainen et al. (1977) carried out studies on the Tvärminne area (SW Finland, northern Baltic Sea), but they did not separate the *Zostera* infauna from the leaf fauna.

Abundance and biomass of leaf fauna are not linearly correlated with the standing stock or leaf area of *Zostera*, as recorded elsewhere in fresh water by Cooper and Crowder (1979), and in marine areas by Orth et al. (1984) and Mattila et al. (1999). The standing stock and leaf area at localities 1–7 are relatively similar (Fig. 10.3), whereas the abundance and biomass for these localities vary quite remarkably (Fig. 10.4). However, moving into the Baltic all these biological parameters are reduced.

Zostera localities with high exposure were mainly found in the Baltic (see Sect. 10.2). With increasing exposure the *Zostera* beds change from detritus sinks to detritus sources (Fonseca et al. 1983; Fonseca and Fischer 1986; Boström and Bonsdorff 2000). Hence, due to such differing current regimes (Gambi et al. 1990), the amounts of organic matter available as food resources or building materials (amphipod tubes) for the leaf fauna differ among the regions studied.

Consequently, dividing the leaf fauna into feeding groups (Fig. 10.5), the detritivores (*Erichtonius difformis* and *Corophium insidiosum*) and suspension feeders (juvenile *Mytilus* attached to the leaves) dominate in the detritus-driven systems (localities 1–5) and, from localities 4–9, grazers (isopods, gammarids) take over successively (Figs. 10.5, 10.6). Among the meiofauna, copepodes (harpacticoids) and nematodes (dominated by *Southermia zostera* on the west coast) probably play an important role as detritivores. The leaf fouling on the west coast consists of detritus, filamentous algae and fauna

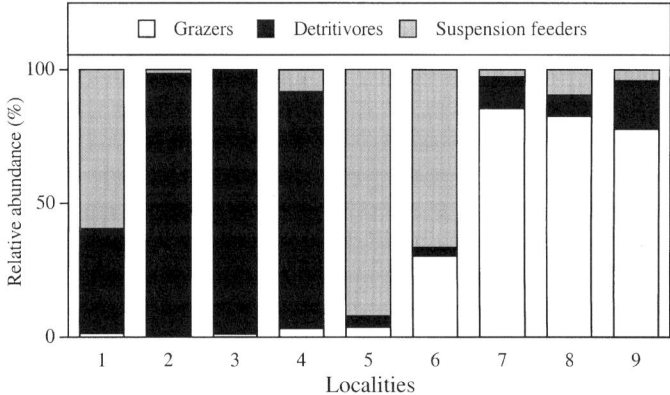

Fig. 10.5. Composition of *Zostera marina* leaf fauna at localities 1–9 based on feeding mode (grazers, detritivores or suspension feeders)

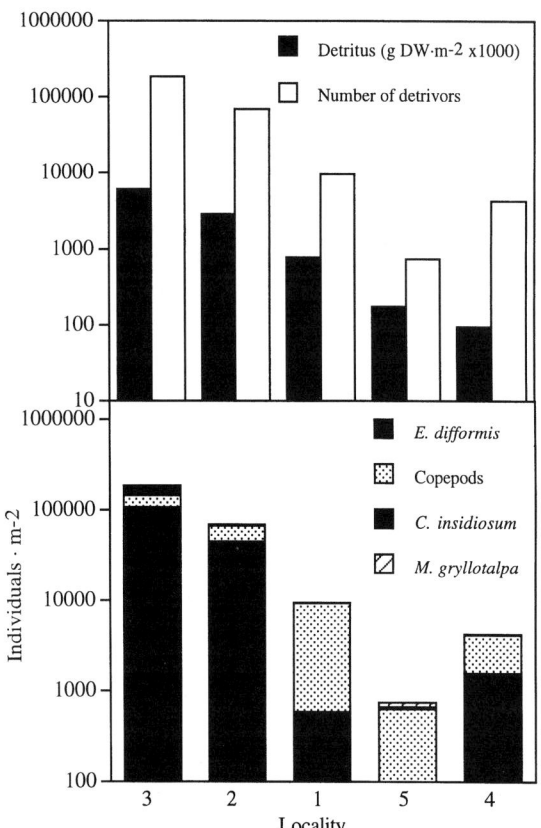

Fig. 10.6. *Zostera marina* localities (1–5) on the Swedish west coast ranked according to the amount of detritus (biomass g DW m^{-2} × 1000) and number of detritivores (individuals · m^{-2}) and the contribution of the dominating detritivores (individuals · m^{-2}) at each locality. Data from Baden et al. (2001)

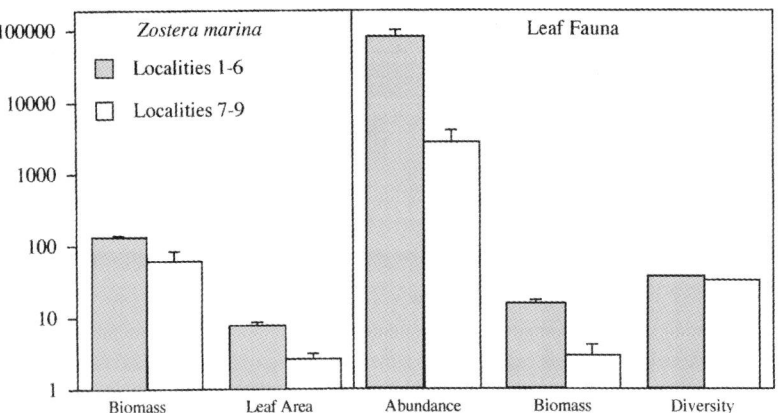

Fig. 10.7. Figure summarizing the *Zostera marina* biomass and leaf area, and leaf faunal abundance, biomass and diversity (number of species) on the Swedish west coast and Kiel (localities 1–6) and along the Swedish east coast (localities 7–9)

in the following mean proportions: 14.7, 3.2 and 9.6 g DW · m^{-2} (Baden et al. 2001). In a RDA analysis, the amphipod fauna was mostly related to the amount of detritus on the leaves, as shown in Fig. 10.6. In the Baltic Sea, the *Zostera* leaves were clean from fouling during the study (August) but tended to accumulate epiphytic algae on the leaves after the growing season (November–May).

A leaf canopy dominated by detritivores is unusual and the potential role of a detrivorous leaf fauna in seagrass has not been investigated. Mobile detritivores on seagrass are gained by increased inputs of organic matter and may through tube building further enhance the effects of eutrophication by binding more detritus to the leaves, increasing the negative effects of shading (Philippart 1995). Most investigations on the fouling communities deal with a leaf fauna community dominated by grazers. One focus in seagrass research has been on the relationship between epiphytes and grazers, thoroughly reviewed by Van Montfrans et al. (1984) and Jernakoff et al. (1996). Epiphytic fouling of seagrass leaves, with subsequent shading effects, increase with increasing amounts of nutrients (bottom-up regulation processes) (Borum 1985; Lapointe et al. 1994; Duarte 1995). An important role for grazers is to reduce the effects of nutrient surplus (top-down regulation processes) (Williams and Ruckelshaus 1993; Hauxwell et al. 1998). Such processes might in turn be affected by the influence of intermediate and top predators, as hypothesized by Heck et al. (2000).

At the exposed localities in the Baltic, fouling does not seem to be a problem to *Zostera* during the growing period. This might be due to physical stress by water movements preventing epiphytes and detritus from growing or accumulating, or because epiphytes (and *Zostera*) are efficiently grazed upon, as observed in the Kiel Bight. Differences in leaf growth provide conflicting evidence, as the leaf turnover rate in the Kattegat area is 14 days (Denmark: Sand Jensen 1977) compared to about 60 days in the Baltic (Roos 2000). This rules out the possibility that the difference in fouling between the Swedish west coast and the Baltic Sea is due to a difference in turnover rate of the leaves.

10.4.3 Couplings Between Leaf Fauna and Infauna

Even though the leaf canopy fauna and the sedimentary infauna may be treated as separate systems, the fauna in the basal part of the leaf canopy and the animals living at the sediment surface most likely interact or migrate vertically. The proportion of animals found among leaves or on the sediment may also vary with season, leaf density, leaf growth, or depend on mobility, life stage or reproductive status of the animals (Orth 1973). Further, sampling precision does not usually allow for a clear-cut distinction between the leaf fauna and the infauna. Consequently, the composition of the leaf fauna is partly determined by the infaunal assemblage.

On the Swedish west coast, detritivores (*Microdeutopus gryllotalpa*, *Erichtonius difformis* and *Corophium insidiosum*) dominate the leaf fauna, but are also abundant among the infauna. In the northern Baltic Sea, these species are absent, but their function is partly maintained by detrivorous chironomid larvae, which occur almost exclusively as leaf fauna (Boström and Bonsdorff 1997). On the west coast, chironomids make up the bulk of the infauna (S.P. Baden, unpublished data), but are also represented in high numbers on the leaves (3.5×10^3 individuals $\cdot$ m^{-2}; Baden et al. 2001). Here, the gastropod and bivalve assemblages are mainly supported by the leaf canopy, while the sediment supports low numbers of molluscs. In the Baltic, mud snails (*Hydrobia* spp.) are the most dominant group in both sediment and upon leaves (see above), thus occupying a broader niche by vertical migrations. Adult blue mussels (*Mytilus edulis*) are abundant on the sediment surface, while juveniles are found in low or moderate numbers on the leaves. On the west coast nematodes (mainly *Southermia zosteraceae*) are represented in high numbers both in the sediment and on the leaves, while the Baltic *Zostera* beds seem to have too poor a detrital settlement in order to support leaf canopy nematodes, though this group is abundant in vegetated sediments (Boström and Bonsdorff 1997).

Orth (1973) found that species abundant among or upon *Zostera* leaves were found in lower numbers in the sediment, and vice versa. Our data suggest that the west coast fauna is not restricted to a specific habitat, since at least detrivorous amphipods, chironomids and nematodes dominate both the sediment and the leaf canopy. In northern Baltic *Zostera* beds such patterns are true only for mud snails (*Hydrobia* spp.), while other groups (e.g. chironomids and nematodes) show more habitat-specific distributions.

10.4.4 Predators

The quantity and composition of leaf fauna and infauna are determined by "bottom-up" as well as "top-down" processes of which the potential importance of "top-down" (predator) regulation is discussed here. Generally, most seagrass meadows contain a richer fauna than the neighbouring unvegetated bottoms in terms of both abundance and biomass. Besides being a three-dimensional habitat with sub-niches allowing more animals to exist, the main hypothesis for the richer fauna is protection against predators. The potential for exploitation of the rich fauna by both intermediate (small fishes and invertebrates) and top (mainly piscivorous fishes) predators is thus a complex balance between the relative sizes of predators and prey, as well as the density and complexity of the leaf canopy and the characteristics of the bottom substrate. This has been the subject of many investigations and has been thoroughly reviewed by Heck and Crowder (1991).

As for the fauna in general, investigations focusing on predators in Scandinavian *Zostera* beds are sparse. Mobile epifauna (intermediate and top predators) has been described in terms of abundance and biomass, but not in terms of food intake. At three localities in the Gullmarsfjord area (including locality 3 in this study), monthly quantitative samples were taken from May–December 1980–1982 (Baden and Pihl 1984). In the Baltic (Åland Islands) the efficiency of predation on the infauna within and outside *Zostera* has been investigated by Mattila (1995), and tested experimentally on epifauna using artificial vegetation (Boström and Mattila 1999). In 1999 different sampling methods for seagrass-associated fish were tested at locality 9 (see Sect. 10.3.3).

In Scandinavian seagrass beds, the main difference is the dominance of intermediate predators on the Swedish west coast compared to the high proportion of larger piscivores in the Baltic. Grass shrimp *(Palaemon adspersus)* and green crab *(Carcinus maenas)* make up about 90% of the intermediate predators on the Swedish west coast, and pipe fish *(Syngnathus typhle)* and black goby *(Gobius niger)* contribute less than 10% to the annual mean of both abundance and biomass (Baden 1990). Piscivorous fish like eel *(Anguilla anguilla)* and eelpout *(Zoarces viviparus)* are few. In the Baltic *Zostera* meadows the fish fauna is, in addition to pipefish, dominated by larger species of limnetic origin such as perch (*Perca fluviatilis*), roach (*Rutilus rutilus*) and ruffe (*Gymnocephalus cernuus*); in addition, flounder (*Platichtys flesus*) is also common (Table 10.2).

For both areas studied in Scandinavia (Table 10.2), the diversity of predators is lower than in a *Zostera* bed off Cape Cod (north eastern USA) that has about 20 decapod and fish species (Heck et al. 1989). Robertson (1984) and Ryer (1988) showed that with increasing complexity of seagrass, larger fish become less important predators, whereas intermediate predators like shrimps and pipefish are favoured by high-density vegetation, as found along the Swedish west coast and shown by Nelson (1979, 1981), Heck and Thoman (1984) and Russo (1987). However, many of the fish species found in *Zostera* are generalists and forage in vegetated habitats on both sheltered soft bottoms (reed belts) and rocky shores (brown and green algal belts), or may shift habitat depending on time of season. Hence, the quantitative role of predators in the Scandinavian *Zostera* food webs is still poorly understood.

Table 10.2. Relative importance (indicated by line thickness of mobile predators in *Zostera marina* meadows along the Swedish west coast and at the Åland Islands (Swedish east coast). *Filled circles* indicate permanent species, *open circles* indicate temporarily visiting species

	Swedish west coast	Åland Islands
Fish		
● Siphonostoma typhle	▬▬▬▬▬▬▬▬▬	▬▬▬▬▬▬▬▬
● Nerophis ophidion	▬▬▬▬▬▬▬▬▬	▬▬▬▬▬▬▬▬
● Spniachia spinachia	▬▬▬▬▬	
● Gobius niger	▬▬▬▬▬	
● Anguilla anguilla	▬▬▬▬▬▬▬▬▬	
○ Pungitius pungitius	▬▬▬▬▬▬▬▬▬	
○ Gasterosteus aculeatus	▬▬▬▬▬▬▬▬▬	
○ Pomatoschistus minutus	▬▬▬▬▬	
○ Zoarces viviparus		▬▬▬▬▬
○ Perca fluviatilis		▬▬▬▬▬
○ Platichtys flesus		▬▬▬▬▬
○ Rutilus rutilus		
Crustaceans		
● Palemon adspersus	▬▬▬▬▬	▬▬▬▬▬
○ Carcinus maenas	▬▬▬▬▬	
○ Crangon crangon	▬▬▬	
Echinoderms		
○ Asterias rubens	▬▬▬▬▬	

10.5 Concluding Remarks

Zostera marina extends from the Swedish west coast into the northern archipelago areas of the Baltic Sea, covering a salinity gradient from about 30 to 5.5 psu. Along this gradient, the density and biomass of vegetation decreases and the main habitat of *Zostera* changes from shallow, sheltered, organically rich substrates to deeper, exposed, organically poor sandy bottoms in the Baltic Sea. Despite such quantitative differences in vegetation, the structuring and diversity-promoting role of *Zostera* in Scandinavian shallow benthic ecosystems remains equally important throughout its salinity range. In sheltered *Zostera* habitats along the Swedish west coast, the leaf fauna is diverse consisting of fouling organisms dominated by detritivores, whereas the leaves at the exposed Baltic localities are relatively clean (partly due to grazing) during the growing season. It is intriguing that *Zostera* beds are restricted to exposed habitats in the low saline Baltic areas. Both indirect and direct effects of salinity may act synergetically and explain this pattern. In the Baltic, freshwater plants dominate on sheltered, shallow, soft bottoms, thus partly restricting *Zostera* to more-exposed, deeper areas. In addition, *Zostera* is dependent on carbon uptake from carbonate instead of carbon dioxide, which is an unlimited resource in fully saline water, but probably a limiting factor in the brackish parts of the Baltic Sea. By growing in exposed areas, *Zostera* can partly compensate for this by increasing access to carbonate through increased shear velocity. Strong water movements probably also determine plant morphology, size and density of the *Zostera* leaf canopy, and thus also the composition of the associated animal assemblages.

Acknowledgements. Erik Bonsdorff, Kenneth L. Heck and Leif Pihl gave valuable comments on earlier versions of this chapter. The authors acknowledge the financial support of the Swedish Council for Forestry and Agricultural (Research Grant 40.0182/99) (SB) and the Academy of Finland (38021) (CB).

References

Adams SM (1976) The ecology of eelgrass, *Zostera marina* (L.), fish communities. I. Structural analysis. II. Functional analysis. J Exp Mar Biol Ecol 22:269–311

Asmus H, Theede H, Neuhoff H-G, Schramm W (1980) The role of epibenthic macrofauna in the oxygen budget of *Zostera* communities from the Baltic Sea. Ophelia Suppl 1:99–111

Baden SP (1990) The cryptofauna of *Zostera marina* (L.): Abundance, biomass and population dynamics. Neth J Sea Res 27(1):81–92

Baden SP, Pihl L (1984) Abundance, biomass and production of mobile epibenthic fauna in *Zostera marina* (L.) meadows, western Sweden. Ophelia 23:65–90

Baden SP, Gerhardt L, Roberts C (2001) The fouling community on eelgrass, *Zostera marina* (L.), along the Swedish West coast. J Sea Res, subm.

Bonsdorff E, Pearson TH (1999) Varation in the sublittoral macrobenthos of the Baltic Sea along environmental gradients: a functional-group approach. Aust J Ecol 24:312–326

Borum J (1985) Development of epiphytic communities on eelgrass (*Zostera marina*) along a nutrient gradient in a Danish estuary. Mar Biol 87:211–218

Boström C (1996) Zoobenthic community structure in *Zostera marina* L. meadows – spatial and temporal variation. MSc thesis (in Swedish, abstract in English), Åbo Akademi Univ, 87 pp

Boström C, Bonsdorff E (1997) Community structure and spatial variation of benthic invertebrates associated with *Zostera marina* L.) beds in the northern Baltic Sea. J Sea Res 37:153–166

Boström C, Bonsdorff E (2000) Zoobenthic community establishment and habitat complexity – the importance of seagrass shoot density and morphology for faunal recruitment. Mar Ecol Prog Ser 205:123–138

Boström C, Mattila J (1999) The relative importance of food and shelter for seagrass-associated invertebrates: a latitudinal comparison of habitat choice by isopod grazers. Oecologia 120:162–170

Brönmark C (1985) Interactions between macrophytes, epiphytes and herbivores: an experimental approach. Oikos 45:26–30

Cooper WE, Crowder LE (1979) Patterns of predation in simple and complex environments. In: Shoud RH, Clopper H (eds) Predator prey systems in fisheries management. Sport Fisheries Institute, Washington, DC

Connolly (1994) Removal of seagrass canopy: effects on small fish and their prey. J Exp Mar Biol Ecol 184:99–110

Dahl E (1973) Ecological range of Baltic and North Sea species. Oikos 15:85–90

Den Hartog C (1970) The seagrasses of the world. Verh K Ned Ak Wet Adf. North-Holland Amsterdam 59:1–275

Duarte CM (1995) Submerged aquatic vegetation in relation to different nutrient regimes. Ophelia 4:87–112

Edgar GJ (1990) Population regulation, population dynamics and competition amongst mobile epifauna associated with seagrass. J Exp Mar Biol Ecol 144:205–234

Edgar GJ, Robertson AI (1992) The influence of seagrass structure on the distribution and abundance of mobile epifauna: pattern and process in a Western Australian *Amphibolis* bed. J Exp Mar Biol Ecol 160:13–31

Elmgren R (1984) Trophic dynamics in enclosed, brackish Baltic Sea. Rapp P-V Reun Cons Int Explor Mer 183:152–169

Fagerholm (1978) The effects of ferry traffic (artificial wave action) on the rocky shore macrofauna in the southern Åland archipelago in the northern Baltic. 2. The *Fucus* zone (a quantitative study). Kieler Meeresforsch 4:130–137

Fonseca MS, Fisher JS (1986) A comparison of canopy friction and sediment movement between four species of seagrass with reference to their ecology and restoration. Mar Ecol Prog Ser 29:15–22

Fonseca MS, Zieman JC, Thayer GW, Fisher JS (1983) The role of current velocalitie in structuring eelgrass (*Zostera marina*) meadows. Estuar Coast Shelf Sci 17:367–380

Gambi MC, Nowell ARM, Jumars PA (1990) Flume observations on flow dynamics in *Zostera marina* (eelgrass) beds. Mar Ecol Prog Ser 61:159–169

Gore RH (1992) The Gulf of Mexico: a treasury of resources in the American Mediterranean. Pineapple Press, Sarasota, Florida, 384 pp

Göthberg A, Röndell B (1973) Ekologiska studier i *Zostera* samhället I norra Östersjön. Inf Sötvattenslab, Drottningholm, p 11

Haahtela I (1969) The Finnish IBP-PM Group: quantitative sampling equipment for the littoral benthos. Int Revue Ges Hydrobiol 54:185–193

Håkansson L, Jansson M (1983) Principles of lake sedimentology. Springer, Berlin Heidelberg New York, 316 pp

Hauxwell J, McClelland J, Behr PJ, Valiela I (1998) Relative importance of grazing and nutrient controls of macroalgal biomass in three temperate shallow estuaries. Estuaries 21(2):347–360

Heck KL Jr, Crowder LB (1991) Habitat structure and predator-prey interaction in vegetated aquatic systems. In: Bell SS, McCoy ED, Mushinsky HR (eds) Habitat complexity: the physical arrangement of objects in space. Chapman and Hall, New York, pp 281–299

Heck KL Jr, Orth RJ (1980) Seagrass habitats: the roles of habitat complexity, composition and predation in structuring associated fish and motile macroinvertebrate assemblages. In: Kennedy VS (ed) Estuarine perspectives. Academic Press, New York, pp 448–464

Heck KL Jr, Thoman TA (1984) The nursery role of seagrass meadows in the upper and lower reaches of Chesapeake Bay. Estuaries 7:70–92

Heck KL Jr, Able KW, Fahay MP, Roman CT (1989) Fishes and decapod crustaceans of Cape Cod eelgrass meadows: species composition, seasonal abundance patterns and comparison with unvegetated substrates. Estuaries 12(2):59–65

Heck KL Jr, Pennock JR, Valentine JF, Coen LD, Sklenar SA (2000). Effects of nutrient enrichment and large predator removal on seagrass nursery habitats: an experimental assessment. Limnol Oceanogr 45:1041–1057

Hellblom F, Björk M (1999) Photosynthetic responses in Zostera marina to decreasing salinity, inorganic carbon content and osmolarity. Aquat Bot 65:97–104

Isaksson I, Pihl L (1992) Structual changes in benthic macrovegetation and associated epibenthic faunal communities. Neth J Sea Res 30:131–140

Jacobs (1982) The annual pattern of the diatoms in the epiphyton of eelgrass (Zostera marina L.) at Roscoff, France. Aquat Bot 8:355–370

Jansson B-O (1978) The Baltic – a system analysis of a semi-enclosed sea. In: Charnock H, Deacon G (eds) Advances in oceanography. Plenum Press, New York, pp 131–183

Jernakoff P, Brearley A, Nielsen J (1996) Factors affecting grazer-epiphyte interactions in temperate seagrass meadows. Oceanogr Mar Biol Ann Rev 34:109–162

Kautsky H, Van der Maarel E (1990) Multivariate approaches to the variation in phytobenthic communities and environmental vectors in the Baltic Sea. Mar Ecol Prog Ser 60:169–184

Kruk-Dowgiallo L (1991) Long-term changes in the structure of underwater meadows of the Puck Lagoon. Acta Ichtyol Piscator 22:77–84

Lapointe BE, Tomasko DA, Matzie WR (1994) Eutrophication and trophic state classification of seagrass communities in the Florida Keys. Bull Mar Sci 54(3):696–717

Lappalainen A, Hällfors G, Kangas P (1977) Littoral benthos of the northern Baltic Sea. IV. Pattern and dynamics of macrobenthos in a sandy-bottom Zostera marina community in Tvärminne. Int Rev Ges Hydrobiol 62:465–503

Leppäkoski E, Bonsdorff E (1989) Ecosytem variability and gradients. Examples from the Baltic Sea as a background for hazard assessment. In: Landner L (ed) Chemicals in the aquatic environment – advanced hazard assessment. Springer, Berlin Heidelberg New York, pp 6–58

Leppäkoski E, Helminen H, Hänninen J, Tallqvist M (1999) Aquatic biodiversity under anthropogenic stress: an insight from the Archipelago Sea (SW Finland). Biodiv Conserv 8:55-70

Lotze H (1998) Population dynamics and species interactions in macroalgal blooms: abiotic versus biotic control at different life-cycle stages. Berichte aus dem Inst für Meereskunde, PhD Thesis. Christian-Albrechts-Universität, Kiel, 134 pp

Luther H (1951) Verbreitung und Ökologie der höheren Wasserpflanzen im Brackwasser der Ekenäs-Gegend in Südfinnland. I. Allgemeiner Teil. Acta Bot Fenn 49: 1–231

Mathiesen I, Nielsen J (1956) Botaniske undersøgelser i Randers fjord og Grund Fjord (Botanical investigations in the fjords of Randers and Grund). Bot Tidskr 53:1–34 (English summary)

Mattila J (1995) Does habitat complexity give refuge against fish predation? Some evidence from two field experiments. In: Eleftheriou A, Ansell AD, Smith CJ (eds) Biology and ecology of shallow coastal waters. Proc 28th Eur Mar Biol Symp. Olsen and Olsen, Fredensborg, pp 261–268

Mattila J, Chaplin G, Eilers MR, Heck KL Jr, O'Neal JP, Valentine JF (1999) Spatial and diurnal distribution of invertebrate and fish fauna of a Zostera marina bed and nearby unvegetated sediments in Damariscotta River, Maine (USA). J Sea Res 41: 321–332

Möller P (1986) Physical factors and biological interactions regulating infauna in shallow boreal areas. Mar Ecol Prog Ser 30:33–47

Möller P, Pihl L, Rosenberg R (1985) Benthic faunal energy flow and biological interaction in some shallow marine soft bottom habitats. Mar Ecol Prog Ser 27:109–121

Nelson WG (1979) An analysis of structural pattern in an eelgrass (*Zostera marina* L.) amphipod community. J Exp Mar Biol Ecol 39:231–264

Nelson WG (1981) Experimental studies of decapod and fish predation on seagrass macrobenthos. Mar Ecol Prog Ser 5:141–149

Nielsen R, Kristiansen A, Mathiesen M, Mathiesen H (1995) Distributional index of the benthic macroalgae of the Baltic Sea area. Acta Bot Fenn 155:1–51

Norkko A, Bonsdorff E (1996) Population responses of coastal zoobenthos to stress induced by drifting algae. Mar Ecol Prog Ser 140:141–151

Ohlsson M, Andersson L (1990) Recent investigation of total carbonate in the Baltic Sea: changes from the past as a result of acid rain? Mar Chem 30:259–267

Orth RJ (1973) Benthic infauna of eelgrass, *Zostera marina*, beds. Chesapeake Sci 14:258–269

Orth RJ, Heck KL, Van Montfrans J (1984) Faunal communities in seagrass beds: a review of the influence of plant structure and prey characteristics on predator-prey relationships. Estuaries 7(4A):339–350

Phillippart (1995) Seasonal variation growth and biomass of an intertidal *Zostera noltii* stand in the Dutch Wadden Sea. Neth J Sea Res 33(2):205–218

Phillips RC, Menez EG (1988) Seagrasses. Smithsonian Contrib Mar Sci 34:1–104

Pihl L, Rosenberg R (1982) Production, abundance and biomass of mobile epibenthic marine fauna in shallow waters, western Sweden. J Exp Mar Biol Ecol 57:273–301

Pihl L, Svenson A, Moksnes P-O, Wennhage H (1999) Distribution of green algal mats throughout shallow soft bottoms of the Swedish Skagerrak archipelago in relation to nutrient sources and wave exposure. J Sea Res 41:281–294.

Pinnerup SP (1980) Leaf production of *Zostera marina* L. at different salinities. Ophelia Suppl 1:219–224

Remane (1934) Die Brackwasserfauna. Verh Dt Zool Ges 36:34–74

Reusch TBH (1994) Factors structuring the *Mytilus*- and *Zostera* community in the western Baltic. An experimental approach. PhD thesis, Christian-Albrechts-Universität, Kiel, 162 pp

Reusch TBH (1998) Differing effects of eelgrass *Zostera marina* on recruitment and growth of associated blue mussels *Mytilus edulis*. Mar Ecol Prog Ser 167:149–153

Reusch TBH, Chapman ARO (1995) Storm effects on eelgrass (*Zostera marina* L.) and the blue mussel (*Mytilus edulis* L.) beds. J Exp Mar Biol Ecol 192:257–271

Reusch TBH, Chapman ARO, Gröger JP (1994) Blue mussels (*Mytilus edulis*) do not interfere with eelgrass (*Zostera marina*) but fertilize shoot growth through biodeposition. Mar Ecol Prog Ser 108:265–282

Reusch TBH, Boström C, Stam WT, Olsen JL (1999a) An ancient eelgrass clone in the Baltic. Mar Ecol Prog Ser 183:301–304

Reusch TBH, Stam WT, Olsen JL (1999b) A micro-satellite based estimation of clonal diversity and population subdivision in *Zostera marina*, a marine flowering plant. Mol Ecol 9:127–140

Roos C (2000) A seasonal study of the production fo eelgrass (*Zostera marina* L.) at two sites in the northern Baltic Sea. MSc-thesis, Åbo Akademi University, 56 pp (in Swedish with English abstract

Robertson AI (1984) Trophic inteactions between the fish fauna and macrobenthos of an eelgrass community in Western Port, Victoria. Aquat Bot 18:135–153

Rumohr H, Bonsdorff E, Pearson TH (1996) Zoobenthic succession in Baltic sedimentary habitats. Arch Fish Mar Res 44:179-214

Russell G (1985) Recent evolutionary changes in the algae of the Baltic Sea. Br Phycol J 20:87–104

Russell G (1987) Salinity and seaweed vegetation. In: Crawford RMM (ed) Plant life in aquatic and amphibious habitats. Blackwell, Oxford, pp 35–52
Russell G (1988) The seaweed flora of a young semi-enclosed sea: The Baltic. Salinity as a possible agent of flora divergence. Helgoländer Meeresunters 42:243–250
Russo AR (1987) Role of habitat complexity in mediating predation by the grey damselfish *Abudefduf sordidus* on epiphytal amphipods. Mar Ecol Prog Ser 36:101–105
Ryer CH (1988) Pipefish foraging and the effects of altered habitat complexity. Mar Ecol Prog Ser 48:37–45
Sand-Jensen K (1977) Effect of epiphytes on eelgrass photosynthesis. Aquat Bot 3:55–63
Schramm W (1996) The Baltic Sea and its transitions zones. In: Schramm W, Nienhuis PH (eds) Marine benthic vegetation – recent changes and the effects of eutrophication. Springer, Berlin Heidelberg New York, pp 131–163
Seed R (1992) Systematics evolution and distribution of mussels belonging to the Genus *Mytilus*: an overview. Am Malacol Bull 9:123–137
Setchell (1929) Morphological and phenological notes on *Zostera marina* L. Univ Calif Publ Bot 14:289–452
Short FT (1983a) The response of interstitial ammonium in eelgrass (*Zostera marina* L.) beds to environmental perturbations. J Exp Mar Biol Ecol 68:195–208
Short FT (1983b) The seagrass, *Zostera marina* L.: plant morphology and bed structures in relation to sediment ammonium in Izembek lagoon, Alaska. Aquat Bot 16:149–161
Sillanpää H (2001) The importance of *Zostera marina* L. meadows for fish in the northern Baltic Sea – a methodological approach. MSc-thesis, Åbo Akademi University, 45 pp
Simenstad CA, Reed DJ, Jay DA, Baross JA, Prahl FG, Small LF (1994) Land-margin ecosystem research in the Columbia River estuary: an interdisciplinary approach to investigating couplings between hydrological, geochemical and ecological processes within an estuarine turbidity maxima. In: Dyer KR, Orth RJ (eds) Changes in fluxes in estuaries: implications from science to management (ECSA22/ERF symposium). Olsen and Olsen, Fredensborg, Denmark, pp 437–444
Snoeijs P (1999) Marine and brackish waters. Acta Phytogeogr Suec 84:187–212
Stevenson J C (1988) Comparative ecology of submersed grass beds in freshwater, estuarine, and marine environments. Limnol Oceanogr 33:867–893
Summerson HC, Peterson CH (1984) Role of predation in organizing benthic communities of a temperate-zone seagrass bed. Mar Ecol Prog Ser 15:63–77
Stoner AW (1980) The role of seagrass biomass in the organisation of benthic macrofaunal assemblages. Bull Mar Sci 30:537–551
Tedengren M (1990) Ecophysiology and pollution sensitivity of Baltic Sea invertebrates. PhD thesis, Stockholm University
Tedengren M, Kautsky N (1986) Comparative study of the physiology and its probable effect on size in blue mussels (*Mytilus edulis* L.) from the North Sea and the northern Baltic proper. Ophelia 25(3):147–155
Tedengren M, André C, Johannesson K, Kautsky N (1990) Genotypic and phenotypic differences between Baltic and North Sea populations of *Mytilus edulis* evaluated through reciprocal transplantations. III. Physiology. Mar Ecol Prog Ser 59:221–229
Tomasko DA, Lapointe BE (1991) Productivity and biomass of *Thalassia testudinum* as related to water column nutrient availability and epiphyte levels: Field observations and experimental studies. Mar Ecol Prog Ser 75:9–17
Väinölä R, Hvilsom MM (1991) Genetic divergence and a hybrid zone between Baltic and North Sea *Mytilus* populations (Mytilidae, Mollusca). Biol J Linn Soc 43:127–148
Van Montfrans J, Wetzel RL, Orth RJ (1984) Epiphyte-grazer relationships in seagrass meadows: consequences for seagrass growth and production. Estuaries 7:289–309

Wallentinus I (1991) The Baltic Sea gradient. In: Matthieson AC, Nienhuis PH (eds) Ecosystems of the world, vol 24. Intertidal and littoral ecosystems. Elsevier, Amsterdam, pp 83–108

Ward LG, Kemp WM, Boynton WR (1984) The influence of waves and seagrass communities on suspended particulates in an estuarine embayment. Mar Geol 59:85–108

Williams SL, Ruckelshaus MH (1993) Effects of nitrogen availability and herbivory on eelgrass (*Zostera marina*) and epiphytes. Ecology 74:904–918

Wium-Andersen S, Borum J (1980) Biomass and production of eelgrass (*Zostera marina* L.) in the Øresund, Denmark. Ophelia Suppl 1:49–55

Worm B, Reusch TBH (2000) Do nutrient availability and plant density limit seagrass colonization in the Baltic Sea? Mar Ecol Prog Ser 200:158–166

11 Energy Flow in Benthic Assemblages of Tidal Basins: Ria Formosa (Portugal) and Sylt-Rømø Bay (North Sea) Compared

M. Sprung, H. Asmus, and R. Asmus

11.1 Introduction

Tidal areas are zones of steep physical and biologically determined gradients. Gradients develop vertically in sediment profiles, horizontally along intertidal transects, temporally with respect to temperature, oxygen, nutrients, light with diurnal, tidal and seasonal periods. It is a special feature of tidal systems that these gradients reveal a high small-scale variability in space and time caused primarily by the hydrodynamics. The energy flow in intertidal areas should be high due to the proximity of sea bottom and surface water, the concomitant good supply of light, and the tidal flow. This leads to a close spatial coupling not only between water column and sea bottom processes, but also between the landward and seaward parts of the coastal system and to a distinct distribution of primary and secondary producers (Jørgensen 1983; Reise 1985; Kjerfve 1994b).

In temperate zones, main primary producers are salt marsh plants at the upper fringe, seagrasses in intertidal and subtidal areas, opportunistic green algae and some tolerant brown and red algae, unicellular or filamentous microalgae on the sediment surface or on biogenic structures, and phytoplankton. Phytoplankton abundance and production is driven by three important processes: remineralization of nutrients, nutrient input by river runoff and the grazing activity of zooplankton and suspension feeders at the bottom; the latter having its largest impact in shallow waters (Dame 1993). Import of particulate organic carbon from the sea may be important as well. It is due to this source that coastal systems tend to be overall heterotrophic, i.e. degradative processes frequently outbalance primary production (Smith and Hollibough 1993; Heip et al. 1995).

The abundance and composition of different primary producers should condition to a large extent the quality and quantity of the macrofauna and hence energy flow through the system in addition to physical factors. However, next to the food availability, the macrofauna is also subject to predation.

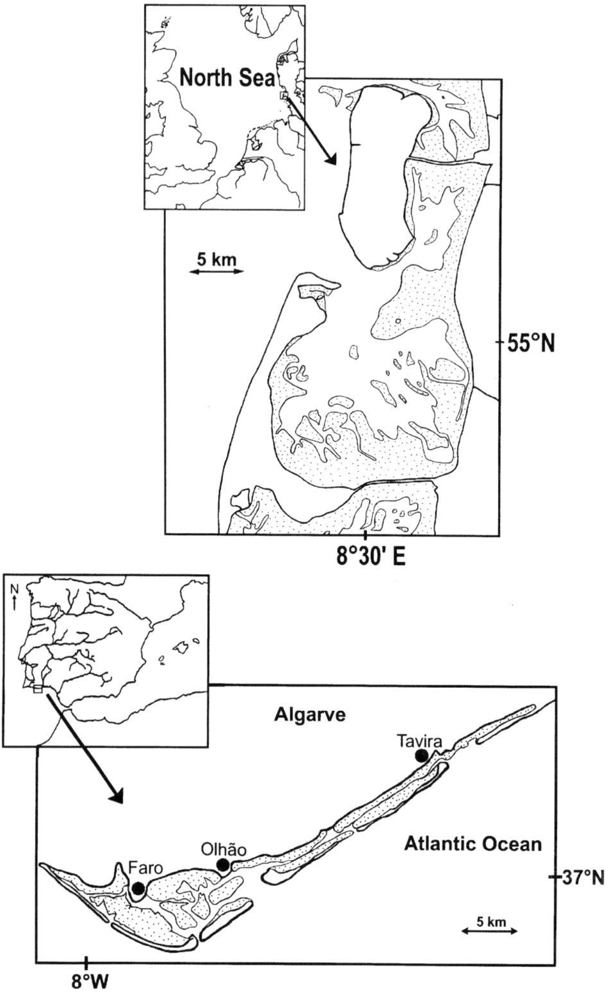

Fig. 11.1. Location of the two tidal basins examined; intertidal areas are shaded

Expressed in ecological terms, macrofauna abundance is not only shaped by bottom-up processes, but by top-down events as well (Reise 1985; Forman et al. 1995).

In this chapter, we attempt to synthesize the knowledge gained during recent years on the energy flow in benthic communities at two intertidal sites, the Ria Formosa lagoon and the Sylt-Rømø Bay (Fig. 11.1). The Ria Formosa lagoon is situated on the Algarve coast (Southern Portugal). Except during heavy rainfall, particularly in winter, it receives no significant freshwater input. It exchanges its water with the Atlantic Ocean by six major inlets. The Sylt-Rømø Bay forms part of the northern Wadden Sea at the German-Danish border. Water exchange is by one large inlet with the comparably shallow North Sea whose water is significantly diluted by riverine input at this site. As in the Ria Formosa lagoon, direct freshwater input into the basin is insignificant. Both systems should be considered leaky lagoon areas according to the classification by Kjerfve (1994a) but are situated in distinct climatic zones. How do macrozoobenthic assemblages reflect primary production? Do more favourable conditions at low latitudes stimulate productivity of the system by more intense nutrient cycling?

11.2 Description of the Sites

Geographical and main environmental characteristics are summarized in Tables 11.1 and 11.2. This information has been extracted from various sources (CCR 1984; C. Andrade 1990; Fidalgo 1996; Gätje and Reise 1998). Published data of the extension of Ria Formosa area types varies considerably. The data used here present the best compromise with priority for the most recent data. The most relevant differences between both systems can be summarized as follows:

1. Water temperature is generally higher in the Ria Formosa, above all in winter (see minimum temperatures in Table 11.1), and shows a lower annual amplitude.

Table 11.1. Main environmental characteristics of the systems

	Ria Formosa	Sylt-Rømø Bay
Geographic coordinates	37°N 8°W	55°N 8°30'E
Total extension	100 km²	404 km²
Intertidal area	67 km²	135 km²
	67%	33%
Salinity	36–38	28–32
Temperature	12–28 °C	−1–20 °C
Monthly insolation (kcal cm^{-2})		
December	6	0.8
June	22	11.5
Tidal range	0.6–2.8 m	1.8 m
Water residence time	0.5–2 days	19–29 days
Range of water nutrient content (μmol l^{-1})		
Nitrate	0.2–4.4	0–56.6
Ammonia	2.4–11.1	0–10.2
Silicate	3.0–29.2	5.5–87.2
Phosphate	0.2–1.6	0–4.9

Table 11.2. Primary and secondary production data

Location				Salt marsh	Seagrass bed	Type of community Mud flat	Sand flat	Mussel bed	Subtidal
Ria Formosa	Area		km²	25	3	19	18		17
	Intertidal (%)			37.3	11.9	28.4	26.9		
	Primary production	Benthic	g C m⁻² a⁻¹	600	350	37	130		103
			t C a⁻¹	15,000	1050	703	2340		1751
		Pelagic	g C m⁻² a⁻¹	0	22	22	22		100
			t C a⁻¹	0	66	418	396		1700
	Intertidal production (%)			75.1	5.6	5.6	13.7		
	Lagoon production (%)			64.0	4.8	4.8	11.7		14.7
	Secondary production		g C m⁻² a⁻¹	3	29	36	17		
			t C a⁻¹	75	87	684	306		
Sylt-Rømø Bay	Area		km²	10	15.5	10	109	0.4	269
	Intertidal (%)			Low	11.5	7.7	80.5	0.3	
	Primary production	Benthic	g C m⁻² a⁻¹	?	606	355	360	36	31
			t C a⁻¹	?	9393	3550	39240	14.4	8339
		Pelagic	g C m⁻² a⁻¹	0	43	43	43	43	218
			t C a⁻¹	0	667	430	4687	17.2	58,724
	Intertidal production (%)			Low	17.4	6.9	75.8	0.05	
	Lagoon production (%)			Low	8.0	3.2	35.1	0.03	53.6
	Secondary production		g C m⁻² a⁻¹	Low	58	56	53	309	
			t C a⁻¹	?	899	560	5777	124	

2. Insolation in the Ria Formosa lagoon is in December about ten times higher caused by the higher angle of the sun with the horizon and longer day time. Although in June day length is shorter than in the Sylt-Rømø Bay, insolation is still twice as high, caused by lower cloudiness.
3. Both systems show a different share of subtidal and intertidal subsystems. At mean low tide two-thirds of the Ria Formosa lagoon is intertidal whereas only one-third of the Sylt-Rømø Bay is uncovered.
4. Although both systems are mesotidal, differences between neap and spring tide are more pronounced in the Ria Formosa. This is probably one reason for the extended Ria Formosa salt marshes, which occupy the fringe between neap and spring high tide. In the lower part *Spartina maritima* dominates, in the upper part shrub-like plants such as *Arthrocnemum* spp., *Suaeda vera* or *Halimione portulacoides*. Originally salt marshes also fringed a considerable part of the Sylt/Rømø Bay but these have been largely reduced by embankments (Reise 1998).
5. Nutrient concentrations in the water column are distinctly lower in the Ria Formosa lagoon. This can be attributed to the shorter residence time of the water and the proximity of the continental shelf. The lagoon water is potentially mixed with a much larger water volume as at the North Sea site during each tide. In the Sylt-Rømø Bay the tidal exchange (with eutrophicated coastal North Sea water!) is lower and keeps nutrient levels high inside the bay.

11.3 Material and Methods

11.3.1 Primary Production

Salt Marsh Plants. Spartina maritima above-ground production has been estimated monthly during an annual cycle by the increment of individually marked shoots at two sites of the Ria Formosa. This production estimate has been assumed representative for the whole salt marsh area.

Microphytobenthos. Production of microphytobenthos has been recorded by the oxygen method in dark and transparent bell jars at monthly intervals at different intertidal sites. This method is described in Asmus and Asmus (1985). For the Sylt-Rømø site additional laboratory experiments with sediment cores were carried out under in situ conditions (Kristensen et al. 1997).

Seagrasses. In the Sylt-Rømø Bay spatial distribution of seagrasses was mapped by true colour remote sensing from air planes (Asmus et al. 1998).

Photographs (1:10000 and 1:25000) were taken monthly during the vegetation period from April to October between 1988 and 1994. Simultaneously, ground truth data were obtained by estimating dry weight and C-content of above-ground biomass of subareas of $0.25\,m^2$. For each seagrass bed 4–5 samples were considered. Production of seagrasses was estimated by multiplying maximum seagrass biomass per area and a P/B_{max} ratio of 10, as calculated for the subarea Königshafen (Asmus 1984). For the Ria Formosa lagoon estimates for the above-ground production of a *Zostera noltii* meadow in the Palemones estuary (S. Spain) have been taken as reference (Pérez Llorens and Niell 1993).

Green Algae. Because of its ephemeral occurrence, monthly increment in biomass has been used as the annual production parameter for the Ria Formosa lagoon (Aníbal 1998). In the Sylt-Rømø Bay, green algae were registered by air plane and by direct mapping in the field as described for *Zostera* beds, but direct production measurements on green algae are missing in this area.

Phytoplankton. Production has been estimated by the oxygen dynamics in dark and transparent bottles incubating for 6 h of daytime.

A comprehensive overview of methods and primary production rates of different compartments of the Sylt-Rømø Bay is given by Asmus et al. (1998).

11.3.2 Secondary Production

Macrobenthic secondary production data for different sites in the Sylt-Rømø Bay have been published in Asmus and Asmus (1985), Asmus (1987) and Asmus et al. (1998), and for intertidal sites of the Ria Formosa lagoon in Sprung (1993, 1994a). Data have been – whenever possible – calculated by the increment summation method between sample intervals of 4–6 weeks, otherwise by $P/\overline{B}$ ratios calculated in most cases with the equation given by Banse and Mosher (1980).

Secondary production of the salt marsh fauna of the Ria Formosa lagoon has been extrapolated from abundance and body weight using $P/\overline{B}$ values reported for similar species groups. The calculations are based on 54 core samples of $113\,cm^2$ collected in spring and autumn 1997 and autumn 1999 in the lower, central and higher marsh. In the Sylt-Rømø Bay, secondary production of the marginal salt marshes was considered to be too low for inclusion.

Feeding types have been attributed to species as published in Sprung (1994a) and Asmus et al. (1998). In cases of more than one feeding mode within species, it has been assumed that the species display these feeding modes during equal time fractions. Hence, secondary production has been split that way. Omnivorous potential has been calculated as the percentage of the production of species in common for two feeding types.

11.4 Production

11.4.1 Primary Production

The quantity and quality of the primary production turn out to be quite different in both systems (Table 11.2).

Salt Marshes. In the Ria Formosa lagoon, salt marshes are the dominant primary producers, due to its areal extent and its high production rate per area (about 60% of the entire primary production). This figure, however, only holds when the extrapolation of the production rate of the *Spartina* zone to other salt marsh habitats is valid. There are good reasons in favour [see e.g. Bouchard and Lefeuvre (1996) for an *Atriplex portulacoides* salt marsh at Mont Sant Michel, France]. *S. maritima* covers about 8 km^2 corresponding to one-third of the salt marsh area.

In the Sylt-Rømø Bay salt marsh vegetation contributes only to a very limited degree to the primary production. It mainly grows above mean high tide level (*Puccinellia maritima, Festuca rubra, Armeria maritima*); only the pioneer stages of *Salicornia stricta* and *Spartina anglica* extend their range 30 cm below mean high tidal level. Therefore, most of the 10 km^2 of these salt marshes are inundated only during storm floods; this is equivalent to 250 to 20 tidal cycles or 125–10 days per year depending on the accretion level.

Seagrass Beds. The principal seagrass species is *Zostera noltii* in both intertidal areas. *Z. marina* is rare in both systems. In the Ria Formosa, it is almost entirely restricted to the subtidal region, whereas in the Sylt-Rømø Bay it occurs only intertidally, while the subtidal population has disappeared since 1932 (Asmus and Asmus 2000). *Cymodocea nodosa* is, next to *Z. noltii*, the principal subtidal species in the Ria Formosa. The subtidal benthic primary production figure is based on a contribution of 5 km^2 subtidal seagrass beds with the same primary production as *Z. noltii* in the intertidal zone. *C. nodosa* is lacking in the Sylt-Rømø Bay due to the biogeographical range of this warm-temperate to subtropical species (Den Hartog 1970).

In the production estimate for the Sylt-Rømø Bay microphytobenthos and epiphyte production is also included. This explains why production per area is nearly twice as high as in the Ria Formosa. In terms of seagrass above-ground production, however, estimates are lower (260 g C m^{-2} a^{-1}) when compared with Ria Formosa sites (350 g C m^{-2} a^{-1}).

Green Algae. These are regular elements in intertidal areas of the inner Ria Formosa during winter. However, they may persist all year round at some spots that are subject to anthropogenic pollution, above all by domestic waste-water input (Aníbal 1998). In the Wadden Sea, they have occasionally been quite common particularly during summer months in certain years (Reise and Siebert 1994; Asmus et al. 1998). Their contribution to the production of the whole system is very difficult to estimate. In the Ria Formosa lagoon it is, with 36 g C m^{-2} a^{-1}, the main source of primary production of the mud flats (Aníbal 1998). For the whole Wadden Sea, green algal production was assessed to be 15 g C m^{-2} per year, when a maximum coverage of 5–10 % of the intertidal area is assumed (Reise et al. 1993). Hence, it has been assumed to be insignificant in comparison with other sources.

Microphytobenthos. This is the principal source of primary production on the sand flats of the Ria Formosa lagoon and on seagrass beds, mud flats and sand flats of the Sylt-Rømø Bay. In the intertidal zone of the Sylt-Rømø Bay, microphytobenthos attains 83 % of the primary production. The contribution for the Ria Formosa sites turned out to be much lower, for sand flats one-third, for mud flats, whose production is dominated by green algae, less than one-tenth of the Sylt-Rømø Bay.

Phytoplankton. Its importance increases with inundation time and water depth. For the Ria Formosa lagoon, estimates are subject to strong local and temporal variation.

These variations are the result of the concomitant action of nutrient availability, water turbidity and action of suspension feeders. The filtering activity of suspension feeders per unit area may not be extreme. Mendonça (1992) registered in the lower intertidal zone a benthic filtering activity in field chambers of 23.1 l m^{-2} h^{-1} (median in a seagrass bed) and 36.1 l m^{-2} h^{-1} (median on a sand flat). However, taking into account the current speed and water depth, it can be calculated that at this particular site about 10 % of the water column will be cleared during a spring tidal cycle and 37 % during a neap tidal cycle by the action of suspension feeders.

An annual average of 100 g C m^{-2} has been assumed as most realistic for phytoplankton production. This estimate is based on an average of 45 g C m^{-2} recorded during an annual cycle in 1990/1991 for intertidal sites, and on ^{14}C incubations by Falcão et al. (1991) in 1985/6 which varied between 0.7–80.7 µg C l^{-1} h^{-1}. Assuming a water depth of 1 m as representative for the whole lagoon would imply that phytoplankton contributes roughly 11 % to total primary production.

In the Sylt-Rømø Bay, phytoplankton is the main primary producer when the subtidal part of the bay is also included, contributing 52 % to total primary production. For the intertidal area the share of phytoplankton in total primary production is about 10 % due to the short inundation time of most tidal flats. This calculation is based on a mean annual production of 159 g C m^{-2} for subtidal areas and 43 g C m^{-2} for intertidal areas.

Detritus Input. This is not directly a source of primary production. However, for the Sylt-Rømø Bay, flume measurements imply a detritus input from subtidal sources, which constitutes an important element for the energy flow through intertidal communities. On average an input of 226 g C m^{-2} a^{-1} has been estimated (Asmus et al. 1998). In the Ria Formosa lagoon, published estimates are still lacking.

11.4.2 Secondary Production

Macrofaunal secondary production per unit area in the Sylt-Rømø system is about three times as high as in the Ria Formosa lagoon in general (Table 11.3, Fig. 11.2). Specifically for the seagrass beds and mud flats, the estimates are about twice as high as in the Ria Formosa lagoon, and for the sand flat three times. No estimates are available for the secondary production of salt marshes in the Sylt-Rømø area, but production is assumed to be very low. For the Ria Formosa lagoon, extrapolations based on abundance and body weight indicate a secondary production in salt marshes which is only one-sixth of comparable intertidal areas. This is mainly due to the fact that the macrofauna is mostly limited to fringes close to the marsh creeks and to the high water line, leaving extended zones in the centre of salt marshes almost free of macrofauna. Mussel beds, the area with the highest secondary production in the Sylt-Rømø Bay, have no equivalent in the Ria Formosa lagoon.

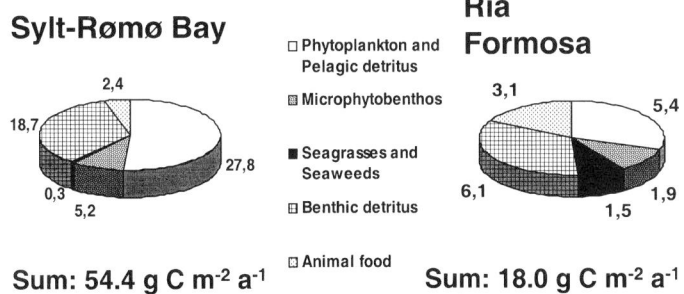

Fig. 11.2. Average secondary production of macrobenthic assemblages (g C m^{-2} a^{-1}) in both systems split into their food sources

Split into feeding types, the most dramatic difference in production rates is at the level of suspension feeders (Table 11.3). Their food is not only phytoplankton, but also pelagic detritus. In terms of an idealized unit area, secondary production of suspension feeders is five times higher in the Sylt-Rømø area and represents the dominant component of the energy flow in this system. The impact of herbivorous macrobenthic organisms on seagrasses and seaweeds is low in both areas. However, geese can remove most of the seagrass biomass in winter in the North Sea area and even show a grazing impact on salt marshes (Olff et al. 1997). For the Ria Formosa no significant impact of herbivorous birds has been documented or observed. Feeders on benthic algae and benthic detritus represent, next to suspension feeders, the most important macrobenthic secondary producers in the Ria Formosa lagoon. In the Sylt-Rømø area, however, their production rates are about three times higher. For carnivorous organisms similar production rates have been estimated; however, with a greater relative share in the Ria Formosa (Fig. 11.2).

Table 11.3. Secondary production (g C m^{-2} a^{-1}) split into different feeding modes; in cases of more than one feeding mode, the secondary production of a species has been split in equal proportions to these feeding modes

Feeding mode (assumed food source)	Community				
	Salt marsh	Seagrass bed	Mud flat	Sand flat	Mussel bed
Ria Formosa					
Suspension feeding (phytoplankton, pelagic detritus)	0.5	9.2	8.8	7.5	
Browsing (microphytobenthos)	0.4	3.2	1.1	4.5	
Deposit feeding (benthic detritus)	1.5	9.5	13.6	3.7	
Herbivores (macrophytes)	0.5	2.4	4.1	0.0	
Predators (animal food)	0.1	4.4	8.1	1.4	
Sylt-Rømø Bay					
Suspension feeding (phytoplankton, pelagic detritus)	?	14.5	16.4	30.0	275.1
Browsing (microphytobenthos)	?	17.9	11.9	2.8	12.6
Deposit feeding (benthic detritus)	?	23.3	25.1	17.4	19.3
Herbivores (macrophytes)	?	0.6	0.8	0.2	1.3
Predators (animal food)	?	2.0	1.7	2.5	0.6

The secondary production has been attributed to feeding types by splitting the production of an organism into its feeding modes when adequate. Hence the degree of omnivory can be indicated as production of a certain feeding type due to organisms in common with the production of another feeding type (Table 11.4). For the Ria Formosa, there appears to be no consistent line.

Table 11.4. Degree of omnivory expressed as percentage share of production realized by the same species, but by another feeding mode. B Browser (grazer); S suspension feeder; D benthic detritus feeder; H herbivore on macrophytes; C carnivore; a share of more than 50 % has been marked in bold

Sylt Rømø Bay

Mud flat

	B	S	D	H	C
B	–	23	34	0	0
S	21	–	21	0	0
D	43	0	–	0	0
H	0	0	0	–	0
C	0	0	0	0	–

Sand flat

	B	S	D	H	C
B	–	8	28	0	0
S	15	–	15	0	0
D	**57**	8	–	0	0
H	0	0	0	–	0
C	0	0	0	0	–

Zostera bed

	B	S	D	H	C
B	–	4	**52**	0	0
S	2	–	1	0	0
D	43	4	–	0	0
H	0	0	0	–	0
C	0	0	0	0	–

Mussel bed

	B	S	D	H	C
B	–	0.25	39	7	3
S	3	–	8	7	3
D	24	0.25	–	7	3
H	2	0.17	4	–	3
C	2	0.17	4	7	–

Ria Formosa

Mud flat

	B	S	D	H	C
B	–	0	0	0	0
S	0	–	**65**	**98**	49
D	0	**100**	–	**100**	**100**
H	0	45	30	–	49
C	0	45	**60**	**98**	–

Sand flat

	B	S	D	H	C
B	–	0	0	0	0
S	0	–	29	34	9
D	0	14	–	10	10
H	0	0	0	–	34
C	0	2	12	34	–

Zostera bed

	B	S	D	H	C
B	–	0	0	0	0
S	0	–	37	**100**	**54**
D	0	41	–	**100**	**73**
H	0	26	25	–	**54**
C	0	26	34	**100**	–

Salt marsh

	B	S	D	H	C
B	–	0	0	0	0
S	0	–	32	0	0
D	0	**100**	–	**100**	19
H	0	0	36	–	0
C	0	0	1	0	–

Sand flat organisms show apparently only a very small degree of omnivory. For the mud flat organisms, omnivory is attributed principally to the capacity of suspension feeding in addition to other feeding modes (particularly manifested by the polychaete *Nereis diversicolor* and the bivalve *Scrobicularia plana*). For organisms in the *Zostera* bed omnivory means to be additionally herbivorous. For the salt marsh organisms, omnivory is linked to the capacity to use benthic detritus next to other food sources (plant material and suspension feeding).

In the Sylt-Rømø area, the main omnivorous capacity for all examined communities refers to a switch between feeding on benthic detritus and benthic microalgae. Three dominant macrofaunal species are involved in this: the gastropod *Hydrobia ulvae*, the lugworm *Arenicola marina* and the bivalve *Macoma baltica*. The latter, and to a minor extent the lugworm, are at times also suspension feeders (see Chap. 4, this volume).

11.5 Energy Flow and Nutrient Cycle

Energy flow from primary to secondary producers, as focused on in this study (Fig. 11.3), constitutes only a part of the larger ecological cycle. In a closed system, primary production will largely depend on the availability of nutrients, which are set free by degradation of organic matter. Macrobenthic secondary producers do not represent the end of the food chain, but are subject to predation.

From this perspective, both systems are distinctly different with consequences elaborated in this chapter. One essential point is that the Ria Formosa is more open in terms of water circulation although at first sight geographically more efficiently separated from the sea than the Sylt-Rømø Bay. However, water exchange is by many inlets on the longitudinal part. Hydrographically, it represents a chain of small tidal basins. Hence, the subtidal water volume is much smaller than the tidal prism, water exchange is more complete and water residence time short (Reise and de Jong 1999). At the same time nutrient supply from the Atlantic water is much lower. Under these conditions vascular plants (salt marsh plants and seagrass) with access to nutrients in the sediment tend to dominate as primary producers. Due to their morphological and biochemical complexity, vascular plants are rarely degraded directly by herbivorous organisms, but mostly as detritus.

The discrepancy between high primary production and low secondary production in the salt marshes of the Ria Formosa implies an export of plant detritus to other lagoon areas. This is contrary to results from studies in North Sea salt marshes, where plant material is mainly degraded in the marsh (Hemminga et al. 1996) or even imported (Murray and Spencer 1997). Stepwise

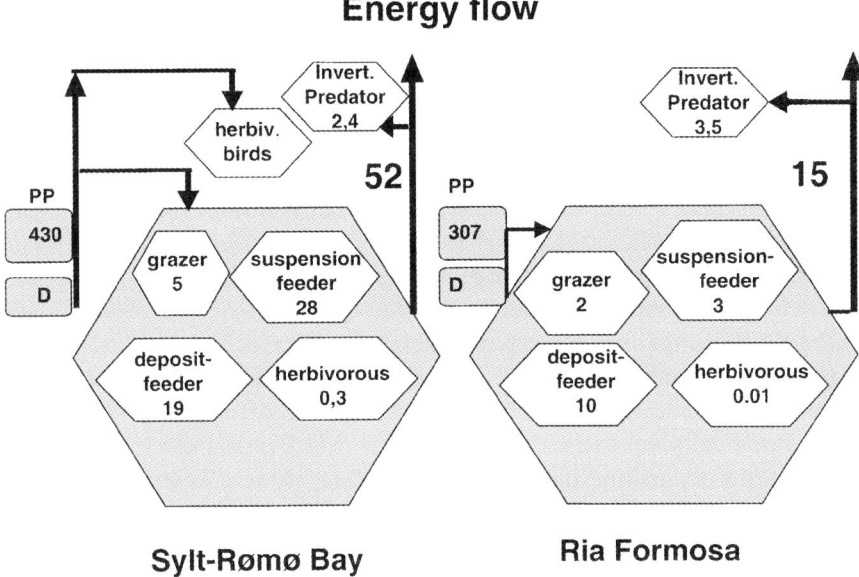

Fig. 11.3. Energy flow (g C m^{-2} a^{-1}) through the benthic systems of Ria Formosa and the Sylt-Rømø Bay

degradation of detritus may result in a more efficient nutrient retention in the system. This is consistent with a strong tendency to detritus feeding in the Ria Formosa.

Browsing microphytobenthos is a significant branch of energy flow in the Sylt-Rømø area, whereas in the Ria Formosa only few secondary producers use this food source (Table 11.3). In seagrass beds of the Sylt-Rømø Bay, microphytobenthic production is controlled by grazing (Schanz et al. 2000).

Equally important for differences in benthic assemblages is the fact that conditions for primary production are very unfavourable during winter in the Sylt-Rømø area. This leads to an increase in nutrient concentration in the water column and stimulates phytoplankton production particularly during spring months. Hence, a significant part of the cycle of biologically relevant nutrients and organic matter is located at the transition of the coastal system to the sea. This cycle between lagoon and ocean is not absent in the Ria Formosa, although it appears not to be as important as in the Sylt-Rømø area. Falcão and Vale (1998) registered during most seasons an export of nitrogen and phosphorus components in transects from the lagoon to the ocean. This seems at first sight incompatible with the high primary production of the lagoon compared to the ocean. However, this cycle may be a ramification of a production/remineralization cycle between the entire lagoon and their vascular plants or of nutrient input to salt marshes from external sources such as ground water or nitrogen fixation from air (White and Howes 1994). As a logic consequence, the diet of suspension-feeding organisms shows a strong gradient in the lagoon from mainly vascular plant detritus in the inner part to mainly phytoplankton in the outer part (Machás and Santos 1999). Mussel beds take advantage of this planktonic food source growing between the coastal system and the sea. They are missing in the Ria Formosa system, probably because this food source is much less developed than in the Sylt-Rømø Bay.

Particulate organic matter (POC) import from the sea is probably one other important source for the mussel beds in the Sylt-Rømø area. Published data for the Ria Formosa on POC exchange with the sea are lacking. As pointed out, it seems more plausible that POC should be in the first range exported from the marshes and seagrass beds to the rest of the lagoon and to the sea in the form of vascular plant detritus. This subject is controversial and has been discussed as an "outwelling" hypothesis in the literature (e.g. Dame et al. 1986). Dame (1982) calculated that detritus transport via *Spartina* leaves should correspond only in the range of 1 % of the net primary production at North Inlet (Atlantic coast of the US). However, leaf shedding is not pronounced in *Spartina alterniflora* on the West Atlantic coast, whereas each shoot of the East Atlantic *Spartina maritima* releases, according to own unpublished observations, between 2 to 4 leaves per month to the system. This may favour outwelling events, particularly from the *Spartina maritima* belt in the lower marsh (Heip et al. 1995).

In both systems, predators can exert a strong top-down effect on secondary production. This includes bird predation, predation by juvenile fish and by invertebrate predators. Bird predation seems at first sight more conspicuous in the Sylt-Rømø area, which serves as a resting area for many migrant bird species (Meltofte et al. 1994; Nehls and Scheiffarth 1998). It has been calculated that, on average, 9 % (or $5.1\,g\,C\,m^{-2}\,a^{-1}$) of the secondary production is removed per year from the tidal flats (Scheiffarth and Nehls 1997; Asmus et al. 1998).

The Ria Formosa is not frequented as a resting place by many migrant bird species. However, considerable numbers of waders can overwinter in this lagoon (Batty 1991). For the Tejo estuary situated some 250 km north of the Ria Formosa, Moreira (1997) calculated that about 12 % of the macrobenthic invertebrate production is removed by birds, more than half of this value by gulls. Predation by juvenile fish is important in both systems, which may be considered nursery areas of many species. This has been particularly investigated, e.g., for *Solea senegalensis* in the Ria Formosa (J.P. Andrade 1990) and *Pomatoschistus microps* in the Sylt-Rømø area (Herrmann et al. 1998).

Invertebrate predators may shape the intertidal macrobenthic community to a considerable extent. In the Ria Formosa the small bivalve *Abra ovata* can be severely decimated from intertidal areas not protected by algal or seagrass cover during late spring/early summer, when recruits of the shore crab (*Carcinus maenas*) colonize these zones at quantities of up to 90 specimens per square metre (Sprung 1994b). Even higher densities of shore crabs have been reported for the Sylt-Rømø area (Scherer and Reise 1981) which may remove 0.4 g C m^{-2} a^{-1} of the macrobenthic secondary production.

In summary, main bottom-up and top-down forces act at different components in the systems. Bottom-up forces have their greatest impact on phytoplankton and microphytobenthos in the Sylt-Rømø Bay, and on macrophytes in the Ria Formosa lagoon. Main top-down forces will be those on microphytobenthos production in seagrass beds and on macrobenthos in the Sylt-Rømø area, and that on phytoplankton (and plankton in general) and on macrobenthos in the Ria Formosa lagoon.

Contrary to our initial assumption, primary production is lower at Ria Formosa sites. This is caused by the lower availability of nutrients, which outbalances the potential of higher annual nutrient turnover. Also secondary production is lower because of lower food availability and high top-down control. Our data are limited in the sense that only two systems with their inevitable geographic peculiarities have been examined and warrant further discussion in a broader context.

Acknowledgements. Natália Dias kindly furnished unpublished data on salt marsh invertebrates. Karsten Reise and Justus v. Beusekom improved the text considerably by their valuable comments.

References

Andrade C (1990) O ambiente de barreira da Ria Formosa (Algarve-Portugal). Tese de Doutoramento, Univ Lisboa, 644 pp

Andrade JP (1990) A importância da Ria Formosa no ciclo biológico de *Solea senegalensis* Kaup 1858, *Solea vulgaris* Ouensel 1806, *Solea lascaris* (Risso 1810) e *Microchirus azevia* (Capello 1868). Tese de Doutoramento, Univ Algarve, 409 pp

Aníbal J (1998) Impacte da macroepifauna sobre as macroalgas Ulvales (Chlorophyta) na Ria Formosa. Tese de Mestrado, Univ Coimbra, 73 pp

Asmus H (1987) Secondary production of an intertidal mussel bed community related to its storage and turnover compartments. Mar Ecol Prog Ser 39:251–266

Asmus H, Asmus R (1985) The importance of grazing food chain for energy flow and production in three intertidal sand bottom communities of the northern Wadden Sea. Helgoländer Meeresunters 39:273–301

Asmus H, Asmus R (2000) Material exchange and food web of seagrass beds in the Sylt-Rømø Bight – How significant are community changes on the ecosystem level? In: Asmus H, Asmus R (eds) Intertidal seagrass beds and algal mats – organisms and fluxes on ecosystem level. Helgol Mar Res 54:137–150

Asmus H, Lackschewitz D, Asmus R, Scheiffarth G, Nehls G, Herrmann JP (1998) Carbon flow in the food web of tidal flats in the Sylt-Rømø Wadden Sea. In: Gätje C, Reise K (eds) Ökosystem Wattenmeer Austausch-, Transport- und Stoffumwandlungsprozesse. Springer, Berlin Heidelberg New York, pp 393–420

Asmus R (1984) Benthische und pelagische Primärproduktion und Nährsalzbilanz. Ber Inst Meereskunde 131:148

Asmus R, Jensen MH, Murphy D, Doerffer R (1998) Primary production of microphytobenthos, phytoplankton and the annual yield of macrophytic biomass in the Sylt-Rømø Wadden Sea. In: Gätje C, Reise K (eds) Ökosystem Wattenmeer Austausch-, Transport- und Stoffumwandlungsprozesse. Springer, Berlin Heidelberg New York, pp 367–391

Banse K, Mosher S (1980) Adult body mass and annual production/biomass relationships of field populations. Ecol Monogr 50:355–379

Batty L (1991) Aspects of the phenology of waders (Charadii) on the Ria Formosa, Portugal. MSc thesis, Univ Wales, Bangor, 123 pp

Bouchard V, Lefeuvre J-C (1996) Hétérogénéité de la productivité d'*Atriplex portulacoides* (L.) Aellen dans un marais salé macrotidal. CR Acad Sci Paris Sci Vie 319:1027–1034

CCR (1984) Caracterização esquemática da reserva da Ria Formosa, Ministério da Administração Interna, Lisboa, 17 pp

Dame RF (1982) The flux of floating macrodetritus in the North Inlet estuarine ecosystem. Estuar Coast Shelf Sci 15:337–344

Dame RF (1993) Bivalve filter feeders in estuarine and coastal ecosystem processes. Springer, Berlin Heidelberg New York

Dame R, Chrzanowski T, Bildstein K, Kjerfve B, McKellar J, Nelson D, Spurrier J, Stancyk S, Stevenson H, Vernberg J, Zingmark R (1986) The outwelling hypothesis and North Inlet, South Carolina. Mar Ecol Prog Ser 33:217–229

Den Hartog C (1970) The sea-grasses of the world. Verhandlingen der Koninklijke Nederlands Akademie van Wetenschappen Afd. Natuurkunde Tweede Reeks Deel 59, no 1. North-Holland Publishing Company, Amsterdam, 275 pp

Falcão MM, Vale C (1998) Sediment-water exchanges of ammonium and phosphate in intertidal and subtidal areas of a mesotidal coastal lagoon (Ria Formosa). Hydrobiologia 373/374:193–201

Falcão MM, Pissarra JL, Cavaco MH (1991) Características quimico-biológicas da Ria Formosa: análise de um ciclo anual (1985–1986). Bol Inst Nac Invest Pescas 16:2–21

Fidalgo L (1996) Aplicação de um sistema de informação geográfico na conservação da biodiversidade. Tese de Mestrado, Univ Algarve

Forman K, Valiela I, Sardá, R (1995) Controls of benthic marine food webs. Scientia Marina 59:119–128

Gätje C, Reise K (1998) Ökosystem Wattenmeer Austausch-, Transport- und Stoffumwandlungsprozesse. Springer, Berlin Heidelberg New York, 567 pp

Heip CHR, Goosen NK, Herman PMJ, Kromkamp J, Middelburg JJ, Soetaert K (1995) Production and consumption of biological particles in temperate tidal estuaries. Oceanogr Mar Biol (annual review) 33:1–149

Hemminga MA, Cattrijsse A, Wielemaker A (1996) Bedload and nearbed detritus transport in a tidal saltmarsh creek. Estuar Coast Shelf Sci 42:55–62

Herrmann JP, Jansen S, Temming A (1998) Fish and decapod crustaceans in the Sylt-Rømø Wadden Sea. In: Gätje C, Reise K (eds) Ökosystem Wattenmeer Austausch-, Transport- und Stoffumwandlungsprozesse. Springer, Berlin Heidelberg New York, pp 81–88

Jørgensen BB (1983) Processes at the sediment-water interface. In: Bolin B, Cook RB (eds) The major biogeochemical cycles and their interactions. SCOPE, pp 477–515

Kjerfve B (1994a) Coastal lagoons. In: Kjerfve B (ed) Coastal lagoon processes. Elsevier Science Publishers, Amsterdam, pp 1-8

Kjerfve B (ed) (1994b) Coastal lagoon processes. Elsevier, Amsterdam

Kristensen E, Jensen MH, Jensen KM (1997) Temporal variations in microbenthic metabolism and inorganic nitrogen fluxes in sandy and muddy sediments of a tidally dominated bay in the northern Wadden Sea. Helgoländer Meeresunters 51:295-320

Machás R, Santos R (1999) Sources of organic matter in Ria Formosa revealed by stable isotope analysis. Acta Oecologica 20:463-469.

Meltofte H, Blew J, Frikke J, Rösner HU, Smit CJ (1994) Numbers and distribution of waterbirds in the Wadden Sea. Results and evaluation of 36 simultaneous counts in the Dutch-German-Danish Wadden Sea 1980-1991. IWRB Publ 34, Water Study Group Bull 74, special issue, 192 pp

Mendonça V (1992) Taxas de filtração de comunidades bentónicas da Ria Formosa. Estágio de Licenciatura em Biologia Marinha e Pescas, Univ Algarve, 43 pp

Moreira F (1997) The importance of shorebirds to energy fluxes in a food web of a South European estuary. Estuar Coast Shelf Sci 44:67-78

Nehls G, Scheiffarth G (1998) Migratory waterbirds in the Sylt-Rømø Wadden Sea. In: Gätje C, Reise K (eds) Ökosystem Wattenmeer Austausch-, Transport- und Stoffumwandlungsprozesse. Springer, Berlin Heidelberg New York, pp 89-94

Olff H, de Leeuw J, Bakker JP, Platerink RJ, van Wijnen HJ, de Munck W (1997) Vegetation succession and herbivory in a salt marsh: changes induced by sea level rise and silt deposition along an elevation gradient. J Ecol 85:799-814

Perez-Llorens JL, Niell FX (1993) Seasonal dynamics of biomass and nutrient content in the intertidal seagrass *Zostera noltii* Hornem. from Palemones River Estuary, Spain. Aquatic Bot 46:49-66

Reise K (1985) Tidal flat ecology. Ecological studies, vol 54. Springer, Berlin Heidelberg New York, 191 pp

Reise K (1998) Coastal change in a tidal backbarrier basin of the northern Wadden Sea: Are tidal flats fading away? Senckenbergiana Maritima 29(1/6):121-127

Reise K, de Jong F (1999) The tidal area. In: de Jong F, Bakker JF, van Berkel CJM, Dankers NMJA, Dahl K, Gätje C, Marencic H, Potel P (eds) Wadden Sea Quality Status Report. Wadden Sea Ecosystem no 9. Common Wadden Sea Secretariat. Wilhelmshaven, Germany, pp 187-190

Reise K, Siebert I (1994) Mass occurrence of green algae in the German Wadden Sea. Dt Hydrogr Z Suppl 1:171-180

Reise K, Asmus R, Asmus H (1993) Ökosystem Wattenmeer – das Wechselspiel von Algen und Tieren beim Stoffumsatz. Biol Unsere Zeit 23:301-307

Schanz A, Polte P, Asmus H, Asmus R (2000) Currents and turbulence as a top-down regulator in intertidal seagrass communities. Biol Mar Medit 7:278-281

Scheiffarth G, Nehls G (1997) Consumption of benthic fauna by carnivorous birds in the Wadden Sea. Helgoländer Meeresunters 51:373-387

Scherer B, Reise K (1981) Significant predation on micro- and macrobenthos by the crab *Carcinus maenas* L. in the Wadden Sea. Kieler Meeresforsch (Sonderh) 5:490-500

Smith SV, Hollibaugh JT (1993) Coastal metabolism and the oceanic organic carbon balance. Rev Geophys 31:75-89

Sprung M (1993) Estimating macrobenthic secondary production from body weight and biomass: a field test in a non-boreal intertidal habitat. Mar Ecol Prog Ser 100:103-109

Sprung M (1994a) Macrobenthic secondary production in the intertidal zone of the Ria Formosa – a lagoon in Southern Portugal. Estuar Coast Shelf Sci 38:539-558

Sprung M (1994b) Observations on the life cycle of *Abra ovata* on an intertidal mud flat in Portugal. J Mar Biol Assoc UK 74:919-925

White DS, Howes BL (1994) Long-term ^{15}N-nitrogen retention in the vegetated sediments of a New England salt marsh. Limnol Oceanogr 39:1878–1892

12 Soft-Bottom Fauna of a Tropical (Banc d'Arguin, Mauritania) and a Temperate (Juist Area, German North Sea Coast) Intertidal Area

H. MICHAELIS and W.J. WOLFF

12.1 Introduction

Numerous studies all over the world have dealt with the ecology of the intertidal soft-bottom fauna. Piersma et al. (1993) reviewed literature data of the zoobenthic biomass of intertidal soft-bottom areas in tropical and temperate latitudes. They encountered a tremendous variety of methods and therefore claimed methodically standardized, comparable studies, in the ideal case performed by identical persons or institutions. A number of studies fulfill this prerequesite, e.g. Reise (1991) (investigating the northern temperate German, the tropical Thailandic and the southern temperate Chilenic coasts), Wolff and Smit (1990) (comparing Mauritania and the Dutch Wadden Sea) and, in a further study (Wolff 1991), also including a part of the Rhine-Scheldt Delta. Reise (1991) and Wolff (1991) suggested a hypothesis of large-scale gradients with a significant increase of macrobenthic species richness but a decline of abundance and biomass from high temperate to low tropical geographical latitudes. Intermediate values from southern Portugal (Sprung 1994) seem to support this view.

On the other hand, concerning the narrow sandy beach habitats, Jaramillo and Lastra (Chap. 3, this Vol.) found beach morphodynamics connected to wave exposure and biological interactions to be the dominant factors governing the structure of macrobenthic assemblages along several thousand kilometers of shorelines from high to low latitudes in different continents.

In the present study, we investigated the soft-bottom fauna of the shallows of the Banc d'Arguin on the tropical Mauritanian coast, and the tidal flats of the island Juist, part of the German Wadden Sea in the temperate North Sea (Fig. 12.1). The criteria used for the comparison are habitat division, species richness of macrozoobenthos and abundance, biomass and trophic structure. Both areas play a prominent role in the network of the East Atlantic Flyway of palaearctic waders (Altenburg et al. 1982; Smit and Piersma 1989; Wolff and Smit 1990). The Banc d'Arguin is a main wintering area and the Wadden Sea

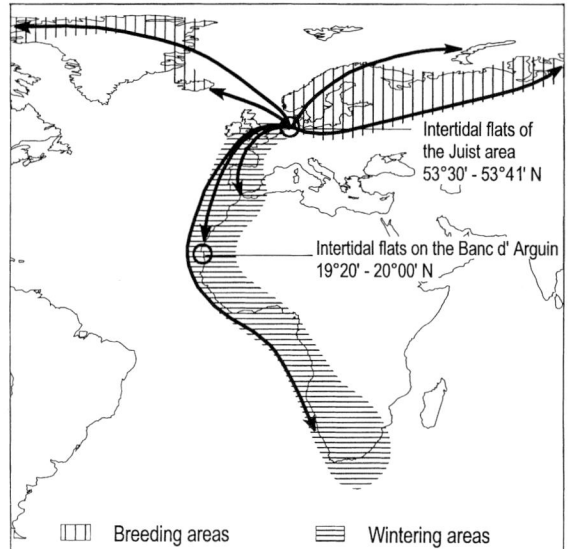

Fig. 12.1. Location of the investigation areas on the Banc d'Arguin (Mauritania) and in the Wadden Sea (Germany) in the system of the East Atlantic Flyway of palaearctic waders (based on a figure of Abrahamse et al. 1976)

is the main resting area during the spring and the autumn migrations. The material of both data sets was gained about the same time with largely identical methods, on the Banc d'Arguin in September 1988 and in the Juist area in the summers of 1987 through 1990. The evaluation of the data and the representation of the results of both studies were also conducted in an identical way.

12.2 Areas, Materials and Methods

The intertidal parts (500 km^2) of the Mauritanian Banc d'Arguin shallows form a complex of sand and mud banks separated by a channel system between 19°20' and 20°N (Fig. 12.2). They are immediately bordered by the Sahara Desert towards the east and by the Atlantic Ocean towards the west. An arid desert climate dominates these coastal reaches. The mean tidal range is about 1.0 m during neap tides and about 2.0 m during spring tides (Wolff et al. 1993a). Seawater temperatures vary seasonally between 20–30 °C (Altenburg et al. 1982; Ens 1985; Wolff and Smit 1990). Salinities amount to 37–42, with, in some semi-enclosed bays, hypersaline concentrations (Wolff and Smit 1990).

The water masses supplied to the Banc d'Arguin from the Atlantic Ocean move across the shallows from north to south becoming more and more isolated from the ocean (Peters 1976; Wolff et al. 1993b). Thus the water acquires high temperature, high salinity and extreme clearness in the tidal-flat area.

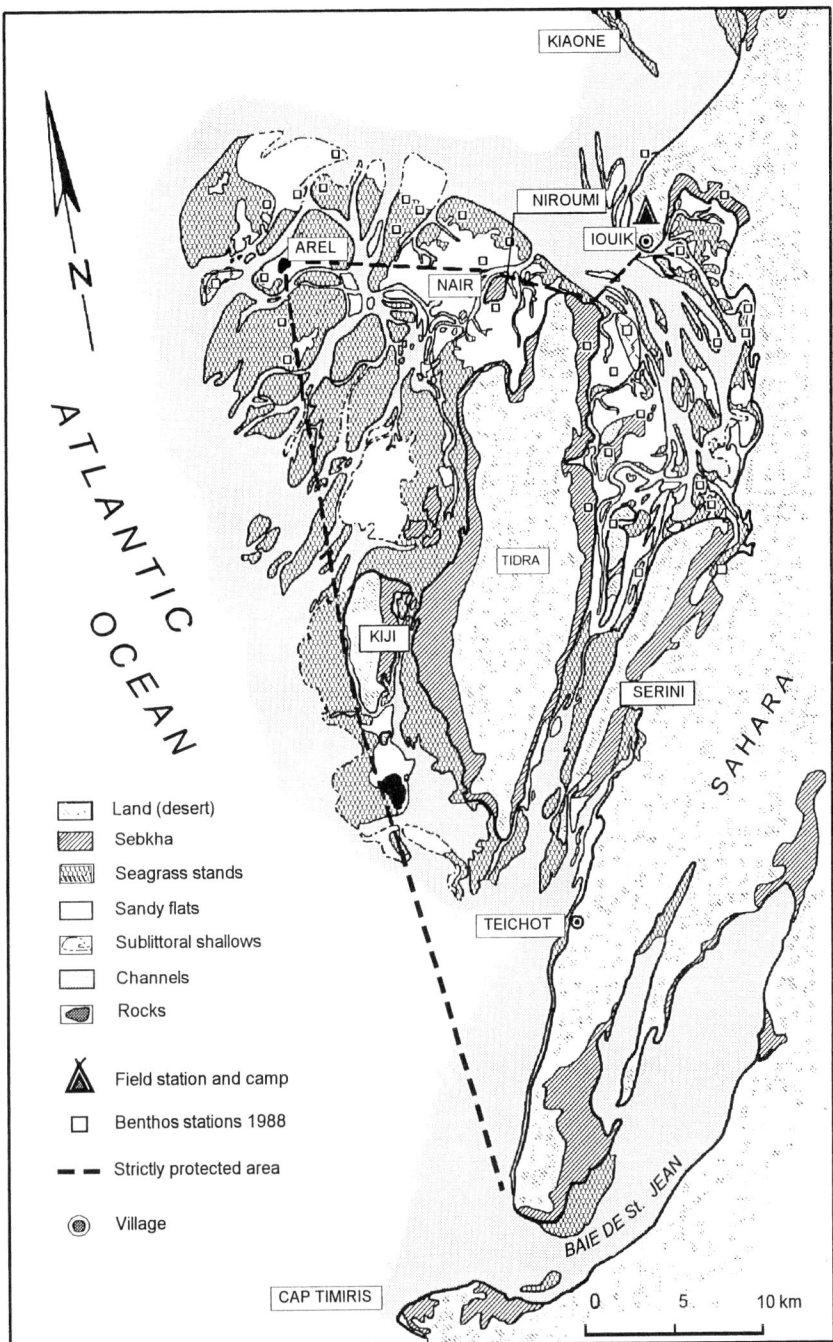

Fig. 12.2. Banc d'Arguin, distribution of the intertidal habitats (based on a map of Altenburg et al. 1982)

Though lying close to the NW-African upwelling area, the Banc d'Arguin water is oligotrophic and has a particularly low plankton and seston content. An influence of the upwelling system is completely lacking (Sevrin-Reyssac 1993; Wolff et al. 1993b).

At the southern North Sea coast, we investigated an intertidal area between the East Frisian island Juist and the mainland (Fig. 12.3) situated between 53°30' and 53°41'N. This area (108 km^2) is part of the Wadden Sea that extends along the Dutch, German and Danish coast and has a coherent intertidal fringe of 4300 km^2 in total. The mean tidal range of the Juist area is 2.5 m. The climate is oceanic within the cool–temperate zone. The seawater temperature – measured at Norderney – shows a long-term mean maximum of 17 °C in July/August and a mean minimum of slightly less than 2 °C in January (Bätje 1986). Severe winters with drift of ice occur frequently. Salinities are widely fluctuating. Euhaline concentrations (>30) prevail during summer and polyhaline ones (<30) during winter (Bätje 1986). High turbidity is a general feature of these waters and eutrophication results in a high production of phytoplankton (Rahmel et al. 1995).

The material from Mauritania comprises benthos and sediment samples from 30 randomly distributed stations (Fig. 12.2). In the Juist area benthos and sediment samples were collected from 104 stations (Fig. 12.3). Altogether, samples, parameters and methods from both areas can be summarized as follows:

- Surface sediment from the top layer (1–5 cm) for analyzing grain-size distribution. Sieving of the fractions >0.06 mm, areometer measurements of the fractions <0.06 mm.
- Macrofauna Banc d'Arguin: At 22 stations 5 core samples covering 177 cm^2 each, 25 cm deep, were taken; at another 8 stations where a patchy distribution was suspected 10 cores covering 78.7 cm^2 each, 25 cm deep, were obtained. Sieving was carried out with 0.6 mm mesh size. The large bivalve *Anadara senilis* was separately collected from 1.5 m^2. Sorting and first-step identification were performed in a live state. The further taxonomic and quantitative evaluation of the conserved material (10% formalin) was carried out in the home laboratories. The biomass was determined as ash-free dry weight (ADW) by incineration at 485 °C (bivalves <5 mm including the shells, >5 mm without the shells; gastropods mainly with the shells).
- Macrofauna Juist area: At 102 stations 7 core samples covering 177 cm^2 each, 25 cm deep, were taken. Samples from two shell bed stations supplied only qualitative information. Sieving was carried out with 1.0 mm mesh size. Examination of all samples (sorting, species identification, counting) was performed in a live state on board or in field laboratories. Then the material was stored in a deep freezer for later determination of biomass as described above. Details are given in Obert and Michaelis (in prep.).

Soft-Bottom Fauna of a Tropical and a Temperate Intertidal Area

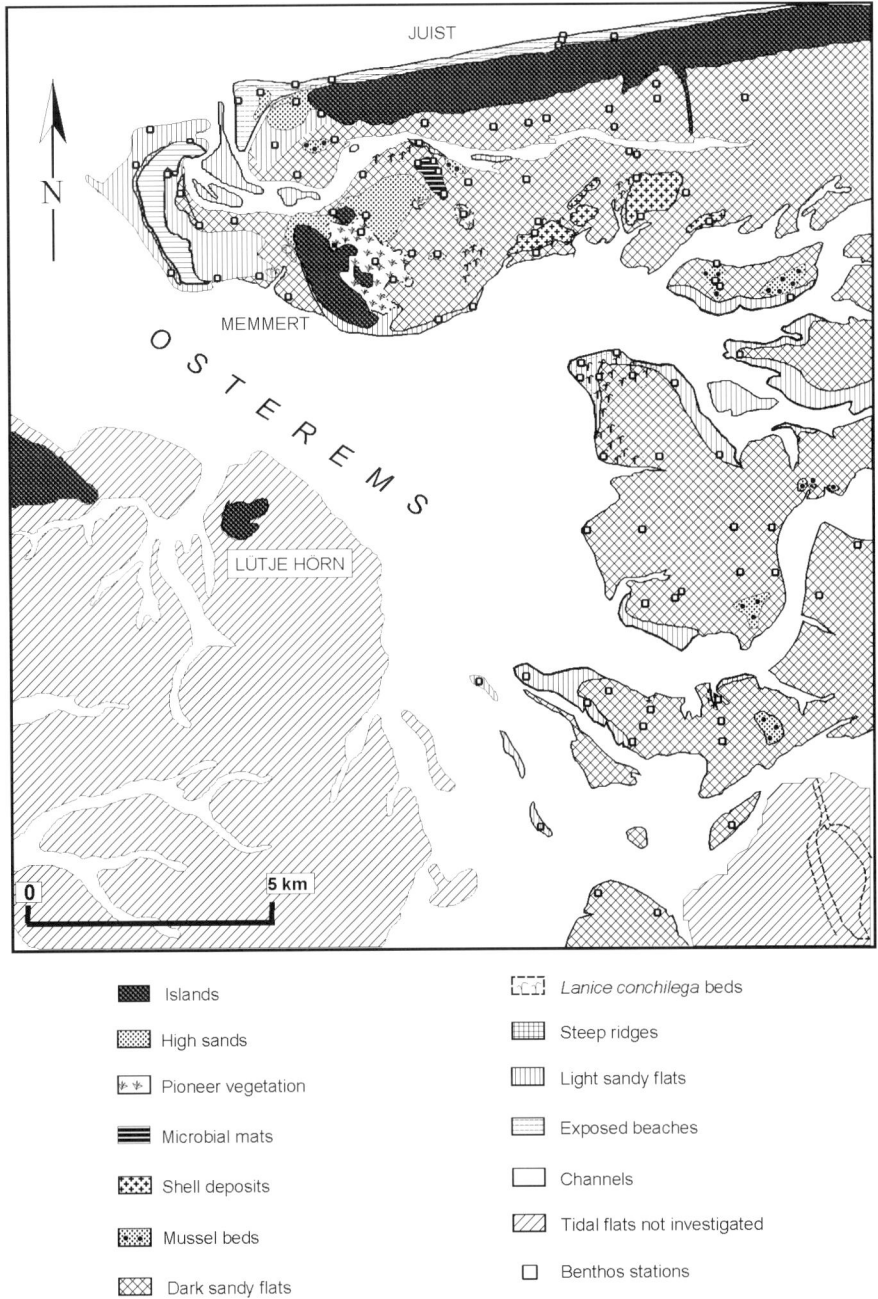

Fig. 12.3. Juist area, distribution of the intertidal habitats

The habitat division on the Banc d'Arguin is based on a map presented by Altenburg et al. (1982). It was actualized in a few details by the observations of September 1988. For the Juist area, a series of aerial photographs (black and white-infrared, 1:15,000) from a low-tide situation in June 1988 was analyzed. In combination with ground truth, all discernable structures were used for characterization and mapping of physiographical units.

12.3 Results

12.3.1 Habitat Division

On the tidal flats of the Banc d'Arguin, habitats had already been distinguished by earlier surveys (Altenburg et al. 1982; Wolff et al. 1993a):

- Stands of seagrasses with the dominant species *Zostera noltii* covering about 80% of the area. Muddy and mixed sediments prevail (Michaelis 1993). Within the *Zostera* stands, minor areas covered by the trichomic alga *Vaucheria* occur.
- Light sandy flats cover about 20% of the intertidal areas. Michaelis (1993) and Wolff et al. (1993a) give data on grain-size distributions and related sediment parameters.

Additional habitats of low quantitative relevance, and therefore not indicated in Fig. 12.2, are exposed beaches and a scant mangrove of *Avicennia africana*. Finally, the sebkha, a supralittoral transition area between the tidal flats and the desert, covers extensive beach plain areas with irregular inundation (Fig. 12.2).

In the Juist area, ten different habitats could be distinguished. The grain-size distribution of the sediments ranges from fine sand to fine-sandy medium sand (Obert and Michaelis, in prep.). The habitats are arranged on a gradient from the wave-exposed seaside to the sheltered positions towards the mainland and behind the island of Juist (Fig. 12.3). The main habitats are light sandy flats forming fringes at the edge of the more exposed sand banks and dark sandy flats covering almost three-quarters of the whole area in the more sheltered positions. The habitat "high sands" are bare sand banks that project to a small degree above the mean high-water level and are irregularly inundated (Reineck 1956).

12.3.2 Faunal Inventories

The inventory of the Banc d'Arguin tidal flats in September 1988 comprises a total of 132 taxa from only 30 sampling stations (Table 12.1). The complete taxonomical and autecological results will be given in a separate publication. The main species groups are (number of species in brackets): Polychaeta (57), Crustacea (32), Mollusca (25), Sipunculida (7), Oligochaeta (3), Echinodermata (3), Porifera (1), Cnidaria (1), Tunicata (1), Acrania (1) and Hexapoda (1).

From 104 sampling stations in the Juist area 78 species were obtained (Obert and Michaelis, in prep.). The main groups are (species numbers in brackets): Polychaeta (30), Crustacea (20), Mollusca (15), Oligochaeta (5), Hexapoda (3), Coelenterata (2), Nemertini (1), Bryozoa (1) and Echinodermata (1). On the tropical Banc d'Arguin as well as in the temperate Juist area polychaetes, crustaceans and molluscs dominate by species number.

12.3.3 Structure and Distribution of Macrozoobenthos Assemblages

Tables 12.2 and 12.3 present mean species numbers, total abundance and total biomass of the individual habitats and of the entire investigated areas on the Banc d'Arguin and in the Wadden Sea. The values for the entire areas were calculated without considering the supralittoral sebkha on the Banc d'Arguin and the supralittoral high sands in the Juist area. On the Banc d' Arguin (Table 12.2) the means of all eulittoral stations amount to 16 species per station, slightly over 2000 individuals m^{-2} and a biomass of 7.6 g ADW m^{-2}. In the latter value the biomass of the large bivalve *Anadara senilis* is not included following the example of the preceding surveys (Altenburg et al. 1982; Wolff 1991; Wolff et al. 1993a). In September 1988 the mean share amounted to 1.7 g ADW m^{-2} which leads to a total biomass of 9.3 g ADW m^{-2}.

In the Juist area (Table 12.3) the resulting mean values from all stations (excluding the mussel beds and the high sands) are 10 species per station, 4180 individuals m^{-2} and 32.7 g ADW m^{-2}. With the mussel beds included, the mean total biomass amounts to 41.2 g ADW m^{-2}.

In the following short descriptions the habitats are arranged from seaward low intertidal through landward high intertidal to supralittoral (see also Tables 12.2 and 12.3 and Figs. 12.4–12.8).

Banc d'Arguin – Exposed Beaches (Table 12.2). The spectrum of 33 species comprises rare species limited to this habitat like the polychaetes *Hyboscolex longiseta* and *Scolelepis squamata*. Three species dominate, viz. Apseudidae sp. 1 and *S. squamata*, dominant in both number and weight, and juvenile *Uca tangeri*, providing the highest share of biomass.

Table 12.1. Macrozoobenthic species of the Banc d'Arguin, September 1988

Porifera
*Porifera sp. 1

Cnidaria
Anthozoa sp.

Mollusca
*Gibbula umbilicalis
Hydrobia cf. ulvae
Turritella torulosa
*Crepidula sp.
Nassarius sp.
Cymbium sp.
*Persicula chudeani
*Clavatula sp.
*Bullaria adansoni
*Haminea orbignyana
*Haminea elegans
*Cychlina grimaldii
*Gastropoda sp.
Arca afra
*Arca subglossa
Anadara senilis
Crenella dollfusi
*Phacoides adansoni
Loripes lacteus
*Cerastoderma edule
*Dosinia cf. lupinus
*Venus rosalina
*Venerupis aurea
Tagelus angulatus
Abra tenuis

Sipunculida
*Sipunculida sp.1
Sipunculida sp. 2
*Sipunculida sp. 3
Sipunculida sp. 4
Sipunculida sp. 5
Golfingia sp.
Sipunculida sp.

Annelida
*Polynoidae sp.
*Eteone siphodonta
Eteone cf. foliosa
*Typosyllis mauretanica
Typosyllis sp. 2
Typosyllis sp. 3
Perinereis cultrifera
*Nereis caudata

Nereis diversicolor
Nereis sp. 1
Nereis sp. 2
Glyceria convoluta
Diopatra neapolitana
Diopatra neapolitana
  capensis
Marphysa sanguinea
Arabella iricolor
*Drilonereis filum
*Drilonereis cf. monroi
*Lumbrinereis heteropoda
Lumbrinereis heteropoda
  difficilis
*Lumbrinereis sp. 1
Nainereis laevigata
Scoloplos chevalieri
Scoloplos sp. 1
*Paraonis lyra
*Paraonidae sp. 1
*Paraonidae sp. 2
*Aedicira cf. belgicae
Aricidea sp.
Nematonereis unicornis
Scolelepis squamata
*Polydora antennata
Cirriformia tentaculata
Cirriformia sp. 1
*Cirriformia sp.
Tharyx dorsobranchialis
Tharyx sp.
Heteromastus filiformis
Capitella capitata
Capitellidae sp. 1
Capitellidae sp. 2
*Capitellidae sp.
Petaloproctus terricola
Euclymene lüderitziana
Euclymene cf. natalensis
Macroclymene monilis
*Axiothella sp.
*Rhodine sp.
Hyboscolex longiseta
*Polycirrus aurantiacus
*Scionella lornensis
Terebella lapidaria
*Trichobranchus glacialis
Bogueidae sp. 1
Hypsicomus cf. capensis
Spirorbis sp.
Polychaeta sp.

Tubificidae sp.
Enchytraeidae sp.
Oligochaeta sp.

Crustacea
Crustacea sp.
*Nebalia cf. bipes
*Tanais dulongii
Apseudidae sp. 1
Anthuridae sp. 1
*Idotea chelipes
Sphaeromatidae sp. 1
*Sphaeromatidae sp. 2
Sphaeromatidae sp. 3
Sphaeromatidae sp. 4
Sphaeromatidae sp. 5
*Isopoda sp.
Cymadusa hirsuta
Urothoe elegans
Urothoe grimaldii
Ampelisca brevicornis
Leucothoe richiardi
*Lysianassa ceratina
*Harpinia cf. pectinatus
Victoriopisa atlantica
Hyale perieri
*Aoridae sp.
Amphipoda sp. 1
Amphipoda sp. 2
*Palaemon cf. elegans
Penaeus cf. kerathurus
Callianassidae sp. 1
Natantia sp. 1
Natantia sp. 2
Uca tangeri
Callinectes sp. 1
*Carcinus sp. 1

Echinodermata
*Ophiactis lymani
Holothurioida sp. 1
Holothurioida sp. 2

Tunicata
*Tunicata sp. 1

Acrania
*Branchiostoma senegalense

Hexapoda
Diptera (pupae)

* Species which were found only in the seagrass stands.

Table 12.2. Intertidal macrozoobenthos of the Banc d'Arguin, view of the habitats

Habitat	Number of stations	Mean values and (SD) of			
		Species number per station	Individuals (m^{-2})	ADW (g m^{-2})	
				Anadara senilis	
				Excluded	Included
Exposed beaches	4	8	3899	6.781	6.781
Sandy flats	5	12	2885	6.085	6.085
Seagrass stands	16	19	1375	8.437	8.664
Zostera/ Vaucheria	1	12	1984	6.120	18.026
Mangrove	1	21	987	6.692	36.436
Sebkha	3	1	121	0.033	0.033
Total area, the sebkha excluded	27	16 (6.9)	2035 (2330)	7.606 (4.175)	9.282 (1.654)

Table 12.3. Intertidal macrozoobenthos of the Juist area, view of the habitats

Habitat	Area (km^2)	Number of stations	Mean values and (SD) of		
			Species number	Individuals (m^{-2})	ADW (g m^{-2})
Exposed beaches	4.41	8	5	1484	6.914
Steep ridges	0.50	4	7	1700	3.760
Light sandy flats	14.04	13	10	1004	10.184
Lanice conchilega beds[a]	2.93	5	11	8341	56.767
Dark sandy flats	80.20	46	13 (3.8)	4212 (5008.7)	51.412 (100.809)
Mytilus edulis beds	1.37	3	9	5649	789.735
Shell deposits	1.79	3	11	2537	19.896
Microbial mats	0.32	4	8	18431	12.380
Pioneer vegetation	2.36	6	5	4270	4.424
High sands	3.87	10	5	5592	3.030
Total area	108.9	102	10 (4.4)	4337 (6127.4)	37.306 (104.485)
Total area without "high sands"	105.0	92	10 (4.3)	4196 (6071.7)	41.156 (109.485)
Total area without "high sands" and "*Mytilus edulis* beds"	103.6	89	10 (4.3)	4180 (6103.6)	32.650 (75.809)

[a] *Lanice conchilega* beds are distributed over parts of the light and dark sandy flats. Therefore, they remain unconsidered in the total area.

Banc d'Arguin – Sandy Flats (Table 12.2). Figure 12.4 shows which of the 41 species play a dominant role. *Scoloplos* sp. 1 is most widely spread. The most abundant species, *Hydrobia* cf. *ulvae* and *Ampelisca brevicornis*, are low in frequency of occurrence but nevertheless belong to the group of four species with the largest biomass.

Banc d'Arguin – Seagrass Beds (Table 12.2). The occurrence of 93 species was ascertained. Numerous species participate in the different aspects of dominance (Fig. 12.5). Nine species reach a frequency of occurrence of 60 % or more, with Anthuridae sp. 1 and *Loripes lacteus* occurring in all samples. Seven species together represent more than 60 % of all individuals with *L. lacteus* taking the largest share. *Marphysa sanguinea* has the highest biomass.

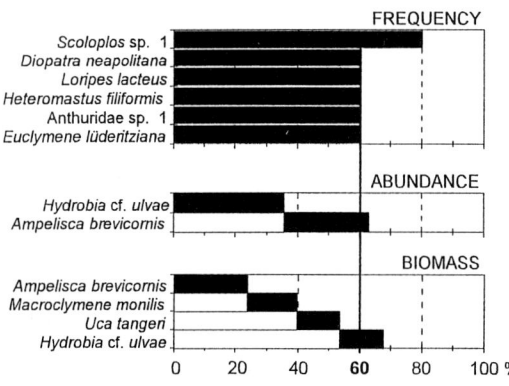

Fig. 12.4. Habitat "sandy flats" on the Banc d'Arguin. Species dominant in frequency of occurrence (% of stations), abundance (% of total) and biomass (% of total)

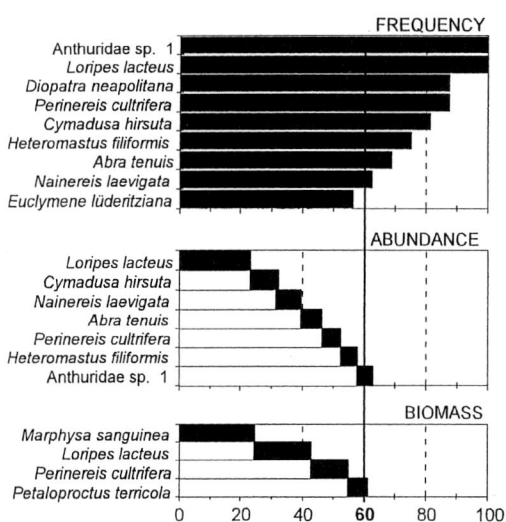

Fig. 12.5. Habitat "seagrass stands" on the Banc d'Arguin. Species dominant in frequency of occurrence (% of stations), abundance (% of total) and biomass (% of total)

In Table 12.1 the species occurring only in the seagrass beds are marked with an asterisk, altogether 53 species. None of them plays a relevant role in terms of number or weight. Their total number is only 176 individuals m^{-2} compared with the 1375 individuals m^{-2} of all species and their total biomass of 1.31 g ADW m^{-2} has only a modest share of the 8.44 g ADW m^{-2} biomass of all species (excluding *Anadara senilis*). *A. senilis* played a minor part in the seagrass beds in 1988, because this bivalve added only 0.23 g ADW m^{-2} to the average biomass (Table 12.2). The most important phytal inhabitants are *Polycirrus aurantiacus, Palaemon* cf. *elegans* and *Idotea chelipes*.

Banc d'Arguin – Zostera/Vaucheria (Table 12.2). None of the 12 species occurring in this habitat revealed a connection to *Vaucheria*.

Banc d'Arguin – Mangrove (Table 12.2). At only one sampling station 21 species were found. *A. senilis* included, the total biomass is higher than in any other habitat (Table 12.2).

Banc d'Arguin – Sebkha (Table 12.2). The supralittoral sebkha is an extremely poor benthic habitat. Two of the investigated stations did not contain any macrofauna. At a third station, *Abra tenuis, Capitella capitata* and juvenile *Uca tangeri* were encountered.

Juist Area – Wave-Exposed Beaches (Table 12.3). Three of the 22 species occurring here are highly specific for wave-exposed locations: *Scolelepis squamata, Haustorius arenarius* and *Pontocrates arenarius*. These species are also dominant in frequency of occurrence (Fig. 12.6). *S. squamata* accounts for about 75 % of the total abundance and biomass. Compared with the other habitats of the Juist area, the exposed beaches belong to the poorest ones.

Juist Area – Steep Ridges in Exposed Positions (Table 12.3). This habitat is closely related to the wave-exposed beaches, but appears in aerial photographs, as well as in the field, as a particularly conspicuous morphological element. Main characteristic is the deeply fallen groundwater level during low

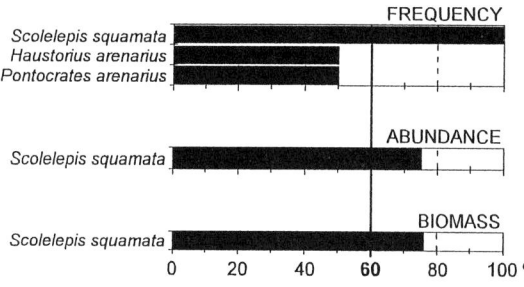

Fig. 12.6. Habitat "exposed beaches" in the Juist area. Species dominant in frequency of occurrence (% of stations), abundance (% of total) and biomass (% of total)

Juist Area – Light Sandy Flats (Table 12.3). The light sandy flats border the low-water line of the tidal flats. The assemblage of 30 species is dominated by *Nephtys hombergi, Macoma balthica, Scoloplos armiger, Arenicola marina* and *Eteone longa* in frequency, by *S. squamata* and *S. armiger* in total abundance, and by *M. balthica, A. marina* and *Cerastoderma edule* in total biomass (Fig. 12.7).

Juist Area – Lanice conchilega Beds (Table 12.3). With abundances of up to 25,000 individuals m^{-2}, the sedentary polychaete *Lanice conchilega* creates a peculiar habitat with its closely placed vertical tubes. *L. conchilega* dominates the whole assemblage with 100% frequency and overwhelming preponderance in total abundance and total biomass. The remaining 29 species rank as merely accompanying members of the assemblage. The *L. conchilega* beds are a rich assemblage in every respect.

Juist Area – Dark Sandy Flats (Table 12.3). This main habitat of the area is inhabited by 56 species. Figure 12.8 demonstrates the dominance aspects. *M. balthica, S. armiger, Heteromastus filiformis, Pygospio elegans* and *Hydrobia ulvae* are dominant in frequency as well as in abundance. *C. edule, Mya arenaria* and *M. balthica* account for 60% of the biomass.

Juist Area – Mytilus edulis Beds (Table 12.3). At the time of investigation, the scattered mussel beds covered only 1.4 km^2 of the area. Exceedingly high biomass – in this case 780 g m^{-2} – is a characteristic of this habitat.

Juist Area – Shell Deposits (Table 12.3). The majority of the 21 species belong to a fouling assemblage. The infaunal species *L. conchilega*, penetrating the

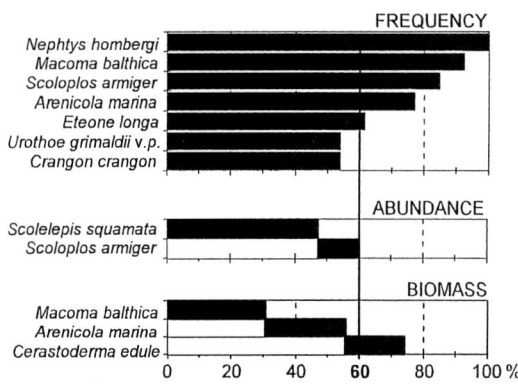

Fig. 12.7. Habitat "light sandy flats" in the Juist area. Species dominant in frequency of occurrence (% of stations), abundance (% of total) and biomass (% of total)

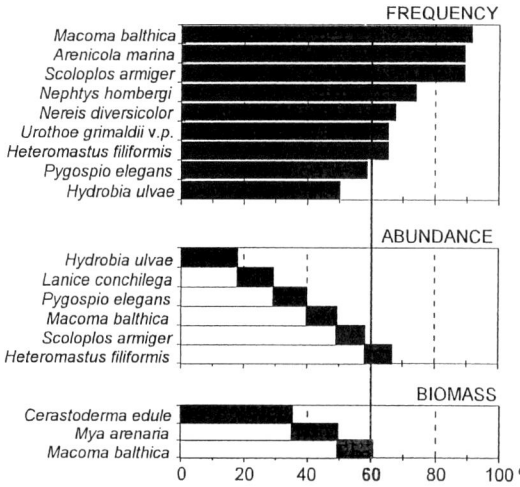

Fig. 12.8. Habitat "dark sandy flats" in the Juist area. Species dominant in frequency of occurrence (% of stations), abundance (% of total) and biomass (% of total)

shell layers with its tubes, supplies 82 % of the total abundance and 71 % of the total biomass.

Juist Area – Microbial Mats (Table 12.3). This habitat is situated immediately below the mean highwater line. Of the 13 species present, *Tubifex costatus*, *P. elegans*, *N. diversicolor*, Enchytraeidae sp., *M. balthica* and *H. ulvae* belong to the animals dominating in frequency, abundance and biomass.

Juist Area – Pioneer Vegetation (Table 12.3). The areas covered by *Salicornia* and *Spartina* occur at the same tidal level as the microbial mats. The benthic fauna is almost identical to that in the microbial mats. However, mean total abundance and mean total biomass are considerably lower, probably due to the dense root system of phanerogamic plants in the sediment.

Juist Area – High Sands (Table 12.3). The faunal assemblage of the supralittoral high sands is closely related to those of microbial mats and pioneer vegetation. This irregularly inundated habitat exhibits a surprisingly dense benthic fauna. However, all species found here are represented only by tiny juvenile stages. From the 19 species encountered, 5 play a dominant role. Only *N. diversicolor* reaches a frequency of 60 %, followed by *P. elegans*, *M. balthica* and *H. ulvae* with 50 % each.

12.4 Discussion

Our study of the tropical Banc d'Arguin and the temperate Juist area reveals similarities as well as considerable differences when habitat division, macrobenthic species richness, standing stock and trophic structure are compared. Whether these criteria are subject to latitudinal gradients (Reise 1991; Wolff 1991) remains a question to be answered by a large-scale or even global review. As regards to the habitats present in both areas a remarkably close correspondence exists between the exposed beaches, because the dominant species, viz. the highly stenotope polychaete *Scolelepis squamata,* is identical. Between the sandy flats of the Banc d'Arguin and the light and dark sandy flats of the Juist area the similarity is confined to the sediment characteristics while macrozoobenthic standing stock and trophic structure show large differences (see below). The seagrass habitat of the Banc d'Arguin is not represented in the Juist area today, although, until the 1970s, it was present to a modest degree, even with the same dominant species *Zostera noltii* (Kastler and Michaelis 1999). The mussel beds, shell deposits, and *Lanice* beds of the Juist area do not have corresponding habitats on the Banc d'Arguin. The mangrove may be considered as a parallel of the northern pioneer zone in a certain sense as, in both habitats, the ground is covered by a pioneer vegetation. Microbial mats very similar to those of the Juist area were also encountered in the high intertidal zone of the Banc d'Arguin but they remained uninvestigated. Finally, the supralittoral "high sands" of the Juist area correspond to the tropical sebkha in certain respects. In the sebkha, an extremely poor benthic fauna was found (Table 12.2). The high sands, on the other hand, are inhabited by a surprisingly rich benthic fauna (Table 12.3). It is assumed that the wet and cool northern climate enables some aquatic animals to survive under these conditions.

A conspicuous difference concerns the spatial distribution of habitats in both areas. In the Juist area – as generally in the Wadden Sea – there exists an arrangement which is evidently governed by the exposure to winds, currents and wave action: The highly exposed, clean sandy habitats face to the west and north and are gradually followed by more sheltered habitats towards the east and the south.

In contrast, in the tidal flat area of the Banc d'Arguin, the distribution pattern of muddy sediments is correlated with the presence of the seagrass beds. They occur over the whole area, even immediately bordering the Atlantic Ocean (Fig. 12.2). Continuous offshore trade winds are certainly part of the explanation.

The differences in species richness of the two areas are evident: 132 species from only 30 stations at the Banc d'Arguin (on average 16 species per station), 78 species from more than 100 stations in the Juist area (on average 10 species per station). The inclusion of preceding surveys (Einsele et al. 1974; Altenburg et al. 1982; Wolff et al. 1993a) would increase the species number of the Banc d'Arguin considerably and the complete spectrum is probably still far from being assessed. A high species richness appears to be a general characteristic of tropical tidal flat areas, as confirmed by studies from Vargas (1987, 1988), Reise (1991), Dittmann (1995) and Baron et al. (1993). On the other hand, the modest richness of the Juist area is a general characteristic of the Wadden Sea confirmed in numerous surveys with many thousands of sampling stations (see Michaelis 1987; Essink et al. 1998; Beukema 1989; Reise 1991 and literature cited therein). Next to differences between tropical and temperate areas in general, an important additional explanation could be salinity conditions. Along the coast of the southern North Sea these are rather variable and considerably low compared with the open sea. Stenohaline marine species could be either completely excluded from the coastal waters or displaced into the sublittoral waters with higher salinities.

Because of differences in body size, the usefulness of the parameter total abundance of individuals is limited without additional information on biomass. Total abundance in tropical and subtropical localities appears to depend considerably on habitat type, season, and other conditions (Alongi 1990). Examples of highly different figures are available which may, however, be only cautiously compared because of different methods of sampling and calculation of mean values. Total abundances of <1000 individuals m^{-2} were found in intertidal seagrass stands of New Caledonia (Baron et al. 1993), numbers of 1000–2000 individuals m^{-2} in sandy and muddy flats of the Andaman Sea (Reise 1991) and in estuarine flats of the NE Australian coast (Dittmann 1995); high numbers from 10,000–20,000 individuals m^{-2} are reported from Costa Rica (Vargas 1987, 1988), and a very high total abundance of 45,000 individuals m^{-2} (annual average) is given by Puttick (1977) for a locality in Langebaan Lagoon, South Africa. Against this background, the total abundance figures for the Banc d'Arguin may be considered as rather modest. The survey of February–April 1986 (Wolff et al. 1993a; Wijnsma et al. 1999) yielded a mean total abundance of about 1400 individuals m^{-2}; in September 1988 (present study) the average of all stations (sebkha excluded) amounted to roughly 2000 individuals m^{-2} (Table 12.2). In the temperate Juist area (where samples were sieved with 1 mm mesh size!), the mean total abundance varies from 1000 (light sandy flats) to 18,000 individuals m^{-2} (microbial mats). The means of both the largest habitat (dark sandy flats) and of all habitats (excluding high sands) lie close to each other around 4000 individuals m^{-2} (Table 12.3). From these figures no major difference between the total abundances at low and high geographical latitudes can be derived, a fact also confirmed by the "comprehensive, but not exhaustive" compilations of Alongi (1990).

In the previous studies the interest was focused on the biomass of the intertidal macrozoobenthos of the Banc d'Arguin, and its role as a food resource of wintering waders (Altenburg et al. 1982; Wolff and Smit 1990; Wolff 1991; Wolff et al. 1993a,b). In the Wadden Sea, biomass has been used as a general means to characterize the productivity of different coastal areas or habitats (e.g. Michaelis 1987; Beukema 1989; Essink et al. 1998; Reise and Lackschewitz 1998). As in the earlier surveys (Altenburg et al. 1982; Wolff and Smit 1990; Wolff 1991; Wolff et al. 1993a), the present study again points to a remarkably low biomass on the Banc d'Arguin tidal flats: A mean value of 7.6 g ADW m^{-2} without *Anadara senilis* and of 9.3 g ADW m^{-2} with the bivalve included (Table 12.4). It may be concluded that seasonal differences are largely lacking and that a low macrozoobenthic standing stock is a general characteristic of this tropical tidal flat area. Its biomass is only a small fraction of the values found in the Juist area: A mean total biomass of 32.7 g ADW m^{-2} without the mussel beds and of 41.2 g ADW m^{-2} with the mussel beds included. Summer values in that range (30–40 g ADW m^{-2} without *Mytilus* beds) are also characteristic for other subareas of the Wadden Sea, as shown by the examples given in Table 12.4.

The vast seagrass beds and the extraordinary clearness of the seawater of the Banc d'Arguin point to a system in which the phytoplankton is insignificant and the primary production is almost exclusively achieved by seagrasses and their periphyton (Wolff and Smit 1990). Sevrin-Reyssac (1993) calls the poverty of plankton in the Banc d'Arguin waters extreme. This is in line with the observation that benthic suspension feeders are quantitatively unimportant in the tidal flats (Wolff et al. 1987, 1993a,b). Though the knowledge on autecology is insufficient, we conclude that the majority of the species belong to the deposit and epistrate feeders. In the Juist area, the largest habitat, i.e. dark sandy flats, is dominated by the suspension feeders *Cerastoderma edule* and *Mya arenaria* followed by the epistrate feeder *Macoma balthica* (see Fig. 12.8). In the habitat "light sandy flats" (Fig. 12.7), the first places in biomass contribution are occupied by the epistrate feeder *M. balthica*, the deposit and suspension feeder *Arenicola marina*, and the suspension feeder *Cerastoderma edule*. Generally, three main sources are considered as being the nutritive basis of primary consumers in the Wadden Sea: Autochtonous phytoplankton production, autochthonous microphytobenthos production, and the import of particulate organic material from the North Sea (van den Hoek et al. 1979).

In contrast to the Wadden Sea, the main portion of primary production on the Banc d'Arguin tidal flats is not directly utilizable as a food source. With the exception of a few sap-sucking isopod species, the benthic invertebrates do not feed on living seagrass and therefore no invertebrate food webs are based on this immensely rich supply. Nevertheless, the seagrass material has to be considered as the main energy source of the Banc d'Arguin system (Wolff and Smit 1990; Wolff et al. 1993b). The leaves, shed after dying, become partly displaced but the major part is deposited and decomposed in situ

Table 12.4. Intertidal macrozoobenthic biomass (g ADW m^{-2}) of the Banc d'Arguin and the Juist area compared with other figures from the Wadden Sea

	Anadara senilis resp. *Mytilus* beds		Source(s)
	Excluded	Included	
Banc d'Arguin			
February 1980	2.9	7.6	Altenburg et al. (1982); Engelmoer et al. (1984)
February–April 1986	8.9	17.0	Wolff and Smit (1990); Wolff et al. (1993a)
September 1988	7.6	9.3	Present investigation
Juist area			
Summers 1987–1990	32.7	41.2	Present investigation
Western Dutch Wadden Sea			
Summer/autumn		38.5	Beukema (1989)
Eastern Dutch Wadden Sea			
Annual means 1985–1995		30–70	Essink et al. (1998)
Randzel, Niedersachsen coast			
Summer 1981	36.0	109.0	Obert (1982)
Jade Bay, Niedersachsen coast			
Summers 1975–77	43.0	51.0	Michaelis (1987)
Sylt-Rømø area (Schleswig-Holstein and Danish coasts)			
Summers 1992–1994	37.5	40.3	Reise and Lackschewitz (1998)

(Hemminga and Nieuwenhuize 1991). Thus, the detritus-cycle is the basis of most food webs in the system (Kikuchi and Peres 1977). Additionally, the periphyton, the microalgae growing on seagrass leaves, plays a certain role as a direct food source that, on the Banc d'Arguin, is especially exploited by the fiddler crab, *Uca tangeri* (Hootsmans et al. 1993).

Finally, we conclude that the low standing stock of macrozoobenthos in all intertidal habitats of the Banc d'Arguin will probably remain a crucial question as long as direct studies on biomass production are lacking. Piersma et al. (1993) show that low biomass is not a general characteristic of intertidal flats in the tropics. There are many arguments in favour of the view that the tidal flat area of the Banc d'Arguin owes its peculiar character to a unique regional constellation of topographical, hydrographical, climatic and biological conditions. To the latter ones belongs an extremely high predation pressure, as 2 million palaearctic waders spend about the same period of time on the Banc d'Arguin flats (500 km^2) as on the intertidal flats of the Wadden Sea (4300 km^2). Wolff and Smit (1990), Wolff (1991) and Wolff et al. (1993a,b) "solved" the problem by postulating sufficiently high production rates of the bottom fauna. The direct proof, however, has not yet been provided.

Acknowledgements. We thank the Mauritanian authorities, in particular the director and staff of the Parc National du Banc d'Arguin, for the permission to carry out the study at the Banc d'Arguin and for their support. We are grateful for assistance in the field to our colleagues, in particular Ineke de Boom and Dirk Michaelis, and in the laboratory work to Hans H. Kramer and Hilmar Hinz. In taxonomical questions we received support from Dr. F. Licher, Dr. Ch. Meyer, Dr. D. Platvoet, Prof. Dr. H. K. Schminke, Dr. M. Türkay, Prof. Dr. J.-W. Wägele and Prof. Dr. W. Westheide. The Mauritanian project was made possible thanks to financial support from the Netherlands Marine Science Foundation. Additional financial support was provided by the Netherlands Research Institute for Nature Management, Deutsche Forschungsgemeinschaft, Worldwide Fund for Nature, and the German Bundesministerium für Bildung, Wissenschaft, Forschung und Technologie.

References

Abrahamse J, Joenje W, van Leeuwen-Seelt N (1976) Waddenzee, natuurgebied van Nederland, Duitsland en Denemarken. Landelijke Vereniging tot Behoud van de Waddenzee, Harlingen

Alongi DM (1990) The ecology of tropical soft-bottom benthic ecosystems. Oceanogr Mar Annu Rev 28:381–496

Altenburg W, Engelmoer M, Mes R, Piersma T (1982) Wintering waders on the Banc d'Arguin, Mauritania. Communication no 6 of the Wadden Sea Working Group, Groningen, 283 pp

Baron J, Clavier J, Thomassin BA (1993) Structure and temporal fluctuations of two intertidal seagrass-bed communities in New Caledonia (SW Pacific Ocean). Mar Biol 117:139–144

Bätje M (1986) Salzgehalt und Wassertemperatur an der ostfriesischen Küste. Jber 1985 Forschungsstelle Küste 37:125–145

Beukema JJ (1989) Long-term changes in macrozoobenthic abundance on tidal flats of the western part of the Dutch Wadden Sea. Helgoländer Meeresunters 43:405–415

Dittmann S (1995) Benthos structure on tropical tidal flats of Australia. Helgoländer Meeresunters 49:539–551

Einsele G, Herm D, Schwarz HU (1974) Holocene eustatic sea level fluctuations at the coast of Mauritania. Meteor Forsch Ergeb, Reihe C 18:43–62

Engelmoer M, Piersma T, Altenburg W, Mes R (1984) The Banc d'Arguin. In: Evans PR, Goss-Custard JD, Hale WG (eds) Coastal waders and wildfowl in winter. Cambridge Univ Press, Cambridge, pp 293–310

Ens B (1985) Tussen Sahara en Siberië. Stichting WIWO, Ewijk, 32 pp

Essink K, Beukema JJ, Madsen PB, Michaelis H, Vedel GR (1998) Long-term development of biomass of intertidal macrozoobenthos in different parts of the Wadden Sea. Governed by nutrient loads? Senckenbergiana Maritima 29:25–35

Hemminga MA, Nieuwenhuize J (1991) Transport, deposition and in situ decay of seagrasses in a tropical mudflat area (Banc d' Arguin, Mauritania). Neth J Sea Res 27: 183–l90

Hootsmans MJM, Vermaat JE, Beijer JAJ (1993) Periphyton density and shading in relation to tidal depth and fiddler crab activity in intertidal seagrass beds of the Banc d' Arguin (Mauritania). Hydrobiologia 258:73–80

Kastler T, Michaelis H (1999) The decline of seagrasses, *Zostera marina* and *Zostera noltii*, in the Wadden Sea of Lower Saxony. Senckenbergiana Maritima 29[Suppl]: 77–80

Kikuchi T, Peres JM (1977) Consumer ecology of seagrass beds. In: McRoy CP, Helfferich C (eds) Seagrass ecosystems, a scientific perspective. Mar Sci 4:147–193

Michaelis H (1987) Bestandsaufnahme des eulitoralen Makrobenthos im Jadebusen in Verbindung mit einer Luftbildanalyse. Jber 1986 Forschungsstelle Küste 38: 13–97

Michaelis H (1993) Food items of the grey mullet *Mugil cephalus* in the Banc d'Arguin area (Mauritania). Hydrobiologia 258:175–183

Obert B (1982) Bodenfauna der Watten und Strände um Borkum – Emsmündung. Jber 1981 Forschungsstelle Küste 33:139–162

Peters H (1976) The spreading of water masses of the Banc d'Arguin in the upwelling area of the northern Mauritanian coast. Meteor Forsch Ergeb A 18:78–100

Piersma T, de Goede P, Tulp I (1993) An evaluation of intertidal feeding habitats from a shorebird perspective: towards relevant comparisons between temperate and tropical mudflats. Neth J Sea Res 31:503–512

Puttick GM (1977) Spatial and temporal variations in intertidal animal distribution at Langebaan Lagoon, South Africa. Trans R Soc S Afr 42:403–433

Rahmel J, Bätje M, Michaelis H, Noack U (1995) *Phaeocystis globosa* and the phytoplankton succession in the East Frisian coastal waters. Helgoländer Meeresunters 49:399–408

Reineck HE (1956) Die Oberflächenspannung als geologischer Faktor in Sedimenten. Senckenbergiana Lethaea 37:265–287

Reise K (1991) Macrofauna in mud and sand of tropical and temperate tidal flats. In: Elliott M, Ducrotoy JP (eds) Estuaries and coasts: spatial and temporal intercomparisons. Olsen and Olsen, Fredensborg, pp 201–216

Reise K, Lackschewitz D (1998) Benthos des Wattenmeeres zwischen Sylt und Rømø. In: Gätje C, Reise K (eds) Ökosystem Wattenmeer, Austausch-, Transport- und Stoffumwandlungsprozesse. Springer, Berlin Heidelberg New York, pp 55–64

Sevrin-Reyssac J (1993) Hydrology and underwater climate of the Banc d' Arguin, Mauritania: a review. Hydrobiologia 258:1–8

Smit CJ, Piersma T (1989) Numbers, mid-winter distribution and migrations of wader populations using the East Atlantic Flyway. In: Boyd H, Pirot JY (eds) Flyways and reserve networks for water birds. International Waterfowl and Wetland Research Bureau special publications 9, Slimbridge, pp 24–63

Sprung M (1994) Macrobenthic secondary production in the intertidal zone of the Ria Formosa – a lagoon in southern Portugal. Estuar Coastal Shelf Sci 38:539–558

van den Hoek C, Admiraal W, Colijn F, De Jonge VD (1979) The role of algae and seagrasses in the ecosystem of the Wadden Sea: a review. In: Wolff WJ (ed) Flora and vegetation of the Wadden Sea. Report 3 of the Wadden Sea Working Group, Leiden, pp 9–118

Vargas JA (1987) The benthic community of an intertidal mud flat in the Gulf of Nicoya, Costa Rica. Description of the community. Rev Biol Trop 35:299-316

Vargas JA (1988) Community structure of macrobenthos and the results of macro-predator exclusion on a tropical intertidal mud flat. Rev Biol Trop 36:287-308

Wijnsma G, Wolff WJ, Meijboom A, Duiven P, De Vlas J (1999) Species richness and distribution of benthic tidal flat fauna of the Banc d'Arguin, Mauritania. Oceanol Acta 22:233-243

Wolff WJ (1991) The interaction of benthic macrofauna and birds in tidal flat estuaries: a comparison of the Banc d' Arguin, Mauritania, and some estuaries in The Netherlands. In: Elliott M, Ducrotoy JP (eds) Estuaries and coasts: spatial and temporal intercomparisons. Olsen and Olsen, Fredensborg, pp 299-306

Wolff WJ, Smit CJ (1990) The Banc d' Arguin, Mauritania, as an environment for coastal birds. Ardea 78:17-38

Wolff WJ, Gueye A, Meijboom A, Piersma T, Sall MA (1987) Distribution, biomass, recruitment and productivity of *Anadara senilis* (L.) (Mollusca: Bivalvia) on the Banc d' Arguin, Mauritania. Neth J Sea Res 21:243-253

Wolff WJ, Duiven AG, Duiven P, Esselink P, Gueye A, Meijboom A, Moerland G, Zegers J (1993a) Biomass of macrobenthic tidal flat fauna of the Banc d' Arguin, Mauritania. Hydrobiologia 258:151-163

Wolff WJ, van der Land J, Nienhuis PH, De Wilde PAWJ (1993b) The functioning of the ecosystem of the Banc d' Arguin, Mauritania: a review. Hydrobiologia 258:211-222

Young DK, Young MW (1982) A comparative study of macrobenthic invertebrates in bare sand and seagrass (*Thalassia testudinum*) in Carrie Bow Cay lagoon. Smithson Contrib Mar Sci 12:115-126

13 Tropical Tidal Flat Benthos Compared Between Australia and Central America

S. DITTMANN and J.A. VARGAS

13.1 Introduction

At all latitudes, tidal flats occur along the shorelines of the world's oceans, covering areas up to thousands of square kilometers (Veenstra 1976; Mathieson and Nienhuis 1991). These soft-sediment environments provide habitat for a species-rich and abundant benthic fauna. As an ecotone between the land and the sea, tidal flats often function as a turntable, both for certain life stages of various organisms, as well as for energy and matter (Reise 1985; Gätje and Reise 1998). From decades of intensive ecological studies, an understanding has grown of the processes that structure benthic communities in temperate tidal flats (Wolff 1983; Reise 1985). Yet, how universal are these processes and do they play a comparable role in tidal flats in tropical environmental settings?

While several studies have compared differences in diversity and community composition of intertidal soft-bottom fauna between latitudes (Warwick and Rushwayuni 1987; Reise 1991; Vargas 1996), we set out here to compare tidal flat benthos within tropical latitudes, seeking generalities or singularities of the fauna and their ecological functions. Based on our respective backgrounds, we derive our examples from tidal flats of tropical Australia and the Pacific coast of Central America. If possible, information was compiled from several tropical tidal flats in Australia and America, but a more detailed analysis is restricted to studies from North Queensland, Costa Rica and El Salvador. These areas belong to different biogeographic regions and are presumably isolated from one another by the East Pacific Barrier (Veron 1995), as the lack of islands sets an oceanic barrier for larval dispersal of nearshore benthic animals. However, this barrier may not be as impenetrable as once believed (Lessios et al. 1998).

To assess generalities versus singularities, we take some general paradigms derived from ecological studies in temperate tidal flats or latitudinal comparisons (see e.g. Alongi 1989) and use them as hypotheses. We test whether these paradigms are applicable to the studied tropical tidal flats and also look for site-specific variation. The following hypotheses are tested and discussed:

I. Species diversity is higher in tropical than in temperate tidal flats, while abundances are lower. Soft sediments are usually home to a great variety of invertebrate species. It has been postulated by Sanders (1968) that diversity increases in the tropics, but Thorson (1957) did not record an increase in infaunal species numbers towards the tropics. While tidal flats were not included in the examples of Sanders and Thorson, several studies have reported very high species numbers from tropical tidal flats (e.g. Macnae and Kalk 1962; Day 1974; Chap. 12). Furthermore, the entire species stock has not always been fully assessed (Reise 1991; Dittmann 1995). Contrary to this latitudinal trend in species numbers, Reise (1991) found lower abundances in tropical tidal flats than in temperate tidal flats.

II. Intertidal benthic fauna shows a zoned distribution following an environmental gradient between the tide marks. Sediment composition and food availability are among known abiotic factors that determine animal distributions in soft sediments around the world (Petersen 1991; Snelgrove and Butman 1994). In most temperate tidal flats, the infauna is distributed along environmental gradients and can be separated into distinct faunal groups (see e.g. Beukema 1976; Whitlatch 1977; Dankers and Beukema 1983). Dittmann (2000) reported a similar distribution of benthic fauna in a tropical tidal flat which followed this general zonation pattern.

III. The majority of macrobenthic organisms in tidal flats are deposit-feeders. Deposit-feeders occur preferentially in sediments with higher mud contents and dominate in many temperate tidal flats (Sanders et al. 1962; Whitlatch 1977). Abundances of suspension-feeders increase towards the lower intertidal (Beukema 1976) and are further determined by tidal range and exchange rates of tidal basins. Although little is known on the biology of most tropical benthic organisms, a rough classification was made on their feeding modes, based on observations and literature records on related species from other parts of the world (e.g. Fauchald and Jumars 1979). As many benthic organisms demonstrate a high versatility in their feeding behavior (see Chap. 4), a classification based on trophic modes always remains tentative.

IV. Small-scale distributions are structured and regulated by interactions between species. Reise (1985) provided a conceptual hierarchy of ecological processes affecting population sizes of tidal flat organisms. From comparing temperate and tropical tidal flats it appeared that unrelated species can have similar ecological functions at different latitudes (Reise 1991; Vargas 1996). Here, we set out to test whether this assumption is also valid within tropical latitudes by comparing species with similar ecological roles, looking especially at those roles implying interactions between species.

We faced several constraints for the comparison: The studies carried out in tidal flats of Australia and Central America had not been designed for a comparative analysis and differed in their respective aims, durations as well as in the methods used. We therefore restrict most of the comparison to the macrofauna size fraction retained on a 500-µm sieve, leaving out the smaller mesofauna (Dittmann 1995). Furthermore, the taxonomy of benthic invertebrates is much better known from Central America than from northern Australia. Because of this and the few available studies on tropical tidal flats in the Pacific, a detailed biogeographic comparison of benthic diversity remains a future endeavor. Throughout the comparison we indicate research demand for tropical tidal flats. Mangroves and seagrass beds, which also occur in the intertidal of tropical coasts, were excluded from the comparison.

13.2 Tropical Tidal Flats in Australia and Central America

Tidal flats can develop where a tidal range of greater than 1 m occurs in combination with an appreciable distance between high and low water line and a prevalence of soft sediments (Veenstra 1976). At the study sites in Australia and Central America compared here, tides were semidiurnal with average tidal ranges of about 2–3 m (Table 13.1). However, tidal ranges can vary from 1 to >9 m along the coast of northern Australia (Lane 1987).

The following sites were used for the comparison (Fig. 13.1). Data on macrobenthos from tidal flats in tropical Australia were available from Wells (1983) for the Bay of Rest in West Australia and from Dittmann (1995, 1996, 2000) for Hinchinbrook Channel and the Haughton River estuary in North Queensland. Further records were available from Alongi (1987, 1988) and Pepping et al. (1999). In Central America, tidal flats in the Estero de Jaltepeque (El Salvador) were surveyed for Polychaeta by Hartmann-Schröder (1959) and her sites were re-sampled by Molina in 1990–1991 (Molina 1992; Molina and Vargas 1994, 1995). In the sheltered location of the Gulf of Nicoya (Costa Rica), Vargas (1987, 1988, 1989, 1996) studied a mudflat at Punta Morales and a further site (Cocorocas) located nearby. Further records were available from foreshore tidal flats in Panama (Kaufman 1976; Lee 1978).

The studied sites differed in their areal extent and comprised different types of tidal flats. While tidal flats on the Pacific coast of Central America often cover a few square kilometers, and are more scattered and isolated coastal habitats, large coherent tidal flats exist in West Australia. The north-

Table 13.1. Environmental conditions in tidal flats of tropical Australia and Central America. The data for Australia are derived from Wells (1983), Alongi (1988) and Dittmann (1995) and for Central America from Hartmann-Schröder (1959), Molina and Vargas (1994, 1995), Vargas (1987, 1996) and Jiménez and Soto (1985)

Parameter	Australia		Central America	
	Bay of Rest, West Australia	North Queensland	Estero de Jaltepeque, El Salvador	Punta Morales, Costa Rica
Climate				
Wet season	Feb–June	Oct–April	May–November	
Annual rainfall (mm)	276	>1000–2000	ca. 1000	ca. 1500
Sediment temperature (°C)	18–47	22–35	20–40	25–41
Tidal range (m)	3	3	2	2.3
Salinity (PSU)	39–48	25–41	14–30	24–36
Sediment				
Median grain size (mm)	0.12	0.12–0.42	–	–
Silt and clay (%)	–	1–27	3–92	21–48
Organic matter (% afdw)	–	2.4	–	2

– = No data available

east of Australia has a more heterogeneous coastline where single tidal flats extend over several square kilometers to hectares. Most of the tidal flats studied in Central America were bordered by mangroves and had an estuarine influence, the Estero de Jaltepeque being a tidal lagoon (Hartmann-Schröder 1959; Molina and Vargas 1994), the Punta Morales mudflat a beach foreshore, and the Cocorocas site an open tidal flat near a river mouth. All Australian sites were located seaward of mangrove forests and were under estuarine influence as well. The Haughton River site was an extensive tidal flat in the mouth of an estuary, while the tidal flats in Hinchinbrook Channel and the Bay of Rest were narrow (several hundred meters) seaward fringes in front of the mangroves.

In spite of these differences in setting, the environmental conditions at the tidal flats compared were similar (Table 13.1) and provided similar soft-sediment habitats for the benthic fauna. In both areas, wet and dry seasons were pronounced, although the west Australian coast receives much less rain than the northeast coast. The sediments ranged from sand to mud, depending on single sampling sites and their location in the intertidal, with soft mud encountered in sheltered locations and in front of mangroves and firm sandflats occurring elsewhere. In tidal flats of North Queensland, the sediment showed

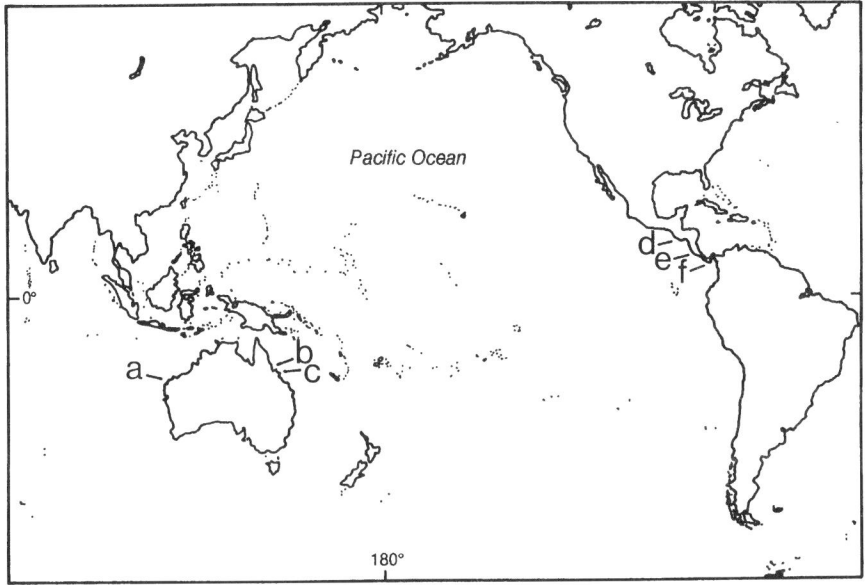

Fig. 13.1. Map of the Indo-Pacific showing the location of the tidal flats compared between Australia (*a–c*) and Central America (*d–f*). *a* Bay of Rest at the northwest Cape of West Australia (Wells 1983), *b* Hinchinbrook Channel (Dittmann 1995 and unpubl.) and *c* Haughton River estuary (Dittmann 1995, 2000) in Queensland, northeast Australia, *d* Estero de Jaltepeque, El Salvador (Hartmann-Schröder 1959; Molina and Vargas 1994, 1995), *e* Punta Morales and Cocorocas, Costa Rica (Vargas 1987, 1988, 1996), *f* foreshore tidal flats in Panama (Kaufman 1976; Lee 1978)

a shoreward fining trend along transects between the high and low water line (Dittmann 2000, unpubl.). The sediment in the Australian tidal flats considered here can be classified as sandy mud to sand (Wells 1983; Dittmann 1995), while muddy sand–sandy mud prevailed in the tidal flats on the Pacific coast of Central America (Vargas 1987; Molina and Vargas 1994).

13.3 Species Diversity and Abundance

13.3.1 Species Richness in Tropical Tidal Flats

The macrofauna species numbers from tropical tidal flats in Australia and Central America confirm a high species diversity for tropical areas, but also show a high degree of variation within as well as between tropical sites on either side of the Pacific (Fig. 13.2). Diversity as measured by the Shannon-Wiener index (H') (base ln) ranged from 1.5–4 within the studied sites. How-

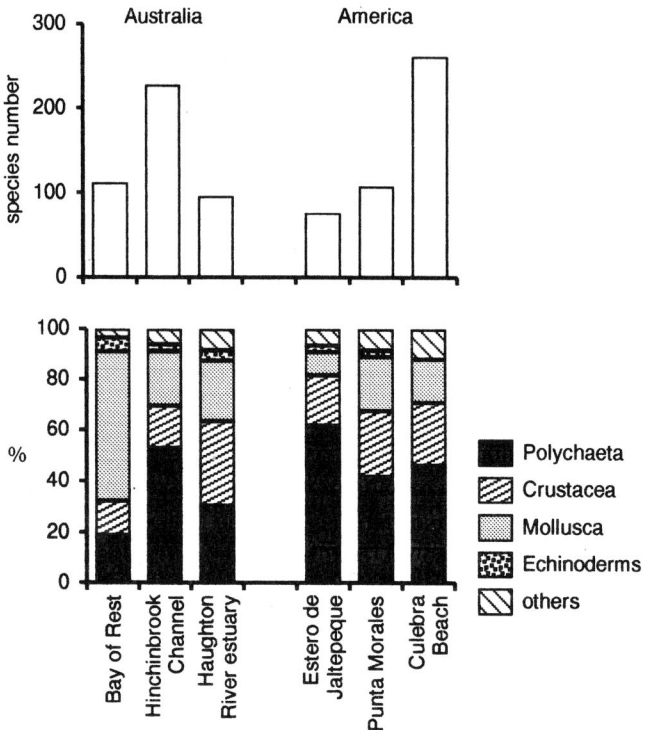

Fig. 13.2. Species numbers and taxonomic composition of macrobenthos from tropical tidal flats in Australia and Central America. Values are taken from Wells (1983) for Bay of Rest, Dittmann (1995) for Hinchinbrook Channel and the Haughton River estuary, Hartmann-Schröder (1959), Molina (1992) and Molina and Vargas (1994) for Estero de Jaltepeque, Vargas (1987, 1988, 1989 and unpubl. records) for Punta Morales, and Kaufman (1976) for Culebra Beach, Panama. "Others" include Oligochaeta, Nemertinea, Brachiopoda, Sipuncula, Echiurida, Hemichordata and Cephalochordata

ever, the species numbers from the tidal flats considered here were not as high as those from tidal flats in East Africa (Macnae and Kalk 1962; Day 1974). For the Haughton River estuary and the Estero de Jaltepeque they were even more comparable to temperate tidal flats (Dankers and Beukema 1983). As species-area curves for the tidal flats in northeast Australia were still increasing at the maximum area sampled (Dittmann 1995), it is expected that the records will increase with further surveys. In Costa Rica, Vargas (1988) found that a higher number of replicate samples gives higher species numbers, as more rare species are recorded.

The majority of the species in tropical tidal flats are represented by single, or a few, individuals only. Dittmann (1995) recorded >50% of the species in Australian tidal flats with only 1–3 individuals while in Costa Rica, 35% of the species accounted for 95 % of the individuals, leaving 2/3 of the species represented with few specimens (Vargas 1987). Such a low species frequency was found in other tropical tidal flats as well (Reise 1991; Wolff et al. 1993a; Pepping et al. 1999). The high number of rare species implies the need for more extensive sampling programs to assess also those species with low population densities. It also poses questions as to the co-existence of the many species and the availability of niches in tropical tidal flats.

13.3.2 Similarity in Taxonomic Compositions

The macrofauna in the tropical tidal flats in Australia and Central America is mainly composed of polychaete, crustacean and mollusc species (Fig. 13.2). Polychaeta account for about half of all species both at the Central American sites and the Hinchinbrook Channel in Australia. At the other Australian sites, molluscs or crustaceans were richer in species.

To compare taxonomic similarity, the taxa in common were counted and the Sørensen index QS (Sørensen 1948) calculated for each taxonomic level (Table 13.2). This comparison gives a first idea on taxonomic similarity at

Table 13.2. Similarity in the taxonomic composition of macrobenthos recorded from several tropical tidal flats in Australia and on the Pacific coast of Central America. Records are based on unpublished species lists of the authors and records by Jones and Morgan (1994), Hartmann-Schröder (1959) and Molina and Vargas (1994). The values are based on the present state of taxonomic descriptions and could change with ongoing taxonomic work and revisions

	Australia	America	Taxa in common	QS[a]
Entire macrobenthos				
Phyla	14	11	11	0.88
Class	17	16	16	0.97
Order	34	29	28	0.89
Family	94	65	53	0.67
Genus	203	135	41	0.24
Species	336	173	0	0
Polychaeta				
Order	14	12	12	0.92
Family	41	32	30	0.82
Genus	103	66	26	0.31
Species	184	86	0	0

[a] QS=Sørensen index ($2 \times C/(A+B)$, where A and B are the taxa in each area and C the number of taxa in common).

various levels, as not all phyla were studied to the same extent and taxonomic work is still continuing. Based on all macrobenthic invertebrates, Table 13.2 shows that dissimilarities start at the family level (QS=0.68) and the two sides of the Pacific had few genera and no species in common. For Polychaeta, the similarity was still very high at the family level, but here, too, no species occurred both in Australia and Central America.

13.3.3 Individual Abundances

Macrobenthic abundances were lower in the Australian than Central American tidal flats, although the maximum values recorded were similar (Fig. 13.3). In general, the variations in abundance were very high both within as well as between sites on either side of the Pacific. This high variability could be the result of patchiness but requires future investigations. Furthermore, abundances were found to vary with season (Vargas 1996), with higher abundances coinciding with the dry season. Long-term records by Vargas (1989) revealed species-specific variations in abundance with moderate changes of some species, while others showed strong oscillations. Thus, much of the variation in abundance recorded in tropical tidal flats cannot be explained until more long-term data are available.

A rank order of abundance was calculated from the macrobenthic surveys in North Queensland and Central America (Table 13.3). Representatives of various taxa were among the most common species, but there was little similarity among the abundant macrobenthic species within or between the Australian and Central American sites. Thus each site had a distinct and different macrobenthic assemblage. This might be explained by the hetero-

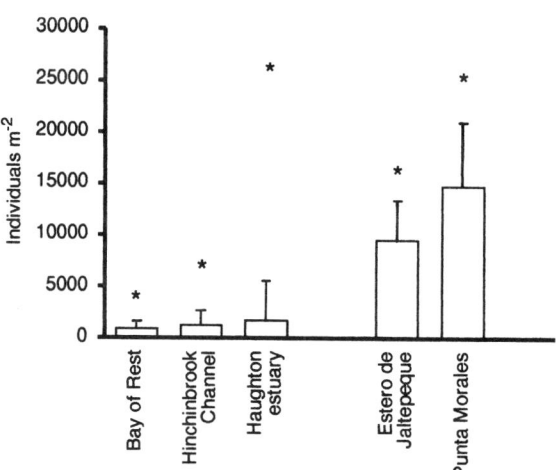

Fig. 13.3. Abundances of macrobenthos (retained on 500-µm mesh size) from tropical tidal flats in Australia (*left columns*) and Central America (*right columns*) (see Fig. 13.1 for details of the sites). Values given are mean abundances with standard deviations. The *asterisks* indicate the highest recorded individual density at each site

Table 13.3. Abundant macrobenthic species (500 µm) in several tropical tidal flats of Australia and Central America. The species are listed in a decreasing rank order of abundance. See Fig. 13.1 for sites and sources. *A* Amphipoda; *B* Bivalvia; *Br* Brachiopoda; *C* Cumacea; *D* Decapoda; *G* Gastropoda; *O* Ostracoda; *Op* Ophiurida; *P* Polychaeta; *S* Sipuncula

Rank order	Site		
Australia	Hinchinbrook Channel	Haughton River Estuary	
1	*Theodoxus oualaniensis* (G, Neritidae)	*Tellina australis* (B, Tellinidae)	
2	*Cerithium* cf. *coralium* (G, Cerithiidae)	Gammaridae indet. (A, Gammaridae)	
3	*Amygdalum glaberrima* (B, Mytilidae)	*Armandia secundariopapillata* (P, Opheliidae)	
4	*Lingula anatina* (Br, Lingulidae)	Haustoriidae indet. (A, Haustoriidae)	
5	*Pitar coxeni* (B, Veneridae)	*Barantolla* sp. (P, Capitellidae)	
Central America	Estero de Jaltepeque	Punta Morales	Cocorocas
1	*Dasybranchus lumbricoides* (P, Capitellidae)	*Coricuma nicoyensis* (C)	*Glottidia audebarti* (Br, Lingulidae)
2	Cumacea indet. (C)	*Mediomastus californiensis* (P, Capitellidae)	*Sipunculus nudus* (S)
3	*Acesta lopezi* (P, Paraonidae)	*Paraprionospio pinnata* (P, Spionidae)	Nereidae sp. 1 (P, Nereidae)
4	*Haploscoloplos elongatus* (P, Orbiniidae)	*Cyprideis pacifica* (O)	*Amphipholis geminata* (Op)
5	*Sigambra ocellata* (P, Pilargiidae)	*Pinnixia valerii* (D, Pinnotheridae)	*Dosinia dunkeri* (G, Veneridae)

geneous nature of the tropical tidal flats that differed in areal extent, sediment gradients and megabenthic fauna (see below). With the complex animal-sediment relations in soft-sediment systems, the varying occurrence of macrobenthic organisms can result from the availability of certain substrate and/or interactions with macro- and megabenthic species providing microhabitats and therefore functioning as ecosystem engineers.

This longitudinal comparison of species diversity and abundances showed a high degree of variation within tropical tidal flats. Species numbers were higher in some, but not all of the studied tropical tidal flats compared with their temperate counterparts, which agrees with a concept by Alongi (1989). Low macrobenthic abundances in tropical tidal flats support a previous comparison by Reise (1991). Species composition and abundances were site-specific and varied as much within the tropics as between latitudes.

13.4 Community Structure and Distribution

13.4.1 Spatial Zonation Along Environmental Gradients

The macro- and megafauna occur in several distinct assemblages throughout tropical tidal flats in Australia (Dittmann 1995). These distribution patterns are related to environmental parameters and biotic interactions and are in agreement with zoned distributions found for benthic fauna in temperate tidal flats (Dittmann 2000). The spatial distribution of life forms and benthic assemblages is less well known from Central America. Therefore, the comparison depicted in Fig. 13.4 remains preliminary until further tropical tidal flats are surveyed with the same sampling design. So far there are indications of some similarities in the spatial distributions of life forms, e.g. the occurrence of surface-deposit-feeding tube-worms in the mid-intertidal muddy sandflats and sand dollars and hermit crabs in the lower intertidal sandflats. Brachiopods occur in the mid-intertidal: *Glottidia audebarti* at Cocorocas, Costa Rica, and *Lingula anatina* in Hinchinbrook Channel, Australia. Furthermore, there are site-specific variations in habitat ranges of benthic organisms that partly result from varying habitat diversities in each tidal flat. Macnae and Kalk (1962) have already pointed out that the predominance of benthic associations is related to the extent of their habitat.

While no temporal variation in spatial zonations and community composition was found by Dittmann (2000) in an Australian tidal flat, Vargas (1988, 1989) recorded within-site differences in community composition over time. These were mainly due to the varying abundances of the single species described above.

13.4.2 Trophic Groups

In most tidal flats studied in Central America and Australia, about two-thirds of the macrobenthic organisms were deposit feeders (Fig. 13.5). A distinction between surface- and subsurface-deposit-feeders showed that surface-deposit-feeders were more abundant in the Haughton estuary and the

Tropical Tidal Flat Benthos Compared Between Australia an Central America

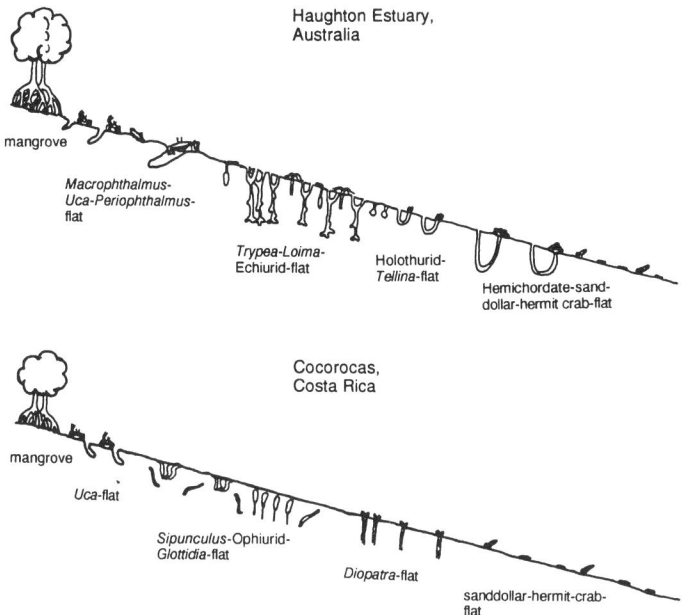

Fig. 13.4. Spatial zonation in two tropical tidal flats from Australia and Central America showing the distribution of prominent macrobenthic organisms from the high towards the low tide level

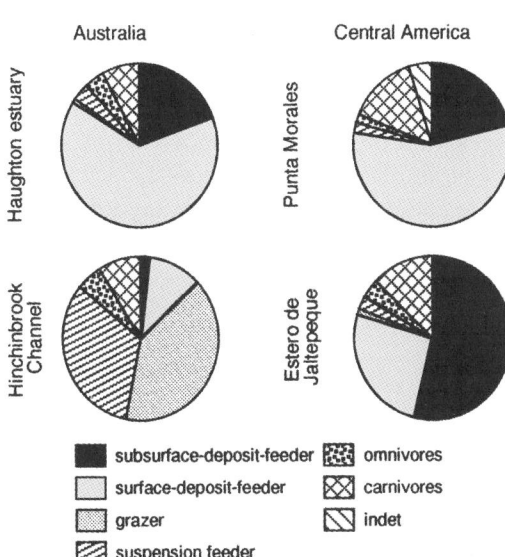

Fig. 13.5. Percentage composition of trophic groups (% of individuals) from tropical tidal flats in Australia and Central America (see Fig. 13.1 for details of the sites). With limited knowledge of the biology of most tropical benthic organisms, some classifications are tentative. Please note that the food items of surface deposit feeders and grazers overlap

mudflat at Punta Morales (Vargas 1987; Dittmann 1995, 2000). Snails, grazing on the microphytobenthos at the sediment surface, dominated in the tidal flats in the Hinchinbrook Channel (Fig. 13.5). Suspension-feeders were scarce at most studied sites. However, suspension-feeding bivalves and brachiopods were abundant in the Hinchinbrook Channel (Fig. 13.5) and Cocorocas, and bivalves accounted for most of the biomass in tropical tidal flats on the west coast of Australia (Wells 1983; Pepping et al. 1999).

The dominance of deposit-feeders appears to be a general characteristic of tropical tidal flats in Australia and Central America. Deposit-feeders also dominated in tidal flats of the Banc d'Arguin in west Africa (Wolff et al. 1993b) and in temperate tidal flats (Sanders et al. 1962; Whitlatch 1977); therefore, no latitudinal variation could be detected for the composition of trophic types (Dittmann 2000). The abundance of deposit-feeders in the tropical tidal flats of Australia and Central America is probably sustained by detritus derived from mangrove litter and the microphytobenthic layer.

13.4.3 Species Interactions

To assess whether related or unrelated species take similar ecological roles within tropical tidal flats, we followed the conceptual hierarchy on ecological processes of Reise (1985) and looked at the respective species in the tropical tidal flats of Australia and Central America (Table 13.4). This table shows that on either side of the Pacific many related species (genus or family level) occur with similar ecological roles, while not many unrelated species with similar roles were listed. Tropical tidal flats provide similar niches and, following a diversification on either side of the Pacific, macrobenthic species take comparable ecological roles. In a few cases, no counterpart for a specific role has been found, indicating some singularities of ecological functions in specific areas. The structural and functional similarity seen between tropical tidal flats is therefore to a large extent a result of related species occupying similar niches and only to a smaller degree due to unrelated species taking analogous ecological roles (Vargas 1987; Reise 1991).

13.4.3.1 Promotive Interactions

A prominent example of promotive interactions in tidal flats is the provision of microhabitats in the burrows of larger macrofauna (Reise 1985). An example for this type of accommodation on associated infauna was reported by Dittmann (1996) for burrows of callianassid shrimps in tidal flats of northeast Australia. These shrimps are abundant in the muddy sandflat of the mid-intertidal and their promotive effect is a major process structuring the

Table 13.4. Ecological roles and prominent macrobenthic species taking these roles in tropical tidal flats in Australia and on the Pacific coast of Central America. *SL* Similarity level: *I* related species (on genus or family level) with similar ecological roles; *II* unrelated species (on genus or family level, or species with uncertain taxonomic classification) with similar ecological roles; *III* no (known) counterpart for a species-specific ecological role. *A* Amphipoda; *Br* Brachiopoda; *Ch* Chondrichtyes; *D* Decapoda; *E* Echinodermata; *Ec* Echiura; *G* Gastropoda; *H* Hemichordata; *P* Polychaeta; *S* Sipuncula

Ecological role	SL	Australia	Central America
Bioturbators/sediment reworking/ Biogenic structures: Burrowers	I	*Uca* spp. (D, Ocypodidae) *Lingula anatina* (Br, Lingulidae) *Mediomastus* sp., *Notomastus* sp., *Barantolla* sp. (P, Capitellidae) *Lumbrineris* spp. (P, Lumbrineridae)	*Uca* spp. (D, Ocypodidae) *Glottidia audebarti* (Br, Lingulidae) *Mediomastus* spp. (P, Capitellidae) *Lumbrineris* spp. (P, Lumbrineridae)
	II	*Trypea australiensis* (D, Callianassidae) Enteropneust indet. (H, Enteropneusta) Echiuride indet. (Ec)	*Alpheus* sp. (D, Alpheidae) Enteropneust indet. (H, Enteropneusta) *Thalassema steinbecki* (Ec)
	III	*Paracaudina* sp. (E, Holothuroidea) *Macrophthalmus latreillei* (D, Ocypodidae)	*Sipunculus nudus* (S)
Biogenic structures: tube-builders	I	*Prionospio* spp. (P, Spionidae) *Owenia* sp. (P, Oweniidae)	*Prionospio* spp., *Paraprionospio pinnata*, *Carazziella citrona* (P, Spionidae) *Owenia* spp. (P, Oweniidae)
	II	*Loimia* cf. *medusa* (P, Terebellidae)	*Diopatra ornata* (P, Onuphidae)
	III		*Corophium* sp. (A)
Small-scale sediment disturbance	I	*Polinices* spp., *Natica* spp. (G, Naticidae) *Nassarius* spp. (G, Buccinidae) *Uca* spp. (D, Ocypodidae)	*Natica* spp. (G, Naticidae) *Nassarius* spp. (G, Buccinidae) *Uca* spp. (D, Ocypodidae)
	II	*Arachnoides placenta* (E, Echinoidea) stingrays (Ch, Elasmobranchii)	*Encope stokesi* (E, Echinoidea) stingrays (Ch, Elasmobranchii)
	III	*Mictyris longicarpus* (D, Mictyridae) *Scopimera inflata* (D, Ocypodidae)	

infaunal community at this site (Dittmannn 1998, 2000). Therefore the ecological role of *Trypea australiensis* equals the one of the lugworm *Arenicola marina* in temperate tidal flats. Promotive effects were also recorded for *Callianassa japonica* in southern Japan (Tamaki and Suzukawa 1991). Callianassid shrimps are documented from intertidal habitats of the Pacific coast of Colombia and from the Gulf of Mexico, where polyclad worms, pinnotherid crabs, caridean shrimps and alpheid shrimps were associated with the burrows (Felder et al. 1993; Manning and Felder 1995; Felder and Manning 1997, 1998). In Costa Rica, alpheid shrimps build similar burrows to *T. australiensis*, but as yet they have not been investigated for associated fauna, nor have the burrows of fiddler crabs (*Uca* spp.).

Burrows of other crustaceans and of brachiopods were also seen to accommodate higher infaunal numbers, but this effect varied both with burrow type and season (Dittmann 1996). Tube-worms are also known to affect infaunal distributions (Zühlke et al. 1998) and, in the subtidal of the Gulf of Nicoya, Maurer and Vargas (1984) found a higher diversity of benthic fauna around tubes of *Spiochaetopterus costarum*.

Although promotive interactions have only been studied in a few tropical tidal flats of Australia so far, it is suspected that larger macrobenthic organisms exert comparable promotive effects in the tropics as in temperate latitudes.

13.4.3.2 Repressive Interactions

Predation is seen as a prevailing process regulating the benthic fauna in tidal flat communities in temperate latitudes (Reise 1985). Several predatory species (e.g. crabs, nemerteans, snails, birds, fish and juvenile prawns) occur in tropical tidal flats and effects on macro- and meiobenthic prey have been reported (Dittmann 1993; Christy et al. 1998). The experimental exclusion of soldier crabs resulted in a fivefold increase in meiofauna numbers in a tropical tidal flat of northeast Australia (Dittmann 1993). Using predator exclusion cages in a mudflat near Punta Morales, Vargas (1988) concluded that macro-predators (fish, birds, crabs and snails) have only a negligible effect on benthic community structure in this tropical tidal flat. He found almost no changes in abundance in the exclusion cages, but did see changes in the rank order of abundance of dominant species.

At present, the findings on the relevance of predation in tropical benthic communities are somewhat contradictory. There is little evidence for a regulating effect of epibenthic predators on infaunal assemblages in the tropics. However, it cannot be excluded that multiple predatory effects (e.g. Ambrose 1984) exist in tropical tidal flats, as the predatory effects of endobenthic predators have not yet been addressed. The report by Christy et al. (1998) of a nemertean preying on fiddler crabs is an indication for further predatory effects in tropical sediments. Furthermore, macrobenthic predators and deposit feeders can affect bivalve recruitment. Vargas (1988) found higher recruitment rates within the exclusion cages and Dittmann (1996) found higher recruitment of bivalves in shrimp exclusions, similar to the competitive exclusion reported by Peterson (1977). Predatory snails could also account for the low bivalve numbers in many tropical tidal flats (Lee 1978; own observations).

Competition may have significant effects on community structure in tidal flats, although this process is generally less important in soft sediments compared to rocky shores (Reise 1985; Peterson 1991). In tropical tidal flats, competition has not yet been investigated, but observations in northeast Australia showed intraspecific density regulation in terebellid polychaetes and echiurids. There also seem to be adult-juvenile interactions in *M. longicarpus*, where juveniles gathered in separate groups from adult soldier crabs, and in *T. australiensis*, where juveniles were found in different areas of the tidal flat. Tamaki and Ingole (1993), however, found intraspecific facilitation between juvenile and adult *Callianassa japonica* in tidal flats in Japan. Interspecific competition could explain the absence of tube-worms in colonies of fiddler crabs or the inhibition of feeding activities of *T. australiensis* when soldier crabs were active on the sediment surface. All these cases require further investigations.

13.5 Conclusions

This comparison of tropical tidal flat benthos between Australia and Central America has revealed a high degree of faunal similarity despite the different biogeographic areas. Species numbers were high, although not as high as expected for tropical sites. Furthermore, comparing several tidal flats in each area showed a high degree of variation in species diversity. Few species were abundant and abundances were generally low and very variable with higher abundances recorded in tidal flats from the Pacific coast of Central America than Australian sites. In all tidal flats compared here, polychaetes, crustaceans and molluscs accounted for most of the species. The rank order of abundance revealed site-specific species compositions and dominance structures of macrobenthic organisms. Large-scale spatial distributions were related to environmental gradients in Australian tidal flats and indications exist for a similar zonation of life forms in a tidal flat in Costa Rica. Deposit feeders dominated the benthic fauna numerically, while suspension feeders were rare. Related species often had similar ecological roles and only few cases were identified where an ecological counterpart did not exist in both Central

American and Australian tidal flat communities. Examples for promotive and repressive interactions were found in the tidal flats on either side of the Pacific, although further research is required to assess the relevance of predation and competition in these areas.

The great variety of benthic assemblages in tropical tidal flats within and between regions illustrated by this comparison can result from site-specific variations in habitat diversity as well as the presence or absence of certain "ecosystem engineers". Larger macrobenthic organisms accommodate other infauna and biotic interactions are probably more important for benthic community structures in the tropics than is often assumed (Alongi 1987, 1990).

Among the generalities of tropical tidal flat benthos are similarities in life forms of benthic fauna and similar ecological roles taken by related species. However, future studies are needed to assess the full range of structural and functional similarities or dissimilarities in benthic assemblages of tropical tidal flats before more generalizations can be drawn. Singularities include site-specific variations in species diversity and abundance. In the Indo-West-Pacific, some specific ecological roles exist (e.g. small-scale sediment disturbance and predation by soldier crabs; Dittmann 1993) which have no equivalents in the east Pacific. Verifying the paradigms derived from studies in temperate tidal flats, this comparison has revealed a high degree of similarity of ecological processes at all latitudes.

Acknowledgements. We thank H. Dean and W. Wolff for their helpful comments on the manuscript. The studies of the first author in northeast Australia were supported by the German Science Foundation (DFG III 02-Di 396/1-2) and carried out at the Australian Institute of Marine Science.

References

Alongi DM (1987) Intertidal zonation and seasonality of meiobenthos in tropical mangrove estuaries. Mar Biol 95:447–458
Alongi DM (1988) Microbial-meiofaunal interrelationships in some tropical intertidal sediments. J Mar Res 46:349–365
Alongi DM (1989) Ecology of tropical soft-bottom benthos: a review with emphasis on emerging concepts. Rev Biol Trop 37:85–100
Alongi DM (1990) The ecology of tropical soft-bottom benthic ecosystems. Oceanogr Mar Biol Annu Rev 28:381–496
Ambrose WG (1984) Role of predatory infauna in structuring soft-bottom communities. Mar Ecol Prog Ser 17:109–115
Beukema JJ (1976) Biomass and species richness of the macro-benthic animals living on the tidal flats of the Dutch Wadden Sea. Neth J Sea Res 10:236–261
Christy JH, Goshima S, Backwell PRY, Kreuter TJ (1998) Nemertean predation on the tropical fiddler crab *Uca musica*. Hydrobiologia 365:233–239

Dankers N, Beukema JJ (1983) Distributional patterns of macrozoobenthic species in relation to some environmental factors. In: Wolff WJ (ed) Ecology of the Wadden Sea. Balkema 1:69–103

Day JH (1974) The ecology of Morrumbene estuary, Moçambique. Trans R Soc S Afr 41:43–97

Dittmann S (1993) Impact of foraging soldier crabs (Decapoda: Mictyridae) on meiofauna in a tropical tidal flat. Rev Biol Trop 41:627–637

Dittmann S (1995) Benthos structure on tropical tidal flats of Australia. Helgoländer Meeresunters 49:539–551

Dittmann S (1996) Effects of macrobenthic burrows on infaunal communities in tropical tidal flats. Mar Ecol Prog Ser 134:119–130

Dittmann S (1998) Spatial and temporal patterns of platyhelminth assemblages in intertidal sediments of northeast Australia. Hydrobiologia 383:41–47

Dittmann S (2000) Zonation of benthic communities in a tropical tidal flat of northeast Australia. J Sea Res 43:33–51

Fauchald K, Jumars PA (1979) The diet of worms: a study of polychaete feeding guilds. Oceanogr Mar Biol Annu Rev 17:193–284

Felder DL, Manning RB (1997) Ghost shrimps of the genus *Lepidophthalmus* from the Caribbean region, with description of *L. richardi*, new species, from Belize (Decapoda: Thalassinidea: Callianassidae). J Crust Biol 17:309–331

Felder DL, Manning RB (1998) A new ghost shrimp of the genus *Lepidophthalmus* from the Pacific coast of Colombia (Decapoda: Thalassinidea: Callianassidae). Proc Biol Soc Wash 111:398–408

Felder DL, Rodrigues SA (1993) Reexamination of the ghost shrimp *Lepidophthalmus louisianensis* (Schmitt, 1935) from the northern Gulf of Mexico in comparison to *L. siriboia*, new species, from Brazil (Decapoda: Thalassinidea: Callianassidae). J Crust Biol 13(2):357–376

Gätje C, Reise K (1998) Ökosystem Wattenmeer. Austausch-, Transport- und Stoffumwandlungsprozesse. Springer, Berlin Heidelberg New York

Hartmann-Schröder G (1959) Zur Ökologie der Polychaeten des Mangroven-Estero-Gebietes von El Salvador. Beitr Neotrop Fauna 1(2):69–183

Jiménez JA, Soto RS (1985) Patrones regionales en la estructura y composición florística de los manglares de la Costa Pacífica de Costa Rica. Rev Biol Trop 33:25–37

Jones DS, Morgan GJ (1994) A field guide to crustaceans of Australian waters. Reed, Chatswood

Kaufman H (1976) A quantitative investigation of the fauna of two sandy beaches of Pacific Panama. MSc Thesis, George Washington Univ

Lane B (1987) Shorebirds in Australia. Nelson Publ, Melbourne

Lee H (1978) Seasonality, predation and opportunism in high diversity soft-bottom communities in the Gulf of Panama. PhD Thesis, Univ North Carolina, Chapel Hill

Lessios HA, Kessing BD, Robertson DR (1998) Massive gene flow across the world's most potent marine biogeographic barrier. Proc R Soc Lond B 265:583–588

Macnae W, Kalk M (1962) The fauna and flora of sand flats at Inhaca Island, Moçambique. J Anim Ecol 31:93–128

Manning RB, Felder DL (1995) Description of the ghost shrimp *Sergio mericeae*, a new species from south Florida, with reexamination of *S. guassutinga* (Crustacea: Decapoda: Callianassidae). Proc Biol Soc Wash 108:266–280

Mathieson AC, Nienhuis PH (eds) (1991) Intertidal and littoral ecosystems. Ecosystems of the world, vol 24. Elsevier, Amsterdam

Maurer D, Vargas JA (1984) Diversity of soft-bottom benthos in a tropical estuary: Gulf of Nicoya, Costa Rica. Mar Biol 81:97–106

Molina OA (1992) Estructura del macrobentos en el manglar del Estero de Jaltepeque, El Salvador. Tesis, Magister Scientiae, Univ Costa Rica, 84 pp

Molina OA, Vargas JA (1994) Estructura del macrobentos del estero de Jaltepeque, El Salvador. Rev Biol Trop 42:165–174

Molina OA, Vargas JA (1995) Poliquetos (Annelida: Polychaeta) del estero de Jaltepeque, El Salvador, una comparación 1959–1991. Rev Biol Trop 43:195–205

Pepping M, Piersma T, Pearson G, Lavaleye M (1999) Intertidal sediments and benthic animals of Roebuck Bay, Western Australia. NIOZ Report 1999–1993. Netherlands Institute for Sea Research, Den Burg, The Netherlands

Peterson CH (1977) Competitive organization of the soft-bottom macrobenthic communities of southern Californian lagoons. Mar Biol 43:343–359

Peterson CH (1991) Intertidal zonation of marine invertebrates in sand and mud. Am Sci 79:236–249

Reise K (1985) Tidal flat ecology. Springer, Berlin Heidelberg New York

Reise K (1991) Macrofauna in mud and sand of tropical and temperate tidal flats. In: Elliott M, Ducrotoy J-P (eds) Estuaries and coasts: spatial and temporal intercomparisons. Olsen and Olsen, Fredensborg, pp 211–216

Sanders HL (1968) Marine benthic diversity: a comparative study. Am Nat 102:243–282

Sanders HL, Goudsmit EM, Mills EL, Hampson GE (1962) A study of the intertidal fauna of Barnstable Harbor, Massachusetts. Limnol Oceanogr 7:63–79

Snelgrove PVR, Butman CA (1994) Animal-sediment relationships revisited: cause versus effect. Oceanogr Mar Biol Annu Rev 32:111–177

Sørensen TA (1948) A method to establish groups of equal amplitude in plant sociology based on similarity of species content, and its application to analyses of the vegetation on Danish commons. Biol Skr 5:1–34

Tamaki A, Ingole B (1993) Distribution of juvenile and adult ghost shrimps, *Callianassa japonia* Ortmann (Thalassinidea), on an intertidal sand flat: intraspecific facilitation as a possible pattern-generating factor. J Crust Biol 13:175–183

Tamaki A, Suzukawa K (1991) Co-occurrence of the circolanid isopod *Eurydice nipponica* Bruce & Jones and the ghost shrimp *Callianassa japonica* Orthmann on an intertidal sand flat. Ecol Res 6:87–100

Thorson G (1957) Bottom communities (sublittoral or shallow shelf). Geol Soc Am 67:461–534

Vargas JA (1987) The benthic community of an intertidal mud flat in the Golf of Nicoya, Costa Rica. Description of the community. Rev Biol Trop 35:299–316

Vargas JA (1988) Community structure of macrobenthos and the results of macropredator exclusion on a tropical intertidal mud flat. Rev Biol Trop 36:287–308

Vargas JA (1989) Seasonal abundance of *Coricuma nicoyensis* Watling and Breedy, 1988 (Crustacea: Cumacea) on a tropical mud flat. Rev Biol Trop 31:207–211

Vargas JA (1996) Ecological dynamics of a tropical intertidal mudflat community. In: Nordstrom KF, Roman CT (eds) Estuarine shores: evolution, environments and human alterations. Wiley, Chichester, pp 355–371

Veenstra H (1976) Struktur und Dynamik des Gezeitenraumes. In: Abrahamse J, Joenje W, van Leeuwen-Seelt N (eds) Wattenmeer. Wachholtz Verlag, Neumünster

Veron JEN (1995) Corals in space and time. the biogeography and evolution of the Scleractinia. UNSW Press, Sydney

Warwick RM, Ruswahyuni (1987) Comparative study of the structure of some tropical and temperate marine soft-bottom macrobenthic communities. Mar Biol 95:641–649

Wells FE (1983) An analysis of marine invertebrate distributions in a mangrove swamp in northwestern Australia. Bull Mar Sci 33:736–744

Whitlatch RB (1977) Seasonal changes in the community structure of the macrobenthos inhabiting the intertidal sand and mud flats of Barnstable Harbor, Massachusetts. Biol Bull 152:275–294

Wolff WJ (ed) (1983) Ecology of the Wadden Sea, vols 1–3. Balkema, Rotterdam

Wolff WJ, Duiven AG, Duiven P, Esselink P, Gueye, A, Meijboom A, Moerland G, Zegers J (1993a) Biomass of macrobenthic tidal flat fauna of the Banc d'Arguin, Mauritania. Hydrobiologia 258:151–163

Wolff WJ, van der Land J, Nienhuis PH, de Wilde PAWJ (1993b) The functioning of the ecosystem of the Banc d'Arguin, Mauritania: a review. Hydrobiologia 258:211–222

Zühlke R, Blome D, van Bernem KH, Dittmann S (1998) Effects of the tube-building polychaete *Lanice conchilega* (Pallas) on benthic macrofauna and nematodes in an intertidal sandflat. Senckenbergiana Marit 29:131–138

Part IV

Structural Dynamics and Trophic Supplies to Sedimentary Shores

14 Recovery Dynamics in Benthic Communities: Balancing Detail with Simplification

S.F. THRUSH and R.B. WHITLATCH

14.1 Introduction

Attempting to match pattern with dynamic processes has a long history in ecology (Kitching 1937; Watt 1947). For marine soft-sediment macrobenthic communities, the disturbance mosaic model (Johnson 1970, 1973) has provided a dynamic framework in which to describe patterns of spatial heterogeneity and biodiversity. The model describes the role of local disturbance events in producing patches containing benthic assemblages with different compositions at different successional stages. The perspective provided by this conceptual model emphasises that benthic communities are complex and dynamic. Spatial heterogeneity created by local disturbance events can account for resource patchiness (Thistle 1981; Van Blaricom 1982), communities with mixed trophic structure (Probert 1984) and ubiquity of opportunistic species. Thus, local disturbance events frequently play a central role in influencing the structure and function of benthic communities.

Field experiments that involve monitoring macrobenthic recolonisation in previously defaunated sediments have provided insights into the nature of benthic succession and the relative importance of various biotic and abiotic factors affecting the recovery process. These experimental studies are often interpreted and generalised within the patch-dynamic conceptual framework developed by Johnson (1970, 1973). This information is also often used to help understand or predict the ecological consequences of much larger scale disturbance events. Practical and ethical considerations limit the scales over which field experiments are feasible (Kareiva and Andersen 1988; Schneider 1994; Thrush et al. 1996a, 1997a, 2000). Since experimentally defaunated sediment patches used in experiments are typically restricted to spatial scales of centimetres to metres, we must be cautious about scaling-up.

Scaling-up is not as straightforward as directly extrapolating from results at smaller scales. Non-linear processes are most likely to be important in organising the shift from one range of scales to another. Not only can broad-scale large and slow processes control smaller scale events but also nature can surprise us and the latter can affect the former (Holling 1996). However, despite the scale dependence of their results, field experiments are still powerful tools in developing a mechanistic understanding and revealing the importance of life and natural history characteristics of component species. For example, general predictions of the consequences of habitat disturbance by commercial fishing, based on the results of field experiments and natural history characteristics (e.g., changes in biodiversity, density of epifauna and large and long-lived species), have been tested and largely validated by broad-scale surveys (Thrush et al. 1998). The power of inference drawn from such broad-scale surveys is greatly increased by the ability to make *a priori* predictions. Thus, while direct scaling-up may be difficult, useful information can be generated and used in an iterative process of prediction and testing to assess broad-scale effects. Elsewhere we have discussed various strategies that are available to improve our ability to scale-up experimental results (Thrush et al. 1997b, 1999, 2000).

The opportunity of working in different locations is always exciting and most field ecologists are quick to identify both the differences and similarities in ecological patterns and process. Thus comparative ecology is an intuitively useful way of discerning generality. But how do we move from intuition to quantification and theory? In this chapter we have intentionally focused on disturbance-recovery experiments that have been conducted with our colleagues in New Zealand or New England (USA). We also attempt to develop a conceptual framework to generalise across localities to measure the dynamic response of benthic communities to disturbance and thus assess their resilience. In order to generalise using experimental results some knowledge of how recovery processes vary with location or with the size of the area disturbed is vital. Both issues need to be addressed if we are to make benthic ecology more predictive. These issues are particularly important when addressing human impacts on marine benthos because we are often concerned with extensive, broad-scale and chronic changes. Usually we do not have pre-disturbance data, adequate knowledge of site history or availability of adequate control sites. Potentially, broad-scale degradation of benthic ecosystem health requires the development of appropriate empirical tools to identify broad-scale changes in recovery. If we can develop a way to generalise across localities it may be possible to measure the dynamic response of benthic communities to disturbance and thus assess their resilience to natural and anthropogenic disturbance.

14.2 Searching for Generality Part I

In attempting to make comparisons between studies conducted in different locations it is important to identify some common and meaningful processes or measures. Meta-analysis provides some statistical tools for the development of quantitative syntheses of separate studies (e.g., Gurevitch and Hedges 1993, 1999; Arnqvist and Wooster 1995; Englund et al. 1999; Osenberg et al. 1999). However, sophisticated meta-analysis will be meaningless unless the variables used to compare studies are ecologically meaningful.

Even when only a qualitative synthesis is possible, we must be aware of potential confounding variables when making comparisons. Disturbance is a very nebulous term (see Pickett and White 1985 for definitions) and disturbance events can occur at various intensities, frequencies and extents, all of which are likely to have important consequences for the subsequent ecological recovery (Zajac et al. 1998). Potential problems can also arise with operational definitions of succession/recovery end points for a defaunated patch. Birth and mortality rates of the benthos within the patch plus their flux across the patch boundary (i.e., immigration and emigration) determine the path of recovery. Also important are the broader scale temporal changes in the benthic community of the surrounding undisturbed sediments, because these provide the pool of potential colonists for patch recovery. We must also assess recovery by comparison with community structure in the adjacent sediments rather than against some theoretical baseline. Confidence intervals for density estimates in and out of experimental plots may overlap extensively during recovery (indicating no significant difference between experimental plots and ambient sediments) but subsequently once again separate. Assuming an adequate sample size is used, this difference can result from recruitment events; changes in spatial variance, (particularly when densities are low); or the merging of experimental patterns with broader scale temporal patterns (such as a seasonal decline in population density). Thus, it is important that similarities in density and assemblage structure in experimental plots and ambient sediments persist over time before recovery is assumed. Unfortunately, this is frequently not the case given the limited time over which many experiments are sampled.

The "classical" succession models for marine soft-sediment macrobenthic communities were developed by Pearson and Rosenberg (1978) and Rhoads et al. (1978). These conceptual models are very similar, although the former was focused on recovery processes in space and the latter on time. Both conceptual models were developed using a combination of broad-scale survey work and detailed smaller scale experimentation and observation. These models have become the cornerstones of the assessment of environmental impacts on macrobenthic communities and the framework against which disturbance/recovery dynamics are frequently assessed. Essentially, close to the disturbance, small, fast growing and rapidly colonising opportunistic species reach high densities. Moving further from the disturbance a transition zone occurs which is still dominated by opportunistic species although not at such high densities. Some larger and more mobile species are also found. At the end of the successional trajectory, a diverse assemblage dominated by large and slow growing deeper burrowing organisms is found. Important over this trajectory is habitat modification by species that influence the success of other colonists (e.g., Rhoads 1974). When comparing locations it is important to remember that Pearson and Rosenberg (1978) recognised both the contribution of adult life-stages as early colonisers and the hydrodynamic regime in controlling the recovery processes. Their later research also emphasised the importance of food supply in affecting broad-scale variations in benthic successional processes (Pearson and Rosenberg 1987). These factors can account for many of the exceptions to the predictions of these classical successional models.

In order to find generality between different locations and/or times we seek to identify processes that might influence broad-scale differences (i.e. those that operate over large scales) and use these to explain the variability between studies. Hierarchy theory is often advocated as an approach to help deal with scale (Allen and Starr 1982, but see Schneider 1994). A hierarchical approach to understanding the relative importance of environmental processes, life history and species interactions was developed by Zajac and Whitlatch (1985). Although this model was not originally designed to be spatially explicit, more recent work has refined this framework to better understand how recovery processes may vary with increasing scale of disturbance (Zajac et al. 1998). The hierarchy of Zajac and Whitlatch (1985) dictates the importance of environmental factors in broad-scale comparisons: with increasing spatial extent we expect to encompass a greater range of environmental conditions. While this type of conceptualisation warrants further investigation, hierarchical approaches can be problematic in that it is often difficult to isolate processes to individual scales (Thrush 1991). Furthermore, interactions between processes classified at different levels in the hierarchy can be important determinants of the recovery processes.

14.3 Some General Mechanisms Influencing Recovery

14.3.1 Seasonality

Many environmental factors potentially important to the recovery process vary with season. For example, water temperature may influence the seasonal availability of larval and juvenile recruits. The strength of temperature effects may be expected to vary with latitude and climate (stronger seasonal variation in air and coastal water temperatures are found on continental land masses rather than islands). In the wet tropics, strong seasonal effects can be generated by seasonal changes in freshwater inputs (Alongi 1990). Zajac and Whitlatch (1982a,b) demonstrated that rates of recovery increased in experiments conducted in spring and summer compared to those conducted at other times of the year. Similarly, Ford et al. (1999) demonstrated rates of recolonisation were slow in winter compared to summer and that seasonal differences in recovery rate were linked to the density of the dominant taxa that were typically more abundant in the summer. For long-lived species, Zajac and Whitlatch (1989) demonstrated that the demographic state of the population at the time of disturbance had an important influence on its recovery. Other factors, such as benthic primary production or wave climate, may also influence recovery dynamics and vary with season or other long-term cyclical atmospheric/oceanographic processes (e.g., El Niño Southern Oscillation).

14.3.2 Hydrodynamics

Hydrodynamic conditions can have a profound influence on benthic recovery processes (e.g., Eckman 1983; Jumars and Nowell 1984; Butman 1987; Hall 1994; Paterson and Black 1999). Water depth and flow characteristics influence benthic food quality and quantity and the supply of colonists to disturbed patches. Water depth will also influence the type of waves that expend energy on the seabed. On sand flats, small wind-waves that generate 1–2 cm high sand ripples can be particularly important in resuspending animals and sediments so that tidal currents can transport them (Bell et al. 1997). Hydrodynamic processes associated with the mobility of macrofauna have a major influence on the scale of disturbance needed to detect differences in the relative importance of adult, juvenile and larval colonists. Even in quiescent mudflat environments tidal flows may be sufficient to transport post-settlement macrofauna to defaunated plots (Thrush and Roper 1988). Whitlatch et al. (in press) reviewed shallow-water experimental defaunation experiments and found no clear evidence for differences in the rate of recovery dependant on habitat factors (deep/shallow and sand/mud) that are likely to reflect hydrodynamic differences. While this finding may seem counter-intuitive, it probably reflects problems with comparing experimental studies that have used different disturbance techniques and have been conducted in habitats possessing different physical and biological attributes.

14.3.3 Mobility

The presence of adults and post-larval stages has been recognised in many disturbance studies (Pearson and Rosenberg 1978; Santos and Simon 1980; Thrush 1986; Smith and Brumsickle 1989), but the relative importance of these life stages to recovery processes has only recently been acknowledged. Modes of colonisation vary between species, within species associated with different life stages, and with environmental conditions (Zajac 1991a,b; Commito et al. 1995; Shull 1997). Gunther (1992) provided a conceptual model of the relative importance of different life stages in recolonisation dependent on the size of the area disturbed. Due to variation in the dispersal and mobility of different life stages, the model predicts dispersal of adults will occur over smaller spatial scales than post-larvae, which, in turn, will be restricted to smaller scales than larvae. Whitlatch et al. (1998, in press) used demographic models to investigate the impact of colonisation by different life stages of the spionid polychaete *Polydora cornuta* on the rate of population recovery in a disturbed patch in cyberspace. The model revealed that life-stage effects on patch recovery time occurred in the following order of importance: larvae>juveniles>adults. However, the life stages did interact: for example, when larval recruitment was at its maximum, colonisation by juveniles still improved patch recovery time. Essentially, the model functions to balance density of colonists with their potential to contribute to reproduction within the disturbed patch. Some support for a scale- and life stage-dependent model of recolonisation has also been presented by Whitlatch et al. (1998) for another spionid polychaete, *Boccardia syrtis*. However, different patterns of colonisation of defaunated patches by *Boccardia* of different sizes did not persist throughout the recovery process, further highlighting the role of hydrodynamics in affecting the recovery process (Thrush et al. 1996a).

14.3.4 Opportunistic Responses

The classical opportunistic response is defined as short-term high abundance of colonists immediately following a disturbance. The absence of this type of response in many experimental studies has been noted (Zajac and Whitlatch 1991; Whitlatch et al., in press). Zajac and Whitlatch (1991) illustrated how opportunistic "boom and bust" abundance could be influenced by demographics. Some species that do achieve high densities in disturbed patches are highly mobile and quite effective at utilising patchy/ephemeral resources (e.g., Van Blaricom 1982; Oliver and Slattery 1985). These types of species operate over broader scales than individual patches and their demographics are more influenced by the landscape of patches rather than tied to events within a single patch. This type of response is often associated with subtidal pits created by feeding predators. A number of studies of recolonisation of intertidal pits have failed to indicate any resource exploitation; rather recolonisation occurred through the passive movement of animals from the surrounding sediments (Levin 1984; Savidge and Taghon 1988; Thrush et al. 1991). This intertidal pattern is probably due to tidal currents and wind-waves mobilising surficial sediments and quickly transporting sediment and organisms into pits. Demonstration of opportunistic responses requires opportunists that are present in the ambient species pool, as well as a high level of resources within the disturbed area that are not overwhelmed by fluxes across the patch boundary.

14.3.5 Biotic Interactions

The role of early colonists ameliorating sediment biogeochemical conditions for subsequent species is important in the classic succession model for marine benthos (Rhoads 1974). Connell and Slatyer (1977) categorised the interactions between species that influence the successional process as facilitation (i.e. positive effect), tolerance (i.e. no effect), and inhibition (i.e. negative effect). Gallagher et al. (1983) concluded that tube-builders usually facilitate recruitment (see also Noji and Noji 1991). However, manipulative experiments conducted by Whitlatch and Zajac (1985) contrasted with those of Gallagher et al. (1983) and failed to demonstrate a consistent role for a variety of opportunistic species, including tube-builders. Whitlatch and Zajac (1985) concluded that the difference in the outcome of their experiment relative to Gallagher et al. (1983) was a result of the initial densities used in the two experiments, thus indicating that facilitation, tolerance and inhibition in biotic interactions are likely to be density dependent. Other experiments have demonstrated density- and location-dependent variation in the strength and direction of effects of one species on another during the recovery processes (Thrush et al. 1992, 1996b, 1997b). Connell et al. (1987) recognised that the strengths and directions of interactions between colonists may vary over the successional processes. In the multi-species context, identifying the importance of facilitation, tolerance, or inhibition during the successional process will be complicated. But, given that in studying succession we are interested in the patterns that emerge through time from one location, should we simply relegate the role of species interactions to fine-scale detail that is not relevant to any broad-scale comparisons?

Biotic interactions will be particularly important when the mechanisms underlying species interactions involve habitat modification within the disturbed patch. Particularly important will be situations where colonists modify the biogeochemistry or the physical stability of the sediments with the patch. Modification of the biogeochemical environment will be important when conditions in the disturbed area are "hostile" (e.g., high concentrations of sulphides, ammonia or contaminants). Hydrodynamic processes can also ameliorate the hostile conditions. Hydrodynamic conditions capable of mobilising sediments can ultimately preclude the development of complex benthic communities with high densities of large, long-lived species (Hall 1994); however, the hydrodynamic limits to species distributions in soft-sediment habitats have yet to be determined. It is important to recognise that some macrofauna can play significant roles in stabilising sediments and modifying habitats, thus ameliorating the effects of hydrodynamic forces on the seabed. For example, Thrush et al. (1996a) demonstrated very slow and scale-dependent rates of recolonisation of defaunated patches of sand flat in an area previously dominated by tube-building polychaetes. The observed scale-dependent recovery dynamics were attributed to a decrease in sediment stability with increasing area without tube-mat. High sediment instability resulted in high rates of emigration from the experimental plots (Thrush et al. 1996a). Cranfield et al. (1999) provide a good example of extensive biogenic reefs existing in an area with strong tidal flows and frequent storms. These important biotic-hydrodynamic interactions are often not considered in assessments of the sensitivity of benthic habitats to disturbance.

14.4 Searching for Generality Part II

Despite variations in the details of the successional process that have been revealed by field experiments (e.g., variations in the strength of an opportunistic response, variations in the interactions between colonists), we can make some generalisations. In Table 14.1 we have categorised processes based on whether they occur within the disturbed patch (intrinsic) or outside the patch (extrinsic). Extrinsic factors may play a dominant role in small-scale experiments because of their short duration and small plot sizes that limit spatial variation in intrinsic factors. For larger scale disturbances, the importance of intrinsic factors may well increase (Zajac et al. 1998). The split into intrinsic and extrinsic processes influencing recovery highlights one of the problems with the development of numerical models to describe macrobenthic successional dynamics. Although models of various intrinsic pro-

Table 14.1. General factors influencing macrobenthic succession

Intrinsic Within the patch	Extrinsic Outside the patch
• Changes to sediment biogeochemistry	• Site history (especially frequency of disturbance events)
• Changes to sediment topography	• Intensity of disturbance (especially its influence on remaining residents)
• Changes to resources relative to ambient conditions	• Spatial extent of disturbance
• Colonist site selection, mobility and survivorship (adults, juveniles, larvae)	• Hydrodynamic conditions (influences sediment stability, supply of colonists and food)
• Colonist demographics	• Availability of colonists (influenced by timing of disturbance, distance colonists have to travel, ambient community and population composition/demography)
• Species interactions (adult-adult; adult-juvenile; predation, interference)	• Timing of disturbance (with respect to environmental characteristics)
• Species-dependent habitat modifications (bioturbation, sediment stabilisation/destabilisation)	• Location (intertidal, subtidal, inner estuary, habitat mosaic)

cesses (e.g., bioturbation; Wheatcroft and Butman 1997) or microflow patterns (e.g., Eckman and Nowell 1984) and extrinsic processes (e.g., benthic productivity; Herman et al. 1999) or hydrodynamic particle dispersion (Keough and Black 1996) exist, these processes are not explicitly linked to enable predictions of the importance of different processes in different locations.

It also should be acknowledged that interactions between intrinsic and extrinsic factors can play a very significant role in accounting for variation in rates and strength of processes during succession. For example, adult/juvenile interactions for the tellinid bivalve *Macomona liliana* can vary in strength and direction from place to place. However, variations in the strength and direction of this intrinsic process can be largely explained by differences between locations in wave climate (an extrinsic process) (Thrush et al. 2000). Interactions between processes operate over different spatial scales that make it difficult to simplify. Inevitably, in ecology, the "devil is in the detail" and this makes it difficult to predict specific outcomes for specific events.

Nevertheless, one of the most consistent patterns to emerge from field experiments, irrespective of the recovery time, is that the disturbed patches recover to contain assemblages very similar to those in the adjacent ambient community (e.g., Bonsdorff 1989). We know of no examples where experimental disturbances in soft sediments have resulted in the development of an assemblage that is distinctly different from that in the adjacent undisturbed sediment. This implies that for small-scale experiments the local pool of colonists and their mobility control the recolonisation process. It also emphasizes the importance of site history in affecting the recovery trajectory.

Unfortunately, most recovery experiments are conducted in the intertidal and very shallow sublittoral where large and long-lived epifauna are often rare. Organisms capable of creating biogenic reefs over soft sediments are likely to be particularly important because they influence sediment stability and facilitate the development of structurally complex benthic communities. Yet these organisms often have a low potential for fast recolonisation. Removal of these organisms, for example in heavily fished areas, may radically influence the benthic community composition achieved as a successional endpoint (e.g., Dayton et al. 1970, 1995; Reise 1982; Riesen and Reise 1982; Cranfield et al. 1999).

14.5 Critical Scales of Disturbance and Recovery Dynamics

Experimentally disturbed patches and many larger-scale disturbances recover to contain benthic assemblages very similar to those in adjacent undisturbed sediments because of the mobility of macrobenthic species. However, it is important to qualify this statement because larger burrowing or biogenic reef-forming organisms that are not very mobile are usually rare, making their influence on statistical comparisons of abundance and community composition weak. Even so, it is clear that mobility is a fundamental process influencing macrobenthic succession. This process can potentially provide a framework to improve our ability to predict rates of recovery and differences in community composition over a range of spatial scales and intensities of disturbances.

To identify when local vs. broader-scale processes drive recovery dynamics, we use a framework developed by Horne and Schneider (1994, 1997) and Schneider et al. (1997). Essentially the aim is to identify critical thresholds in the relative importance of competing processes, i.e. thresholds are reached when the ratio of two processes (with the same units of measurement) equals one. Here, we conceptualise how the potential source of colonists available to colonise a disturbed patch on a sand flat may vary depending on the scale or intensity of disturbance (Fig. 14.1). Our hypothetical sand flat is located in an embayment that is typical of many with a well-mixed water mass driven by tidal flows and wind-waves. Disturbed patches within the sand flat will be colonised predominantly via passive and active movement of organisms in the adjacent sediments, the dominant process here is sediment bed-load

Recovery Dynamics in Benthic Communities: Balancing Detail with Simplification 307

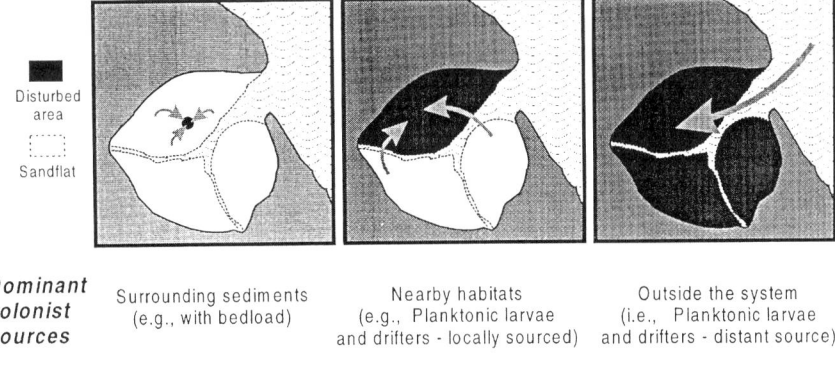

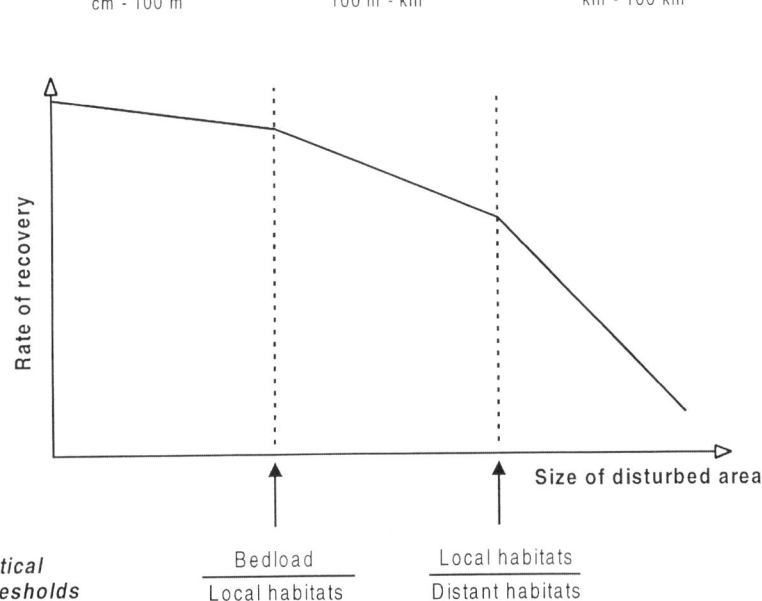

Fig. 14.1. Animal mobility and the potential colonist pool set critical thresholds in recovery rates

transport. Disturbance to the whole sand flat results in colonisation from sand flats elsewhere within the embayment's habitat matrix and is dominated by locally produced planktonic larvae and post-larval drifters. When all sand flat habitats in the embayment are disturbed, colonists must originate from distant populations via broadly dispersing planktonic larvae, drifters or alien invaders. We predict similar patterns when the scale of disturbance is constant but the frequency of disturbance is elevated so that areas adjacent to the disturbed patch are themselves not fully recovered, resulting in a similar broad-scale degradation. Over this range of spatial scales, the relative importance of these different sources of colonists influencing rates of recovery and community composition in the formerly disturbed area sets the space and time scales over which source-sink dynamics can function.

Critical thresholds will vary depending on extrinsic factors (Table 14.1) and the level of biological resolution. For example, the location of critical thresholds on the rate of recovery vs. size of the area disturbed curve will vary with hydrodynamic conditions at the disturbed location, as will the types of organisms used to quantify recovery. For a population, variation can occur due to differences in life history characteristics (Levin et al. 1987). At the community level, emergent patterns will depend on the size and location of animals within the sediment, their resistivity to transport, and modes of reproduction and larval development (Tamaki 1987; Commito et al. 1995; Shull 1997).

This framework predicts that, for species with restricted mobility, rates of recovery will be determined by the proximity of individuals to the disturbed area. Recovery rates will change down the estuary associated with both habitat change and hydrodynamic conditions. Thus, hydrodynamically isolated habitats should not contain species with low mobility when the frequency or spatial extent of disturbance is high. This is consistent with the observed differences in the abundance of oligochaetes in restored and natural salt marshes (Levin et al. 1998). At the community level, more biologically diverse parts of the estuary will take longer to recover, as diverse communities will be composed of species with a wider range of mobilities and reproductive strategies. In contrast, simple low diversity communities, in estuaries at least, are typically dominated by mobile and quick growing species.

This framework emphasises the need to address the source of immigrants to disturbed patches and the role they play in the recovery processes (Whitlatch et al. in press). However, the critical information concerns scales of mobility: for planktonic larvae this could be examined using hydrodynamic models (Keough and Black 1996). It is also important that settlement patterns are assessed because modification of flow over the benthic boundary layer due to physical or biogenic structures can result in a water mass immediately adjacent to the sediment that is not well characterised by modelled bulk flow patterns. For post-larvae, the situation is likely to be even more complex. We know little about the scales of post-settlement movement of most macrofauna, yet this is critical to understanding the important small-scale details and making broad-scale predictions.

This framework can also be used to predict the consequences of chronic degradation/disturbance to macrobenthic communities. As the intensity and/or frequency of natural or anthropogenic disturbance increases, we would expect communities to become increasingly dominated by small, rapidly growing mobile species. Larger, longer-lived and less mobile species would become increasingly rare. Although effects may be more apparent in the upper reaches of the estuary, they are probably more significant in the channels and sand flats because of the inherently higher diversity of these habitats.

14.6. Recovery: a Useful Tool for Assessing Broad-Scale and Cumulative Effects?

Our comparisons of recovery of macrobenthic communities highlight some of the difficulties that require resolution in order to use "recovery time" (resilience) as an indicator of ecosystem health. There is growing concern about chronic broad-scale degradative change in natural ecosystems and how to collect appropriate data to monitor and assess trends. The merging of socio-economic, policy and scientific concepts and concerns over global environmental change has focused attention on the concept of ecosystem health (Rapport et al. 1998). This concept has been strongly debated in the scientific literature because of the difficulties of measurement and inference (e.g., Calow 1992; Suter 1993). Nevertheless, assessment techniques have been proposed and the time taken to recover from disturbance is usually considered a key factor in determining environmental health (Shraderfrechette 1994; Mageau et al. 1995). Developing consistent experimental methodologies to assess resilience would be reasonably straightforward; the trick is comparing responses across systems that vary in environmental characteristics (particularly hydrodynamic conditions). Site history will influence the supply of available colonists, both by direct influences on benthic communities and in the broad-scale spatial arrangement of the habitat matrix. Macrobenthic communities in degraded ecosystems typically are dominated by small, rapidly growing and highly mobile species, which result in the rapid recovery of small-scale disturbances (e.g., Pearson and Rosenberg 1978; Rhoads et al. 1978). Most importantly, we must be aware of the important problem of "shifting baselines" which plagues comparisons when information on site history is unavailable (Leppakoski 1975; Dayton et al. 1998). Measuring the resilience of macrobenthic communities may well result in a counter-intuitive assessment of ecosystem health, because degraded benthic communities will recover from disturbances quickly and we may not have sufficient knowledge to define site history.

14.7 Searching for Generality Part III: The Need To Improve the Information Base

Simply describing the role of processes operating over broad scales may not increase the predictive power of population and community ecology; formal statistical analyses and testing of concepts and predictions are necessary. The most rigorous empirical syntheses can be achieved by formal meta-analysis based on experiments of identical design conducted at different locations in space and/or time, combined with measurements of key environmental parameters that are likely to constrain local biotic interactions. Unless the appropriate co-variables have been measured, we cannot explain what is causing variation in effects from place to place. This limits our ability to make broad-scale predictions about the relative importance of local ecological processes. We have identified a number of variables (Table 14.1) that can be factored into the design of a broad-scale comparative assessment of variation in macrobenthic succession dependent on location.

Understanding natural history is essential, especially when the variability in ecological structure and function can be directly or indirectly ascribed to small-scale processes. We have placed particular emphasis on the importance of mobility and local pools of potential colonists as critical factors influencing variability in response between locations. Generally, however, we lack fundamental data on many intrinsic factors (e.g., within-patch demographics, habitat selectivity or sediment geotechnical characteristics) or how they interact with broader scale processes. While single studies may provide examples of possible mechanisms it is difficult to assess their generality. Careful consideration of the balance between gaining more detailed information versus more general information is always necessary. A coherent programme of the assessment of macrobenthic succession that assesses the relative importance of the intrinsic and extrinsic processes outlined in Table 14.1 could provide important resources enabling us to improve our ability to predict recovery and document chronic degradative change within estuarine and coastal ecosystems.

14.8 Conclusions

Natural disturbance events frequently play a central role in influencing spatial and temporal variation in the structure and function of benthic communities. The development of a patch-dynamic conceptual framework for macrobenthic communities has been aided by manipulative disturbance/recovery experiments. Manipulative experiments are also often used to help understand or predict the ecological consequences of much larger scale disturbance events. Field experiments examining the recovery process reveal that biotic interactions during succession are highly variable but that some important generalisations can be reached based on the functional aspects of the organisms, in particular the provision of complex habitat structure either above or below the sediment/water interface. There is some evidence for scale-dependence limiting our ability to directly scale-up manipulative experiments. We noted the importance of interactions between intrinsic and extrinsic processes that limit the use of simple hierarchical models. However, our review highlights the importance of mobility in influencing colonisation. We suggest that assessing the relative importance of biological and physical processes that influence the mobility of potential colonists over different spatial scales is a critical factor in helping us to scale-up from small manipulative experiments. By conceptualising how the potential source of colonists may vary depending on the scale, intensity and location of disturbance we emphasise the importance of differences in mobility within species, dependent on life-history stage, and between species in determining critical thresholds in recovery rates. Assessing the value of location-dependent recovery rates for different life stages, species or assemblages based on critical thresholds is one way of defining common units to enable us to make quantitative comparisons between locations. This cannot be achieved with the information currently available. We suggest that a coherent programme of the assessment of macrobenthic succession would provide important resources enabling us to improve our ability to predict recovery and document chronic degradative change within estuarine and coastal ecosystems.

Acknowledgements. We thank Judi Hewitt, Vonda Cummings, Alf Norkko, Joanne Ellis, Roman Zajac and Erik Bonsdorff for comments on earlier versions of this manuscript. We thank Karsten Reise for the opportunity to discuss this perspective on coastal comparisons on Sylt. The authors gratefully acknowledge the Foundation for Research Science and Technology (NZ) and the National Science Foundation (USA) for their continued support in helping us to extend the understanding of marine benthic ecosystems.

References

Allen TFH, Starr TB (1982) Hierarchy perspectives for ecological complexity. Univ Chicago Press, Chicago
Alongi DM (1990) The ecology of tropical soft-bottom benthic ecosystems. Oceanogr Mar Biol Annu Rev 28:381–496
Arnqvist G, Wooster D (1995) Meta-analysis: synthesizing research findings in ecology and evolution. TREE 10:236–240

Bell RG, Hume TM, Dolphin TJ, Green MO, Walters RA (1997) Characterisation of physical factors on an intertidal sandflat, Manukau Harbour, New Zealand. J. Exp Mar Biol Ecol 216:11–32

Bonsdorff E (1989) Infaunal colonization and its dependence on environmental variation – experimental evidence from the northern Baltic Sea. In: Ryland JS, Tyler PA (eds) Reproduction, genetics and distribution of marine organisms. 23rd European marine biology symposium. Olsen and Olsen, Fredensborg, pp 349–356

Butman CA (1987) Larval settlement of soft-sediment invertebrates: the spatial scales of pattern explained by active habitat selection and the emerging role of hydrological processes. Oceanogr Mar Biol Annu Rev 25:113–165

Calow P (1992) Can ecosystems be healthy? Critical consideration of concepts. J Aquatic Ecosyst Health 1:1–5

Commito JA, Currier CA, Kane LR, Reinsel KA, Ulm IM (1995) Dispersal dynamics of the bivalve *Gemma gemma* in a patchy environment. Ecol Monogr 65:1–20

Connell JH, Slatyer RO (1977) Mechanisms of succession in natural communities and their role in community stability and organisation. Am Nat 111:1119–1144

Connell JH, Nobel IR, Slayter RO (1987) On mechanisms producing successional change. Oikos 50:136–137

Cranfield HJ, Michael KP, Doonan IJ (1999) Changes in the distribution of epifaunal reefs and oysters during 130 years of dredging for oysters in Foveaux Strait, southern New Zealand. Aquat Cons Mar Freshwater Ecosyst 9:461–483

Dayton PK, Robillard GA, Pain RT (1970) Benthic faunal zonation as a result of anchor ice formation in McMurdo Sound, Antarctica. In: Holdgate MW (ed) Antarctic ecology. Academic Press, London, pp 244–258

Dayton PK, Thrush SF, Agardy TM, Hofman RJ (1995) Environmental effects of fishing. Aquat Conserv Mar Freshwater Ecosyst 5:205–232

Dayton PK, Tegner MJ, Edwards PB, Riser KL (1998) Sliding baselines, ghosts, and reduced expectations in kelp forest communities. Ecol Apps 8:309–322

Eckman JE (1983) Hydrodynamic processes affecting benthic recruitment. Limnol Oceanogr 28:241–257

Eckman JE, Nowell AR (1984) Boundary skin friction and sediment transport about an animal-tube mimic. Sedimentology 31:851–862

Englund G, Sarnelle O, Cooper SD (1999) The importance of data-selection criteria: meta-analyses of stream predation experiments. Ecology 80:1132–1141

Ford RB, Thrush SF, Probert PK (1999) Macrobenthic colonisation of disturbances on an intertidal sandflat: the influence of season and buried algae. Mar Ecol Prog Ser 191:163–174

Gallagher ED, Jumars PA, Trueblood DD (1983) Facilitation of soft-bottom benthic succession by tube builders. Ecology 64:1200–1216

Gunther C (1992) Dispersal of intertidal invertebrates: a strategy to react to disturbances of different scales? Neth J Sea Res 30:45–56

Gurevitch J, Hedges LV (1993) Meta-analysis: combining the results of independent experiments. In: Scheiner SM, Gurvitch J (eds) Design and analysis of ecological experiments. Chapman and Hall, New York, pp 378–398

Gurevitch J, Hedges LV (1999) Statistical issues in ecological meta-analyses. Ecology 80:1142–1149

Hall SJ (1994) Physical disturbance and marine benthic communities: life in unconsolidated sediments. Oceanogr Mar Biol Annu Rev 32:179–239

Herman PMJ, Middelburg JJ, VandeKoppel J, Heip CHR (1999) Ecology of estuarine macrobenthos. Adv Ecol Res 29:195–231

Holling CS (1996) Surprise for science, resilience for ecosystems, and incentives for people. Ecol Apps 6:733–735
Horne JK, Schneider DC (1994) Analysis of scale-dependent processes with dimensionless ratios. Oikos 70:201–211
Horne JK, Schneider DC (1997) Spatial variance of mobile aquatic organisms: capelin and cod in Newfoundland coastal waters. Philos Trans R Soc Lond B 352:633–642
Johnson RG (1970) Variation in diversity within benthic marine communities. Am Nat 104:285–300
Johnson RG (1973) Conceptual models of benthic marine communities. In: Schopf TJM (ed) Models in paleobiology. Freeman and Cooper, San Francisco, pp 148–159
Jumars PA, Nowell ARM (1984) Fluid and sediment dynamic effects on marine benthic community structure. Am Zool 24:45–55
Kareiva P, Andersen M (1988) Spatial aspects of species interactions: the wedding of models and experiments. In: Hastings A (ed) Community ecology, lecture notes in biomathematics. Springer, Berlin Heidelberg New York, pp 38–54
Keough MJ, Black KP (1996) Predicting the scales of marine impacts: understanding planktonic links to populations. In: Schmitt RJ, Osenberg CW (eds) Detecting ecological impacts; concepts and applications in coastal habitats. Academic Press, San Diego, pp 199–234
Kitching JA (1937) Studies in sublittoral ecology. III. Recolonisation at the upper margin of the sublittoral region: with a note on the denudation of Laminarian forests by storms. J Ecol 25:482–495
Leppakoski E (1975) Assessment of degree of pollution on the basis of macrozoobenthos in marine and brackish-water environments. Acta Acad Aboensis Ser B 35:1–90
Levin LA (1984) Life history and dispersal patterns in a dense infaunal polychaete assemblage: community structure and response to disturbance. Ecology 65:1185–1200
Levin LA, Caswell H, DePatra KD, Creed EL (1987) Demographic consequences of larval development mode: planktotrophy vs. lecithotrophy in *Streblospio benedicti*. Ecology 68:1877–1886
Levin LA, Sinicrope Talley T, Hewitt J (1998) Macrobenthos of *Spartina foliosa* (Pacific cordgrass) salt marshes of southern California: community structure and comparisons to a Pacific mudflat and a *Spartina alterniflora* (Atlantic smooth cordgrass) marsh. Estuaries 21:129–144
Mageau MT, Costanza R, Ulanowicz RE (1995) The development and initial testing of a quantitative assessment of ecosystem health. Ecosyst Health 1:201–213
Noji CI-M, Noji TT (1991) Tube lawns of spionid polychaetes and their significance for recolonization of disturbed benthic substrates. A review. Meeresforschung 33: 235–246
Oliver JS, Slattery PN (1985) Destruction and opportunity on the sea floor: effects of Gray whale feeding. Ecology 66:1965–1967
Osenberg CW, Sarnelle O, Goldberg DE (1999) Meta-analysis in ecology: concepts, statistics, and applications. Ecology 80:1103–1104
Paterson DM, Black KS (1999) Water flow, sediment dynamics and benthic biology. Adv Ecol Res 29:155–188
Pearson TH, Rosenberg R (1978) Macrobenthic succession in relation to organic enrichment and pollution of the marine environment. Oceanogr Mar Biol Annu Rev 16:229–311
Pearson TH, Rosenberg R (1987) Feast and famine: structuring factors in marine benthic communities. In: Gee JHR, Giller PS (eds) Organisation of communities past and present. Blackwell, Boston, pp 373–395

Pickett STA, White PS (1985) The ecology of patch disturbance and patch dynamics. Academic Press, Orlando

Probert PK (1984) Disturbance, sediment stability and trophic structure of soft-bottom communities. J Mar Res 42:893-921

Rapport DJ, Costanza R, McMichael AJ (1998) Assessing ecosystem health. TREE 13: 397-402

Reise K (1982) Long-term changes in the macrobenthic invertebrate fauna of the Wadden Sea. Neth J Sea Res 16:29-36

Rhoads DC (1974) Organism-sediment relations on the muddy seafloor. Oceanogr Mar Biol Annu Rev 12:263-300

Rhoads DC, McCall PL, Yingst JY (1978) Disturbance and production on the estuarine seafloor. Am Sci 66:577-586

Riesen W, Reise K (1982) Macrobenthos of the subtidal Wadden Sea: revisited after 55 years. Helgolander Meeresuntersuch 35:409-423

Santos SL, Simon JL (1980) Marine soft-bottom community establishment following annual defaunation: larval or adult recruitment. Mar Ecol Prog Ser 2:235-241

Savidge WB, Taghon GL (1988) Passive and active components of colonization following two types of disturbance on an intertidal sandflat. J Exp Mar Biol Ecol 115:137-155

Schneider DC (1994) Quantitative ecology: spatial and temporal scaling. Academic Press, San Diego

Schneider DC, Walters R, Thrush SF, Dayton PK (1997) Scale-up of ecological experiments: density variation in the mobile bivalve *Macomona liliana* Iredale. J Exp Mar Biol Ecol 216:129-152

Shraderfrechette KS (1994) Ecosystem health: a new paradigm for ecological assessment? TREE 9:456-457

Shull DH (1997) Mechanisms of infaunal polychaete dispersal and colonization on an intertidal sandflat. J Mar Res 55:153-179

Smith CR, Brumsickle SJ (1989) The effect of patch size and substrate isolation on colonization modes and rate in an intertidal sediment. Limnol Oceanogr 34:1263-1277

Suter GW (1993) A critique of ecosystem health concepts and indexes. Environ Tox Chem 12:1533-1539

Tamaki A (1987) Comparison of resistivity to transport by wave action in several polychaete species on an intertidal sand flat. Mar Ecol Prog Ser 37:181-189

Thistle D (1981) Natural physical disturbances and the communities of marine soft bottoms. Mar Ecol Prog Ser 6:223-228

Thrush SF (1986) Spatial heterogeneity in subtidal gravel generated by the pit-digging activity of *Cancer pagurus*. Mar Ecol Prog Ser 30:221-227

Thrush SF (1991) Spatial patterns in soft-bottom communities. TREE 6:75-79

Thrush SF, Roper DS (1988) Merits of macrofaunal colonization of intertidal mudflats for pollution monitoring: preliminary study. J Exp Mar Biol Ecol 116:219-233

Thrush SF, Pridmore RD, Hewitt JE, Cummings VJ (1991) Impact of ray feeding disturbances on sandflat macrobenthos: do communities dominated by polychaetes or shellfish respond differently? Mar Ecol Prog Ser 69:245-252

Thrush SF, Pridmore RD, Hewitt JE, Cummings VJ (1992) Adult infauna as facilitators of colonization on intertidal sandflats. J Exp Mar Biol Ecol 159:253-265

Thrush SF, Hewitt JE, Pridmore RD, Cummings VJ (1996a) Adult/juvenile interactions of infaunal bivalves: contrasting outcomes in different habitats. Mar Ecol Prog Ser 132:83-92

Thrush SF, Whitlatch RB, Pridmore RD, Hewitt JE, Cummings VJ, Maskery M (1996b) Scale-dependent recolonization: the role of sediment stability in a dynamic sandflat habitat. Ecology 77:2472-2487

Thrush SF, Cummings VJ, Dayton PK, Ford R, Grant J, Hewitt JE, Hines AH, Lawrie SM, Legendre P, McArdle BH, Pridmore RD, Schneider DC, Turner SJ, Whitlatch RB, Wilkinson MR (1997a) Matching the outcome of small-scale density manipulation experiments with larger scale patterns: an example of bivalve adult/juvenile interactions. J Exp Mar Biol Ecol 216:153–170

Thrush SF, Schneider DC, Legendre P, Whitlatch RB, Dayton PK, Hewitt JE, Hines AH, Cummings VJ, Lawrie SM, Grant J, Pridmore RD, Turner SJ (1997b) Scaling-up from experiments to complex ecological systems: Where to next? J Exp Mar Biol Ecol 216:243–254

Thrush SF, Hewitt JE, Cummings VJ, Dayton PK, Cryer M, Turner SJ, Funnell G, Budd R, Milburn C, Wilkinson MR (1998) Disturbance of the marine benthic habitat by commercial fishing: impacts at the scale of the fishery. Ecol Apps 8:866–879

Thrush SF, Lawrie SM, Hewitt JE, Cummings VJ (1999) The problem of scale: uncertainties and implications for soft-bottom marine communities and the assessment of human impacts. In: Gray JS, Ambrose W, Szaniawska A (eds) Biogeochemical cycling and sediment ecology. Kluwer, Dordrecht, pp 195–210

Thrush SF, Hewitt JE, Cummings VJ, Green MO, Funnell GA, Wilkinson MR (2000) The generality of field experiments: interactions between local and broad-scale processes. Ecology 81:399–415

Van Blaricom GR (1982) Experimental analysis of structural regulation in a marine sand community exposed to oceanic swell. Ecol Monogr 52:283–305

Watt AS (1947) Pattern and process in the plant community. J Ecol 35:1–22

Wheatcroft RA, Butman CA (1997) Spatial and temporal variability in aggregated grain-size distributions, with implications for sediment dynamics. Cont Shelf Res 17: 367–390

Whitlatch RB, Zajac RN (1985) Biotic interactions among estuarine infaunal opportunistic species. Mar Ecol Prog Ser 21:299–311

Whitlatch RB, Lohrer AM, Thrush SF, Pridmore RD, Hewitt JE, Cummings VJ, Zajac RN (1998) Scale-dependent recolonization dynamics: life-stage-based dispersal and demographic consequences. Hydrobiologia 375/376:217–226

Whitlatch RB, Lohrer AM, Thrush SF (in press) Scale-dependent recovery of the benthos: effects of larval and post-larval stages. In: Aller R, Aller J, Woodin SA (eds) Organism-sediment interactions. Belle W Baruch library in marine science, vol 21. Univ South Carolina, Columbia, South Carolina

Zajac RN (1991a) Population ecology of *Polydora ligni* (polychaeta, spionidae). 1. Seasonal variation in population characteristics and reproductive activity. Mar Ecol Prog Ser 77:197–206

Zajac RN (1991b) Population ecology of *Polydora ligni* (Polychaeta, Spionidae). 2. Seasonal demographic variation and its potential impact on life history evolution. Mar Ecol Prog Ser 77:207–220

Zajac RN, Whitlatch RB (1982a) Responses of estuarine infauna to disturbance. I. Spatial and temporal variation of initial recolonization. Mar Ecol Prog Ser 10:1–14

Zajac RN, Whitlatch RB (1982b) Responses of estuarine infauna to disturbance. II. Spatial and temporal variation in succession. Mar Ecol Prog Ser 10:15–27

Zajac RN, Whitlatch RB (1985) A hierarchical approach to modelling soft-bottom successional dynamics. In: Gibbs P (ed) Proceedings of the 19th European Marine Biological Symposium. Cambridge Univ Press, Cambridge, pp 265–276

Zajac RN, Whitlatch RB (1989) Natural and disturbance-induced demographic variation in an infaunal polychaete, *Nephtys incisa*. Mar Ecol Prog Ser 57:89–102

Zajac RN, Whitlatch RB (1991) Demographic aspects of marine soft sediment patch dynamics. Am Zool 31:808–820

Zajac RN, Whitlatch RB, Thrush SF (1998) Recolonisation and succession in soft-sediment infaunal communities: the spatial scale of controlling factors. Hydrobiologia 376:227–240

15 Population Dynamics of Benthic Species on Tidal Flats: the Possible Roles of Shorebird Predation

J. van der Meer, T. Piersma, and J.J. Beukema

15.1 Introduction

In late summer millions of shorebirds leave their breeding grounds in the Arctic and move to tidal mudflats in the temperate and tropical zones. These mudflats can either be used as stopover sites during migration, or as overwintering areas. Shorebirds necessarily have high feeding rates and the densities of these salient predators can be high. Not surprisingly, many researchers have asked the question what the impact is of shorebirds on their intertidal invertebrate prey, and whether this impact may differ, for example, between temperate and tropical regions (Piersma and Beukema 1993). Generally, two approaches have been adopted to answer these questions. First, estimates of annual consumption by shorebirds were compared with annual invertebrate production estimates. Second, experiments were carried out using exclosure cages. We argue that neither of the two methods is very relevant in answering the questions posed. We suggest a more appropriate, but time-consuming, approach based on long-term observations.

15.2 Production-Consumption Comparisons

A widely applied approach in assessing the impact of migrating or overwintering shorebirds on intertidal invertebrates is the estimation of the amount of annual invertebrate production that is on average consumed by shorebirds. The published accounts that we were able to find (Table 15.1) suggest that consumption/production ratios vary around a value of 0.3 (range 0.12–0.52), a value that does not seem to consistently differ between temperate and tropical regions (Fig. 15.1). Yet, two things should be kept in mind. First, it may be seriously questioned whether the reliability of these published data allows any conclusions at all. Second, even if such conclusions as made above are warranted, one may wonder how relevant the results actually are.

Preferably, production is directly estimated by, for example, the mortality-summation method (Crisp 1984). This method, however, requires knowledge of the age-structure of the population and the changes therein. As this knowledge is rarely available, in most cases published species-specific production/biomass ratios (P/B ratio) were used instead, in combination with species biomass estimates. Only in one study (Baird and Milne 1981) were estimates based on direct methods and then only for one or two species (Table 15.1). Predation causes mortality of the prey and may thus enhance production and turnover rate. Thus, predation itself may increase the P/B ratio, and a certain amount of circularity occurs when consumption/production ratios are used to study the impact of predation assuming constant P/B ratios.

Table 15.1. Annual consumption by shorebirds C (g AFDM m^{-2}), macrozoobenthos biomass B (g AFDM m^{-2}) and annual zoobenthic production P (g AFDM m^{-2}) for various intertidal areas

Area	Region[a]	C	B	P	Method[b]	Source(s)
Ythan	W	39.7	59.8	111.4	ms; *P/B*	Baird and Milne (1981)
Tees	W	17.1	44.4	38.7	*P/B*	Evans et al. (1979)
Wadden Sea	W	5.0	26.6	29.6	*P/B*	Beukema (1976, 1981); Smit (1981); De Wilde and Beukema (1984)
Grevelingen[c]	W	8.3	46.8	56.1	*P/B*	Wolff et al. (1976); Wolff and De Wolf (1977)
Oosterschelde	W	12.3	74.3			Meire et al. (1994)
Westerschelde	W	4.5	14.6			Meire et al. (1989); Stuart et al. (1989)
Tagus	S	4.9		40.8	guess	Moreira (1997)
Banc d'Arguin	A	14.0	14.5	27.0	guess	Wolff and Smit (1990); Wolff (1991); Wolff et al. (1993)
Maputo	A	2.1				De Boer and Longamane (1996)
Berg River	A	26.7	19.4	87.6	*P/B*	Summers (1977); Baird et al. (1985)
Langebaan	A	6.4	9.3	32.0	*P/B*	Velasquez et al. (1991); Kalejta (1992)

[a] W, Western Europe; S, southern Europe; A, Africa.
[b] Production is estimated by the mortality-summation method (ms), through published P/B ratios (*P/B*) or guessed.
[c] The figures presented here refer specifically to the intertidal area and can be derived from the original published figures.

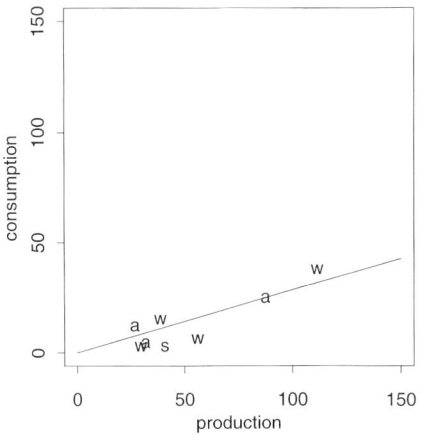

Fig. 15.1. Annual consumption by shorebirds C (g AFDM m^{-2}) versus annual zoobenthic production P (g AFDM m^{-2}) for various intertidal areas (see Table 15.1). W Western Europe, S southern Europe and A Africa. *Line* indicates the average C/P ratio, which equals 0.28

Another reason why the reliability of most of the published data should be doubted is the short time span over which measurements of bird numbers and benthic biomass were collected. The observation period usually did not exceed a single year. However, even if the data were reliable (direct methods applied over a long period), one might wonder whether knowledge of the kind that 30% of the overall production is consumed by birds actually tells us much about the impact of bird predation on their prey populations. It may say something about the amount of depletion of the prey populations, but what does it mean for the dynamics and regulation of the prey population, or for the structure of the benthic community as a whole? Not much, we would say.

The question to what extent production of invertebrate biomass corresponds to consumption by birds can be stated in terms of population ecology. That is, what is the overall contribution of bird predation to prey mortality? However, a much more interesting question in population dynamics is: what sort of relationship exists between prey mortality rate, on the one hand, and densities of prey and predators, on the other? Not the mortality level as such, but the density dependence of mortality rate is fundamental. Studying this density dependence either requires long-term observations or experiments at scales that are generally beyond reach.

15.3 The Balgzand Area: a Long-Term Study

Observational field studies on populations of both invertebrate prey species and shorebird species have been carried out for about 30 years at the Balgzand, a 45-km² tidal flat area in the westernmost Wadden Sea. We selected the cockle *Cerastoderma edule* (L.) and the Baltic tellin *Macoma balthica* (L.), which are among the most important food sources for two shellfish-eating shorebird species occurring in the Wadden Sea, the knot *Calidris canutus* L. and the oystercatcher *Haematopus ostralegus* L. (Hulscher 1982, 1983; Piersma et al. 1993).

Between 1970 and 1998, 12 randomly selected transects of 1-km length were sampled each year in late winter (February-March) and late summer (August). One transect near low-tide level was not taken into account in the present review, because this plot rarely emerges and is hardly available to foraging shorebirds. Along each transect 50 cores were taken, 0.019 m² each in winter and 0.009 m² in summer. Numerical and ash-free dry biomass density values were determined for each age-class separately. More details of the sampling procedure are given in Beukema (1974, 1988, 1993). In the remaining part of this chapter, the term 'recruit' refers to juveniles (i.e. zero-year class) at the first summer sampling in August. We assumed that the cockle is sedentary (and hence there is no net migration in or out of the area) after recruitment. The Baltic tellin is sedentary after the first winter. These assumptions, which enable the estimation of adult mortality rates and production, were confirmed by surveys with plankton nets suspended in tidal streams in the Wadden Sea (Beukema 1989) and by studies on the recolonization of large defaunated areas (Beukema et al. 1999). In both cases no adults, and only post-larvae and juvenile Baltic tellins were observed.

The number of birds that feed on the tidal flats can be estimated from monthly counts at the high-water roosts, available since the mid-1970s (M. Otter, pers. comm.). We assume that birds counted on their daytime roosts represent the population making a living on the Balgzand intertidal. For oystercatcher, in particular, this is probably a robust assumption. Only few (unpublished) data on feeding behaviour and diet at Balgzand are available, but detailed studies have been performed elsewhere in the Wadden Sea (Zwarts and Drent 1981; Hulscher 1982; Zwarts and Blomert 1992; Zwarts et al. 1992; Piersma et al. 1993) and in the laboratory (Hulscher 1976; Piersma et al. 1995).

15.4 Long-Term Variability in Production and Consumption at the Balgzand

For all cohorts (animals "born" in the same year) secondary production (elimination) was estimated for each half-year period by the mortality-summation method (Crisp 1984), i.e. by multiplying the observed decrease in density (averaged over sites) by the average of the individual mass at the start and at the end of the season. The average individual mass was calculated as the average biomass density (averaged over sites) divided by the average numerical density. Production of those individuals that did not survive until the first summer sampling could not be taken into account. For the summers of 1971 and 1972 density and biomass density data were missing and production could not be calculated. For the summers 1973–1979 only data on biomass density were missing, and for each age-class a long-term average individual mass was used instead for that period. Sampling variability of the estimates was estimated by the bootstrap procedure (200 trials; at each trial 11 transects were re-sampled). Efron's first-percentile method was used to obtain 90% bootstrap confidence intervals (CI), i.e. the interval is given by the values that exceed 5 and 95%, respectively, of the generated bootstrap distribution. The among-years variability (not to be confused with the sampling variability) was expressed by the among-years standard deviation (SD).

Monthly high-water roost count data for the knot and oystercatcher were available from July 1975 to June 1998. The missing values for the knot, i.e. 41, and for the oystercatcher, i.e. 39, (out of 276) were imputed by a log-linear model with main effects year and month (Van der Meer et al. 1996). Food demands were based on a field metabolic rate (power) of 4 W for the knot (Wiersma and Piersma 1994). For the oystercatcher, we used a power of 8.8 W, which is slightly higher than the estimate of 7.8 W at thermoneutrality for an oystercatcher weighing 0.52 kg that Zwarts et al. (1996) based on a literature survey. Yet, winter temperatures in the Wadden Sea are slightly lower than the critical temperature of 10 °C below which the costs of thermoregulation increase (Kersten and Piersma 1987). We assumed that the available foraging time per day equals 12 h, the energy density of the bivalves is 22 MJ/kg ash-free dry mass (Beukema 1997) and the gut absorption efficiency is 0.8. The power values are then equivalent to a required food intake rate while foraging of 0.45 and 1.0 mg/s for knot and oystercatcher, respectively.

For each winter and summer season in the period 1973–1998 estimates of production were obtained. For *M. balthica* average production in summer (1.36 g/m^2, SD 0.528, CI 1.13–1.67) and in winter (1.03 g/m^2, SD 0.693, CI 0.71–1.26) were almost equal. In their first winter the zero-year class individuals are too small (about 6 mm) to be profitable as a food source for the two bird species. The older year classes, those that can be eaten, were responsible for about two-thirds of the total winter production (0.69 g/m^2, SD 0.516, CI 0.49–0.85). For *C. edule* average production in winter was 6.53 g/m^2 (SD 4.57, CI 3.97–9.36) and in summer 4.64 g/m^2 (SD 4.78, CI 2.98–6.95). The zero-year class, which is edible for knots, contributed on average about one-fifth (1.31 g/m^2, SD 1.34, CI 0.70–2.07) to total winter production. Production varied considerably from year-to-year in both species (Fig. 15.2). The same holds for the production/biomass ratio, particularly for the cockle (Fig. 15.3).

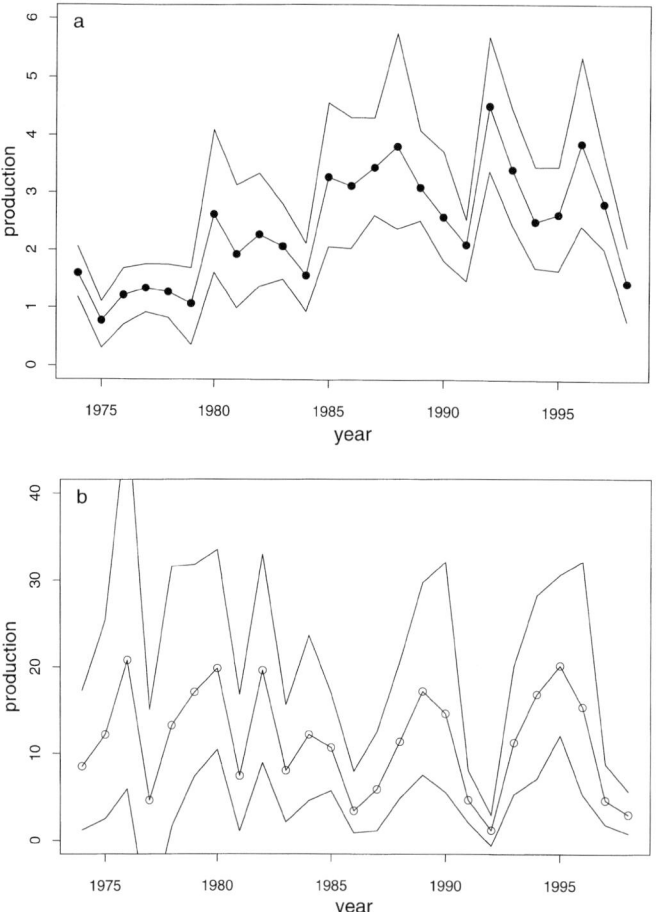

Fig. 15.2. Annual production (g AFDM m^{-2}) versus year (where, for example, 1980 means the period from autumn 1979 until autumn 1980) including 90% bootstrap confidence intervals. **a** *Macoma balthica* and **b** *Cerastoderma edule*

Consumption (estimated by multiplying bird density by their food requirements) by the two bird species largely took place in winter. Annual consumption was estimated as 1.07 g/m^2 (SD 0.590) for knots and 3.35 (SD 0.970) for oystercatchers; winter (September–March) consumption as 0.97 g/m^2 (SD 0.533) for knots and 2.71 (SD 0.874) for oystercatchers. Year-to-year variability in estimated winter consumption (in bird numbers) was higher for the knot than for the oystercatcher, coefficients of variation being 0.55 and 0.32, respectively (Fig. 15.4). Generally, winter production suitable as a food source for knots was mainly determined by the zero-year class of *C. edule*. This prey type accounted for a winter production of on average 1.31 g/m^2 (SD

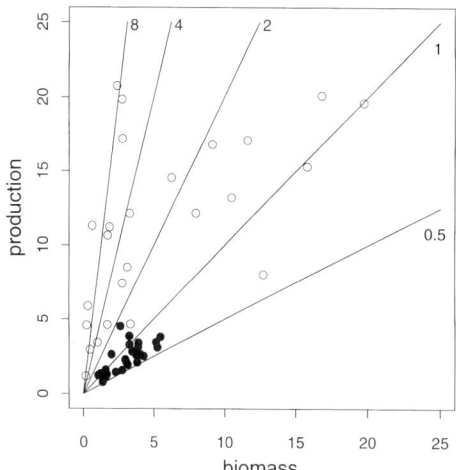

Fig. 15.3. Annual production P (g AFDM m^{-2}) versus biomass density B (g AFDM m^{-2}). *Macoma balthica* (*filled circles*) and *Cerastoderma edule* (*open circles*). Each point refers to a late-winter sample from the period 1973–1997 (for B) and the succeeding year (for P). *Lines* indicate P/B ratios of 0.5, 1, 2, 4 and 8

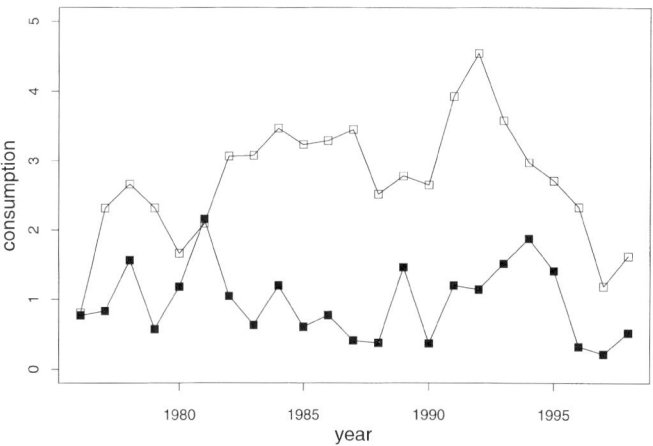

Fig. 15.4. Winter (September–March) consumption (g AFDM m^{-2}) of knots (*filled squares*) and oystercatchers (*open squares*) at Balgzand versus year (where, for example, 1980 means the winter of 1979/1980)

1.34, CI 0.70–2.07) compared with a winter production of 0.23 (SD 0.187, CI 0.12–0.31) for the first and second year class of *M. balthica*. In some winters (such as the winters of 1989 and 1990), however, zero-year classes *C. edule* were very small and their production was close to zero. Production suitable for the oystercatcher (all but the zero-year class of *C. edule* and *M. balthica*) was also mainly determined by *C. edule* production: 5.22 g/m^2 (SD 4.97, CI 3.23–7.50) versus 0.69 g/m^2 (SD 0.516, CI 0.49–0.85). Production that is suitable both for knots and for oystercatchers, i.e. by first and second year class *M. balthica*, was invariably small (0.23 g/m^2, SD 0.187, CI 0.12–0.31).

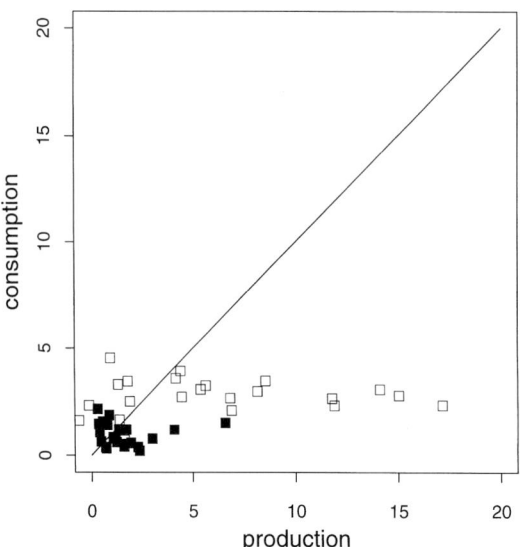

Fig. 15.5. Winter consumption (g AFDM m^{-2}) versus winter production of prey of suitable size (g AFDM m^{-2}) for knots (*filled squares*) and oystercatchers (*open squares*). Each point refers to a winter from the period 1975/1976–1997/1998

When winter production of suitably sized prey is compared with winter consumption it appears that consumption/production ratios varied enormously from year-to-year. In years with low production, the consumption and production roughly matched (Fig. 15.5). Production was much more variable from year-to-year than consumption, and in some years the bulk of the production apparently could not have been consumed by the birds (Fig. 15.5). Average annual production of the two bivalve species suitable as prey for at least one of the two bird species equalled 13.18 g/m² (SD 6.05, CI 9.32–18.42), and was much higher than the average annual consumption (4.21 g/m², SD 1.26).

15.5 Density-Dependent Survival?

In modelling survival we assumed that survival was season-dependent and not age-dependent, except that the first-winter survival was modelled separately. The validity of this assumption and more details of the modelling procedure will be documented elsewhere (Van der Meer et al., in prep.). So, in the model, each summer takes one survival parameter, and each winter two (first-winter and adult) survival parameters. Maximum-likelihood estimates of the parameters were obtained under the assumption of a Poisson-like distribution of the counts. Average survival over the whole period (plus standard deviation to express the among-years variability) was calculated in terms of the average instantaneous death rate z (a^{-1}). Recall that estimates of

Table 15.2. Average instantaneous death rate (a^{-1}) ± among-years SD. Geometric-mean survival percentage (per half-year period) is given in parentheses

	n	Macoma balthica	Cerastoderma edule
First winter	25	1.23 ±0.406 (29%)	1.51 ±0.937 (22%)
Later winters	25	0.146±0.122 (86%)	1.29 ±1.24 (27%)
Later summers	26	0.348±0.161 (71%)	0.597±0.721 (55%)

first-winter survival of *M. balthica* may be biased, because the assumption of no net migration cannot be reliably made for this period.

Annual survival of *M. balthica* was larger and much more constant from year-to-year than survival of *C. edule*. The instantaneous death rate for all individuals older than one year, averaged over the years 1974–1998 ($n=25$), was 0.50 a^{-1} (SD 0.144; CI 0.45–0.54) for *M. balthica*, and 1.91 a^{-1} (SD 1.39; CI 1.79–2.54) for *C. edule*. These averages are equivalent to geometric mean survival percentages of 61 and 15%, respectively. For older *M. balthica* winter survival was higher than summer survival, but summer survival was higher than winter survival for *C. edule* (Table 15.2). For individuals older than 1 year, instantaneous death rate was not linearly density dependent (Fig. 15.6; *M. balthica*, $n=25$, $F=0.83$, $P=0.37$; *C. edule*, $n=25$, $F=0.66$, $P=0.42$) and not related to total predator consumption (Fig. 15.7; *M. balthica*, $n=23$, $F=0.77$, $P=0.39$; *C. edule*, $n=23$, $F=0.30$, $P=0.59$).

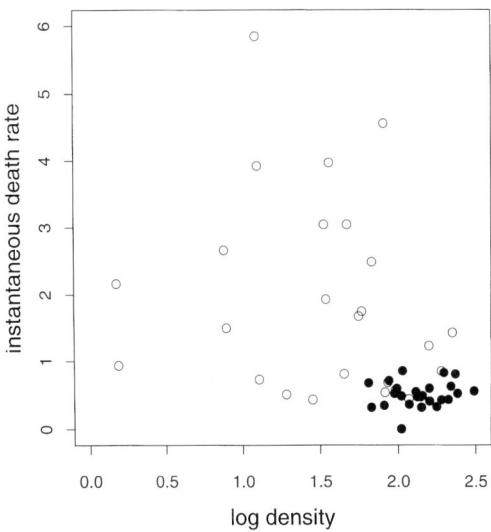

Fig. 15.6. Instantaneous death rate of individuals older than 1 year (a^{-1}) versus log density (m^{-2}). *Macoma balthica* (*filled circles*) and *Cerastoderma edule* (*open circles*). Each point refers to a late-winter sample from the period 1973–1997 (for density) and the succeeding year (for the death rate)

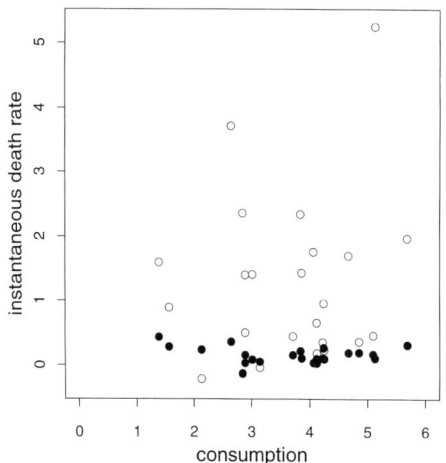

Fig. 15.7. Instantaneous death rate (a^{-1}) versus total shorebird consumption (g AFDM m^{-2}). *Macoma balthica* (*filled circles*) and *Cerastoderma edule* (*open circles*). Each point refers to a winter from the period 1975/1976–1997/1998

The result that mortality was neither related to prey abundance nor to predator numbers was unexpected. One would expect that under constant bird consumption bivalve mortality, i.e. the chance of an individual prey not to survive the year, would be higher in years when prey stocks are low. In years with poor bivalve stocks, bivalve production more or less equalled bird consumption. One might ask what would have been the cause of death of all those bivalves that died, but were not eaten by birds in those years when prey was abundant? In some of those years only a maximum of 10 % of the bivalve production suitable for the birds was actually eaten by them. What type of predator could have eaten the others, or did they die of starvation as a result of food shortage? Were they killed by sulphides (Cadée 1990) or by diseases? If they were eaten, where were all these unknown predators in poor production years when mortality could easily have been attributed to shorebird consumption? We know that densities of other possible shellfish predators, such as gulls *Larus* sp. (Meltofte et al. 1994; Cadée 1995) or shore crabs *Carcinus maenas* (L.) (unpubl. data NIOZ), did not show a numerical response to bivalve densities. If the bivalves died from starvation, is it not remarkable that the negatively density-dependent predation by shorebirds exactly balanced the positively density-dependent starvation?

15.6 Recruitment and the Regulation of Populations

An interesting feature of the bivalve-shorebird system is the spatial and temporal decoupling between the various productive phases in the life histories of both prey and predators and the actual consumption of bivalves by shorebirds. The two bivalve species have a planktonic stage during their first life phase. In spring eggs are spawned into the water and after a few weeks post-larvae leave the water column and settle on the sediments (Beukema 1993). Growth of individual bivalves mainly takes place in spring and summer. Shell growth of *M. balthica* occurs entirely in the period between March and July (Beukema et al. 1985). Most birds arrive from their breeding areas at Balgzand in late summer or autumn, at a time of the year when the productive season of the bivalves is over, and leave again in late winter. Hence for the birds full dinner is served right at the beginning of the annual period that they spend at Balgzand. The impact of bird predation on bivalve populations can thus be stated in terms of the depletion of their prey stocks in the course of winter. After the birds have left, only the surviving bivalves will contribute to the renewal of the population. The important question is thus whether the reduction in prey stocks has any consequences for the renewal of the prey population in the next spring. Stated otherwise: does the reduction in adult stock due to bird predation affect the number of bivalve recruits at the next spat fall?

Recruitment was more variable from year-to-year in *C. edule* compared with *M. balthica*. For *M. balthica* the ratio between the highest and the lowest yearly recruitment equalled 22.9. For *C. edule* the same ratio was almost three times higher: 66.4. Yet, a linear relationship between log recruitment and adult stock at the end of the preceding winter (in terms of log density m^{-2}) was lacking for *M. balthica* ($n=26$, $F=0.19$, $P=0.67$). The relationship in *C. edule* was even negative ($n=26$, $F=6.92$, $P=0.015$). The adult stock was expressed as numerical density (m^{-2}), but the use of alternative expressions like biomass density (g/m^2) did not make much difference. For neither of the two species was recruitment auto-correlated (*M. balthica*, $n=26$, $r_1=-0.13$, $P>0.05$; *C. edule*, $n=26$, $r_1=-0.17$, $P>0.05$), which is a requirement for testing the stock-recruitment relationship.

Adult fecundity is almost unrelated to succeeding recruitment (Honkoop et al. 1999), and the lack of (or a negative) stock-recruitment relationship implies that the survival probability from egg to recruit decreases with increasing adult stock. The log-ratio density of recruits/density of the adult stock at the end of the preceding winter, which expresses this survival probability, indeed significantly decreased with increasing adult stock (Fig. 15.8; *M. balthica*, $n=26$, $F=9.69$, $P=0.005$; *C. edule*, $n=26$, $F=70.5$, $P<0.001$).

This result would mean that strong density-dependent processes take place up to the time of recruitment, that is, up to the time of the sampling survey in the first summer. The increased mortality rate with increasing stocks during this first part of bivalve life may occur in the egg phase, during the early larval stages in the pelagic phase, or during or after the settlement processes. Beukema (1982) found a density-dependent mortality during the first benthic period, between the first of July and late August. Yet, not much

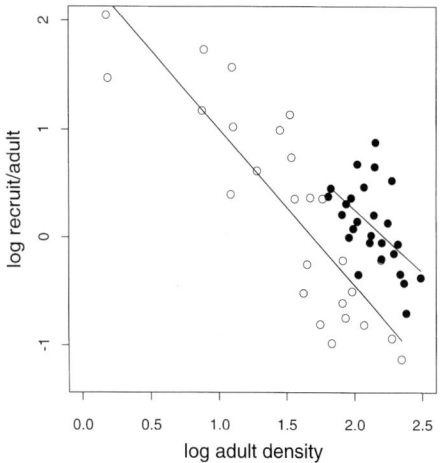

Fig. 15.8. Log of the ratio recruit density/adult density, which indicates survival from egg to recruit versus log adult density. *Macoma balthica* (*filled circles*) and *Cerastoderma edule* (*open circles*). Each point refers to adult density in late winter and recruit density in the succeeding autumn for the period 1973–1998

is known about the underlying causes of the relationships between early mortality and the size of the adult stocks. Does the larger number of eggs produced lead to shortage of food for the young, increased predation risk due to an over-compensatory numerical response of pelagic or benthic predators, or is there a higher infection risk and thus an increased chance to be struck by diseases?

Until now most attention has been paid to the direct or indirect effects of environmental factors on bivalve recruitment. It has been observed that massive recruitment in these bivalves often occurs after a cold winter (Beukema 1982, 1992). Several underlying causes for this phenomenon have been proposed. One example is that metabolic costs in relation to food input might be lower in a severe winter resulting in a better condition in spring (Zwarts 1991; Honkoop and Beukema 1997) and a higher egg production (Beukema 1992; Honkoop and Van der Meer 1997a,b). Yet this explanation has been rejected (Honkoop et al. 1999). Another explanation could be that benthic predators such as crabs and shrimps are severely reduced and arrive later on the tidal flats after a severe winter (Beukema 1992; Beukema et al. 1998). Even if this is all true, such direct or indirect effects of environmental factors do not necessarily explain why mortality in the life stages between eggs and summer spat is negatively density dependent.

The strong density dependence during the first life phase when the bivalves are not suitable as food for birds, and the density-independent mortality during their later life, mean that the dynamics of the bivalve prey populations are only loosely coupled to their avian predators. One other argument rejects the importance of bird predation on the regulation of these bivalve populations. Birds cannot achieve a neutral energy balance when prey density gets too low. For the knot, for example, Piersma et al. (1995) assume that harvestable bivalve density should not get below about half a gram per square meter. The stock-recruitment relationships that we found show that when prey densities are too low to be exploitable, they still are good enough for population renewal. Such prey refugia, as it were, may contribute to the stability of the bivalve populations.

We cannot say much about the impact of bivalve population variability on the regulation of the shorebird populations. Yet, we repeat (Piersma 1987) that a simple comparison between secondary production of a prey species and consumption of a predator (Faldborg et al. 1994) does not allow any conclusion about a possible food limitation of the predators, because higher consumption may simply enhance production.

15.7 The Scale of Population Studies

The natural system that we studied is not a simple closed predator-prey system. Both the bird and bivalve populations are open. As the bivalves have a planktonic phase during their early life stage, recruits at Balgzand may originate from other neighbouring areas. Hence a totally different explanation of the apparent lack (at least in *M. balthica*) of a clear stock-recruitment relationship is related to the open character of the bivalve population. Most of the recruits may originate from neighbouring stocks outside the Balgzand, which are out of phase with the adult stock that we measured. Yet, the high co-variability among adult stocks over a much larger geographical area than the present study area (Beukema et al. 1996) suggests that strong density dependence in the pre-recruitment period is a more likely explanation of the lack of a stock-recruitment relationship than the occurrence of local sub-populations that are dynamically out of phase.

15.8 Exclosure Experiments

The large year-to-year variability in the fraction of the bivalve production that is eaten by shorebirds also sheds some doubts on the generality of so-called exclosure experiments aimed at estimating this fraction by excluding birds from parts of the intertidal flats, if these experiments are only performed in a single or a few years. Such estimates should not be based on short-term observations (Baird et al. 1985). Due to logistic problems these experiments have to be performed over rather short periods (Piersma 1987), so it may come as no surprise that these studies have produced ambiguous results (Raffaelli and Hawkins 1996).

Table 15.3. Duration (months) versus exclosure area (m²) for various published bird exclosure experiments

Duration (months)	Exclosure area (m²)	Source(s)
4–6	2.25–25	Reise (1978)
9–12	0.25–1	Quammen (1981, 1984)
5	2.25	Kent and Day (1983)
1–2	12	Botton (1984)
4–10	4	Raffaelli and Milne (1987)
1	0.72	Wilson (1991)
0.3	2	Székely and Bamberger (1992)
3	9	Kaletja (1993)
10	4	Mercier and McNeil (1994)
1	0.48	Sewell (1996)
1.5–2	1	Botto et al. (1998)

An even more important objection to the exclosure experiments that have been performed hitherto is the use of inappropriate temporal and spatial scales (Raffaelli and Moller 2000). If exclosure experiments aim to assess the impact of bird predation on the benthos at the population or community level, one could argue that the duration should at least exceed the generation time of the prey species. Defining the generation time as the inverse of the instantaneous death rate results, in our case, in a generation time of two years for *M. balthica*. Yodzis (1988) proposed a more objective criterion for determining the length of perturbation experiments (in closed systems) and arrived at twice the summation of the generation times of the predator and its prey species. None of the exclosure experiments that we were able to find exceeded a single year and half of the studies lasted for two months or less (Table 15.3). The spatial scale of a few square meters (most studies presented in Table 15.3 used cages smaller than $4\,m^2$) is probably too small to avoid considerable edge effects, due to animals from outside migrating into the exclosures (Beukema et al. 1999).

15.9 Conclusions

Long-term observational field studies on two bivalve prey species and their main avian predators, carried out for about 30 years in the Wadden Sea, revealed that the death rate of adult bivalves is neither related to prey abundance nor to predator numbers. The long-term data also showed a lack of a stock-recruitment relationship in both bivalve species, which also points to a limited role of bird predation as a regulating factor of the prey populations. If we assume that the bivalve sub-populations in the study area are synchronized with neighbouring sub-populations in terms of adult stock size, the lack of a stock-recruitment relationship implies that strong density-dependent processes take place up to the time of recruitment. Yet, not much is known about the underlying causes of the relationships between early mortality and the size of the adult stocks. Hence, our analyses also point out some avenues for future research. Particularly interesting for understanding the population dynamics of the bivalves are the causes of mortality of bivalves other than predation by shorebirds, the density-dependent factors that operate during the early life phase of the bivalves, and the spatial coupling between adult stock and recruitment.

Still, there are important considerations that we have not touched upon here. In fact, these considerations may be quite central to the evolutionary histories of dynamic interactions between bivalve prey and their shorebird predators. The bivalves living on intertidal flats typically show a wide variety of behavioural and morphological traits that can be interpreted as antipredator adaptations, e.g. deep burying, fast growth, heavy shell armour (Boulding 1984; Vermeij 1987; Zwarts and Blomert 1992; De Goeij and Luttikhuizen 1998). Recent work by our group indicates that the deep-burying response, for example, is relaxed in favour of behaviours enhancing food intake, growth and reproduction when cues indicating the presence of predators are, even temporarily, absent. Although the present analysis has failed to demonstrate any evidence for a tight coupling between population sizes of bivalve prey and shorebird predators, we nevertheless believe that the selection pressure exerted by shorebirds will underlie aspects of the bivalves' population dynamics. Predators like birds and crabs continuously mould the life-history traits of the prey species, as individual prey not expressing traits that prevent predation are constantly weeded out of the population. It is likely that this predation effectively includes selection pressures on life-history traits that bear directly on growth, reproductive investments, and perhaps even on settlement decisions (Beukema 1993). We regard it as our challenge to find out whether in intertidal bivalves like *Macoma* and *Cerastoderma* antipredation traits reflect adaptations to local sets of predators (the predatory environment, as it were) and to explore how such adaptations in turn affect the characteristics of population dynamic processes.

Acknowledgements. We are very grateful to M. Otter for allowing us to use his bird count data. We thank Karsten Reise for giving us a forum to present these ideas. In recent years our work on bird-benthos interactions has been supported by a PIONIER grant to T.P. from the Netherlands Organization for Scientific Research (NWO).

References

Baird D, Milne H (1981) Energy flow in the Ythan estuary. Estuarine Coastal Shelf Sci 13:455–472

Baird D, Evans PR, Milne H, Pienkowski MW (1985) Utilization by shorebirds of benthic invertebrate production in intertidal areas. Oceanogr Mar Biol (Annu Rev) 23: 573–597

Beukema JJ (1974) Seasonal changes in the biomass of the macrobenthos of a tidal flat area in the Dutch Wadden Sea. Neth J Sea Res 8:94–107

Beukema JJ (1976) Biomass and species richness of the macro-benthic animals living on the tidal flats of the Dutch Wadden Sea. Neth J Sea Res 10:236–261

Beukema JJ (1981) The role of the larger invertebrates in the Wadden Sea ecosystem. In: Wolff WJ (ed) Ecology of the Wadden Sea. Balkema, Rotterdam

Beukema JJ (1982) Annual variation in reproductive success and biomass of the major macrozoobenthic species living in a tidal flat area of the Wadden Sea. Neth J Sea Res 16:37–45

Beukema JJ (1988) An evaluation of the ABC-method (abundance/biomass comparison) as applied to macrozoobenthic communities living on tidal flats in the Dutch Wadden Sea. Mar Biol 99:425–433

Beukema JJ (1989) Tidal-current transport of thread-drifting postlarval juveniles of the bivalve *Macoma balthica* from the Wadden Sea to the North Sea. Mar Ecol Prog Ser 52:193–200

Beukema JJ (1992) Expected changes in the Wadden Sea benthos in a warmer world: lessons from periods with mild winters. Neth J Sea Res 30:73–79

Beukema JJ (1993) Successive changes in the distribution patterns as an adaptive strategy in the bivalve *Macoma balthica* (L.) in the Wadden Sea. Helgoländer Meeresunters 47:287–304

Beukema JJ (1997) Caloric values of marine invertebrates with an emphasis on the soft parts of marine bivalves. Oceanogr Mar Biol (Annu Rev) 35:387–414

Beukema JJ, Knol E, Cadee GC (1985) Effects of temperature on the length of the annual growing season in the tellinid bivalve *Macoma balthica* (L.) living on tidal flats in the Dutch Wadden Sea. J Exp Mar Biol Ecol 90:129–144

Beukema JJ, Essink K, Michaelis H, Zwarts L (1993) Year-to-year variability in the biomass of macrobenthic animals on tidal flats of the Wadden Sea: how predictable is this food source for birds? Neth J Sea Res 31:319–330

Beukema JJ, Essink K, Michaelis H (1996) The geographic scale of synchronized fluctuation patterns in zoobenthos populations as a key to underlying factors: climatic or man-induced. ICES J Mar Sci 53:964–971

Beukema JJ, Honkoop PJC, Dekker R (1998) Recruitment in *Macoma balthica* after mild and cold winters and its possible control by egg production and shrimp predation. Hydrobiologia 375/376:23–34

Beukema JJ, Flach EC, Dekker R, Starink M (1999) A long-term study of the recovery of the macrozoobenthos on large defaunated plots on a tidal flat in the Wadden Sea. J Sea Res 42:235–254

Botto F, Iribarne OO, Martinez MM, Delhey K, Carrete M (1998) The effect of migratory shorebirds on the benthic species of three southwestern Atlantic Argentinean estuaries. Estuaries 21:700–709

Botton ML (1984) Effects of laughing gull and shorebird predation on the intertidal fauna at Cape May, New Jersey. Estuarine Coastal Shelf Sci 18:209–220

Boulding EG (1984) Crab-resistant features of shells of burrowing bivalves: decreasing vulnerability by increasing handling time. J Exp Mar Biol Ecol 74:201–223

Cadée GC (1990) Lokale sterfte van kokkels op het wad tijdens een *Noctiluca* bloei. Het Zeepaard 50:119–128

Cadée GC (1995) Birds as producers of shell fragments in the Wadden Sea, in particular the role of the herring gull. Geobios 18:77–85

Crisp DJ (1984) Energy flow measurements. In: Holme NA, McIntyre AD (eds) Methods for the study of marine benthos. Blackwell, Oxford

De Boer WF, Longamane FA (1996) The exploitation of intertidal food resources in Inhaca Bay, Mozambique, by shorebirds and humans. Biol Conserv 78:295–303

De Goeij P, Luttikhuizen PC (1998) Deep-burying reduces growth in intertidal bivalves: field and mesocosm experiments with *Macoma balthica*. J Exp Mar Biol Ecol 228: 327–337

De Wilde P, Beukema JJ (1984) The role of zoobenthos in the consumption of organic matter in the Dutch Wadden Sea. NIOZ Publ Ser 10:145–158

Evans PR, Herdson DM, Knights PJ, Pienkowski MW (1979) Short-term effects of reclamation of parts of Seal Sands, Teesmouth, on wintering waders and shellduck. Oecologia 41:183–206

Faldborg K, Jensen KT, Maagaard L (1994) Dynamics, growth, secondary production and elimination by waterfowl of an intertidal population of *Mytilus edulis* L. Ophelia Suppl 6:187–200

Honkoop PJC, Beukema JJ (1997) Loss of body mass in winter in three intertidal bivalve species: an experimental and observational study of the interacting effects between water temperature, feeding time and feeding behaviour. J Exp Mar Biol Ecol 212: 277–297

Honkoop PJC, Van der Meer J (1997a) Experimentally induced effects of water temperature and immersion time on reproductive output of bivalves in the Wadden Sea. J Exp Mar Biol Ecol 220:227–246

Honkoop PJC, Van der Meer J (1997b) Reproductive output of *Macoma balthica* populations in relation to winter-temperature and intertidal-height mediated changes of body mass. Mar Ecol Prog Ser 149:155–162

Honkoop PJC, Van der Meer J, Beukema JJ, Kwast D (1999) Reproductive investment in the intertidal bivalve *Macoma balthica*. J Sea Res 41:203–212

Hulscher JB (1976) Localisation of cockles (*Cardium edule* L.) by the oystercatcher (*Haematopus ostralegus* L.) in darkness and daylight. Ardea 64:292–310

Hulscher JB (1982) The oystercatcher *Haematopus ostralegus* as a predator of the bivalve *Macoma balthica* in the Dutch Wadden Sea. Ardea 64:292–311

Hulscher JB (1983) Oystercatcher (*Haematopus ostralegus* L.). In: Wolff WJ (ed) Ecology of the Wadden Sea. Balkema, Rotterdam

Kalejta B (1992) Time budgets and predatory impact of waders at the Berg River Estuary, South Africa. Ardea 75:175–187

Kaletja B (1993) Intense predation cannot always be detected experimentally: a case study of shorebird predation on nereid polychaetes in South Africa. Neth J Sea Res 31:385–393

Kent AC, Day RW (1983) Population dynamics of an infaunal polychaete: the effect of predators and an adult-recruit interaction. J Exp Mar Biol Ecol 73:185–203

Kersten M, Piersma T (1987) High levels of energy expenditure in shorebirds; metabolic adaptations to an energetically expensive way of life. Ardea 75:175–187

Meire PM, Seys J, Ysebaert T, Meininger PL, Baptist HJM (1989) A changing delta: effects of large coastal engineering works on feeding ecological relationships as illustrated by birds. In: Hooghart JC, Posthumus WS (eds) Hydroecological relations in the delta waters of the south-west Netherlands. Lakerveld, The Hague

Meire PM, Schekkerman H, Meininger PL (1994) Consumption of benthic invertebrates by waterbirds in the Oosterschelde estuary, SW Netherlands. Hydrobiologia 282/283:525-546

Meltofte H, Blew J, Frikke J, Roesner HU, Smit CJ (1994) Numbers and distribution of waterbirds in the Wadden Sea. CWSS/IWRB/WSG, Wilhelmshaven

Mercier F, McNeil R (1994) Seasonal variations in intertidal density of invertebrate prey in a tropical lagoon and effects of shorebird predation. Can J Zool 72:1755-1763

Moreira F (1997) The importance of shorebirds to energy fluxes in a foodweb of a south European estuary. Estuarine Coastal Shelf Sci 44:67-78

Piersma T (1987) Production by intertidal benthic animals and limits to their predation by shorebirds: a heuristic model. Mar Ecol Prog Ser 38:187-196

Piersma T, Beukema JJ (1993) Foodwebs in intertidal ecosystems: trophic interactions between shorebirds and their invertebrate prey. Neth J Sea Res 31:299-300

Piersma T, Hoekstra R, Dekinga A, Koolhaas A, Wolf P, Battley P, Wiersma P (1993) Scale and intensity of intertidal habitat use by knots *Calidris canutus* in the western Wadden Sea in relation to food, friends and foes. Neth J Sea Res 31:331-337

Piersma T, Van Gils J, De Goeij P, Van der Meer J (1995) Holling's functional response model as a tool to link the food-finding mechanism of a probing shorebird with its spatial distribution. J Anim Ecol 64:493-504

Quammen ML (1981) Use of exclosures in studies of predation by shorebirds on intertidal mudflats. Auk 98:812-817

Quammen ML (1984) Predation by shorebirds, fish, and crabs on invertebrates in intertidal mudflats: an experimental test. Ecology 65:529-537

Raffaelli D, Hawkins S (1996) Intertidal ecology. Chapman and Hall, London

Raffaelli D, Milne H (1987) An experimental investigation of the effects of shorebird and flatfish predation on estuarine invertebrates. Estuarine Coastal Shelf Sci 24:1-13

Raffaelli D, Moller H (2000) Manipulative field experiments in animal ecology: do they promise more than they can deliver? Adv Ecol Res 30:299-338

Reise K (1978) Experiments on epibenthic predation in the Wadden Sea. Helgoländer Wiss Meeresunters 31:55-101

Sewell MA (1996) Detection of the impact of predation by migratory shorebirds: an experimental test in the Fraser River estuary, British Columbia (Canada). Mar Ecol Prog Ser 144:23-40

Smit C (1981) Production of biomass by invertebrates and consumption by birds in the Dutch Wadden Sea area. In: Wolff WJ (ed) Ecology of the Wadden Sea. Balkema, Rotterdam

Stuart JJ, Meininger PL, Meire PM (1989) Watervogels van de Westerschelde. WWE 14. Univ Gent, Gent

Summers RW (1977) Distribution, abundance and energy relationships of waders (Aves: Charadrii) at Langebaan Lagoon. Trans R Soc S Afr 42:483-494

Székely T, Bamberger Z (1992) Predation of waders (Charadrii) on prey populations: an exclosure experiment. J Anim Ecol 61:447-456

Van der Meer J, Duin RNM, Meininger PL (1996) Statistical analysis of long-term monthly oystercatcher counts. Ardea 84A:39-56

Velasquez CR, Kaletja B, Hockey PAR (1991) Seasonal abundance, habitat selection and energy consumption of waterbirds at the Berg River Estuary, South Africa. Ostrich 62:109-123

Vermeij GJ (1987) Evolution and escalation. An ecological history of life. Princeton Univ Press, Princeton

Wiersma P, Piersma T (1994) Effects of microhabitat, flocking, climate and migratory goal on energy expenditure in the annual cycle of knots. Condor 96:257–279

Wilson WHJ (1991) The foraging ecology of migratory shorebirds in marine soft-sediment communities: the effects of episodic predation on prey populations. Am Zool 31:840–848

Wolff WJ (1991) The interaction of benthic macrofauna and birds in tidal flat estuaries: a comparison of Banc d'Arguin, Mauritania, and some estuaries in the Netherlands. In: Elliot M, Ducrotoy JP (eds) Proceedings 19th ECSA Symp Int Symp Ser

Wolff WJ, De Wolf L (1977) Biomass and production of zoobenthos in the Grevelingen Estuary, The Netherlands. Estuarine Coastal Mar Sci 5:1–24

Wolff WJ, Smit CJ (1990) The Banc d'Arguin, Mauritania, as an environment for coastal birds. Ardea 78:17–38

Wolff WJ, Van Haperen AMM, Sandee AJJ, Baptist HJM, Saeijs HLF (1976) The trophic role of birds in the Grevelingen estuary, The Netherlands, as compared to their role in the saline Lake Grevelingen. In: Persoone G, Jaspers E (eds) Proceedings of the 10th European Symposium on Marine Biology. Universa Press, Wetteren

Wolff WJ, Duiven AG, Duiven P, Esselink P, Gueye A, Meijboom A, Moerland G, Zegers J (1993) Biomass of macrobenthic tidal flat fauna of the Banc d'Arguin, Mauritania. Hydrobiologia 258:151–163

Yodzis P (1988) The indeterminacy of ecological interactions as perceived through perturbation experiments. Ecology 69:508–515

Zwarts L (1991) Seasonal variation in body weight of the bivalves *Macoma balthica, Scrobicularia plana, Mya arenaria* and *Cerastoderma edule* in the Dutch Wadden Sea. Neth J Sea Res 28:231–245

Zwarts L, Blomert AM (1992) Why knots *Calidris canutus* take medium sized *Macoma balthica* when six prey species are available. Mar Ecol Prog Ser 83:113–128

Zwarts L, Drent RH (1981) Prey depletion and the regulation of predator density: oystercatchers (*Haematopus ostralegus*) feeding on mussels *(Mytilus edulis)*. In: Jones NV, Wolff WJ (eds) Feeding and survival strategies of estuarine organisms. Plenum, London

Zwarts L, Blomert AM, Wanink JH (1992) Annual and seasonal variation in the food supply harvestable by knots *Calidris canutus* staging in the Wadden Sea in late summer. Mar Ecol Prog Ser 83:129–139

Zwarts L, Ens BJ, Goss-Custard JD, Hulscher JB, Kersten M (1996) Why oystercatchers *Haematopus ostralegus* cannot meet their daily energy requirements in a single low water period. Ardea 84A:269–290

16 Experimental Approaches to Integrating Production, Structure and Dynamics in Sediment Communities

D. RAFFAELLI and M. EMMERSON

16.1 Introduction

The comparative approach to intertidal ecology has a long history, prompted initially by the recurrence of obvious biological patterns worldwide on rocky shores (Stephenson and Stephenson 1949, 1972). The universal zonation patterns found on rocky shores, whereby the same kinds of organisms (lichens, barnacles and kelps) dominate the upper, middle and lower parts, respectively, of shores in many parts of the world, provided a strong indication that the same, strong underlying mechanisms and processes were at work on these shores (Raffaelli and Hawkins 1996). However, the comparative approach has not been so easy for sediment ecologists. Unlike rocky shores, the fauna of sediment beaches is not readily accessible so that recurrent biological patterns are not immediately obvious. Despite this, Dahl (1952) was able to demonstrate a certain universality in the zonation of isopod and amphipod crustaceans in sandy beaches, at least in the north-east Atlantic, whilst in other regions similar zones were occupied by other taxa (e.g. John and Lawson 1991; Santelices 1991; Chap. 3) or by functional groups (Brown and McLachlan 1990; McLachlan 1990). Physical patterns, associated with sediment water content, are often more readily apparent than biological patterns on sandy shores (Salvat 1964, 1967), and these may in turn underpin the major biological patterns described above (McLachlan 1990; Raffaelli et al. 1991). The close association between sediment characteristics and benthic communities has also long been recognised in sublittoral systems through the pioneering work in the North Sea of Thorson (1957, 1966), Glenmarec (1969, 1973), and Petersen (1914–24), where different sediment grades typically support characteristic species, such as the bivalves *Macoma, Tellina, Venus* and *Abra*, the echinoderms *Amphiura* and *Ophiura*, and the amphipod *Haploops* – the so-called Peterson-Thorson communities. Whilst it is now clear that the rigid, categorical association of entire communities with sediment grades (and other physical properties of the habitat) is probably not tenable

(Stephenson et al. 1971; Gray 1981), this comparative approach has been useful in generating hypotheses about likely animal-sediment interactions.

The comparison of biological patterns across different kinds of intertidal and sublittoral sediments and similarities in patterns between biogeographical regions provide indirect evidence of common structuring processes (Reise 1985; Brown and McLachlan 1990; Raffaelli and Hawkins 1996). However, similar patterns may be generated by a variety of alternative mechanisms and it is not always safe to assume causal links between pattern and process. Direct, large-scale comparisons of the processes themselves are preferable, but more difficult because much greater effort is usually required. However, it is now possible to make such comparisons with respect to two functional processes, energy flow and community dynamics.

Describing the flows of material between biological compartments in ecological systems was a major focus of research programmes in the 1960s and 1970s. Many of these were associated with the International Biological Programme (IBP), one of the aims of which was to "*ensure a world-wide study of ... organic production on the land, in fresh waters, and in the seas ...*" (Worthington 1965). The approach adopted by the IBP provided a sensible basis for the comparison of production (and hence food webs) in systems as diverse as rain forests, taiga and, of course, sandy shores and mudflats. Although the role of the IBP largely evaporated in the mid-1970s, the programme did deliver many methodological handbooks (e.g. Vollenwider 1969; Holme and McIntyre 1971) and facilitated a great deal of research on production and energy flow in shallow-water marine systems worldwide (Raffaelli 2000a). This initiative provided the springboard for later detailed studies at many other locations, such as in Gullmar fjord (Pihl 1985) and South-west England (Warwick et al. 1975). The most significant feature of the IBP is that it provided an agreed common methodological framework, mainly through the standard protocols contained in the IBP manuals, and this greatly facilitated the comparative approach.

These comparative studies identified the significance of allochthonous imports of carbon for many sediment systems, where in situ primary production can often be of little importance, except perhaps at the local scale for the most sheltered environments where macrophyte biomass is high (Raffaelli and Hawkins 1996). A second feature to emerge from comparative energy flow studies was the overriding importance of sediment particle size, and hence the morphodynamic state of the shore (Brown and McLachlan 1990), for the retention and processing by heterotrophs of allochthonous organic matter on which higher trophic levels ultimately depend.

In parallel with production studies, intertidal ecologists began to apply the experimental manipulative approach to shores in an effort to identify the important processes which might organise communities. However, in marked contrast to the coordinated approach to energy flow facilitated by the IBP, the experimental manipulative approach has never been coordinated on a global scale and investigations have proceeded idiosyncratically. In particular, the approach has suffered from a lack of agreed protocols for the design and analysis of experiments – the equivalent of the IBP manuals. Although there are now several excellent texts available (e.g. Hairston 1989; Underwood 1997), many of the early experiments were poorly designed, leading to potential confusion in the literature (Raffaelli and Moller 2000). For instance, many published field experiments were not in fact truly replicated (Hurlbert 1984) and many are inappropriately analysed (Underwood 1981), making any formal meta-analysis (Raffaelli and Moller 2000) problematic.

To the authors' knowledge, the only serious attempts at a regional-scale comparative experimental study involving manipulations of populations are that of Bob Paine and his colleagues of the occurrence and significance of keystone predators on rocky shores in different regions of the Pacific (Paine 1966; Paine et al. 1985), although recently there has been a Europe-wide programme on the recruitment dynamics of the barnacle *Semibalanus balanoides* (Jenkins et al. 1999). Shallow-water sediment ecologists have tended not to adopt this comparative approach to field experiments, so that comparisons along the reflective-dissipative gradient and between biogeographical regions can only be made by drawing together a variety of different studies (see reviews in Reise 1985; Raffaelli and Hawkins 1996). Several common processes emerge from manipulative experiments in shallow-water assemblages: (1) keystone predators (species which mediate coexistence amongst a superior competing species and other members of the assemblage) are not a feature of most of the sediment shores studied; (2) in general, top-down effects are weaker than bottom-up effects; (3) small epibenthic consumers, such as crabs and shrimps, tend to have a larger impact than larger fish and shorebirds, except perhaps at the local scale in the case of rays and skates, or where shorebird densities are extremely high; (4) bioturbation, disturbance and stabilisation can have important local-scale effects on the organisation of infaunal communities, and in some cases these effects may scale up to the landscape (see Chap. 7); and (5) biological interactions may have more significance for community organisation on dissipative than on reflective shores. For key references to (1)–(5), the reader should consult Reise (1985), Hall et al. (1994) and Raffaelli and Hawkins (1996).

The energy flow and the manipulative approaches have indicated that similar processes might operate to organise communities in similar ways on similar shores. Whilst both approaches deal with interactions between species, the natures of these interactions (flows of material and dynamic changes in numbers of individuals) are quite different and can result in different perspectives as to which specific processes are important. This point is well illustrated by Paine's (1980) analysis of a simple rocky-shore food where he showed that trophic links along which large amounts of energy flow are often (but not always) functionally insignificant, whilst those links along which virtually no energy flows may be all-important in the organisation of the dynamics of the system. Specifically, sea urchins *Strongylocentrotus* ingested little energy from kelps, because they grazed the settling spores, but removal of urchins resulted in a large increase in the kelp *Hedophyllum*, a competitive dominant (Paine 1980). Surprisingly, there have been few such comparisons carried out on sediment shores, although Raffaelli and Hall (1996) reached a similar conclusion to that of Paine (1980) for a mudflat food web.

There is clearly a pressing need to understand the links between, on the one hand, energy flow and, on the other, the dynamics of community organisation; a need also highlighted by the Second International Food Web Symposium held in Colorado in 1993 (Polis and Winemiller 1996). Here, we describe several areas where the production and organisation of sediment shore communities may well be linked. The investigations described are focused heavily on a well-worked system, the Ythan estuary, Aberdeenshire, UK (see Baird and Milne 1981; Gorman and Raffaelli 1993 for an overview of this system), because few similar studies have been carried out elsewhere. However, we hope that the approaches described here may be taken up by workers in other regions so that between-region comparisons can become possible. Specifically, we discuss the relationships between production subsidies and food chain dynamics, between production and body-size relationships, and between production and biodiversity.

16.2 Effects of Production Subsidies on Food Chain Dynamics

Polis et al. (1997) have suggested that the consumer-resource dynamics in a food web may be profoundly affected by subsidies of material originating from outside the system. For example, terrestrial insects falling into a stream may sustain a high abundance of predatory fish, leading to a reduction in invertebrate grazers and a subsequent benthic algal bloom. The fish can have no effect on the supply rate or production of the subsidy (terrestrial insects) and the dramatic effect of the subsidy on the primary producer is indirect. The most compelling evidence for such effects comes from studies on desert food webs subsidised by carbon from the adjacent oceanic fringe (Polis and Hurd 1996b), but such effects may be more common than realised (Polis et al. 1997). However, experimental tests of the hypothesised effects of subsidies on food web dynamics are lacking (Riley et al. 2000)

By virtue of their open nature, intertidal flats receive large amounts of allochthonous material (subsidies) from the sea, and, in the case of estuarine mudflats, also from the riverine catchment. Intertidal flats, therefore, provide an ideal opportunity to explore the role (if any) that subsidies might play in the dynamics of food webs. Intertidal flats have the additional advantage that, unlike many systems, experimental manipulations can be carried out with relative ease. In this section, we describe the results of a series of field experiments and other research on the Ythan estuary food web that are relevant to the subsidy question. Specifically, we examine the hypothesis that subsidies which increase the abundance of primary producers, invertebrate consumers and their predators, have cascading effects on community organisation, as postulated by Polis et al. (1997).

16.2.1 Effects of Subsidised Primary Producers

The most significant primary producers on the Ythan are the green macroalgae *Enteromorpha*, *Ulva* and *Chaetomorpha*, all of which seem to be nitrogen-limited in this system (Raffaelli 1999). Most of the nitrogen entering the system is imported to the mudflats by the river entering the estuary, being ultimately derived from farmland. Nitrate concentrations arriving at the estuary have increased three-fold (from about 3 mg NO_3 l^{-1} to 8 mg NO_3 l^{-1}) from the 1950s to date (Raffaelli 1999). The biomass and distribution of green macroalgae has similarly increased over the same period, so that 40 % of all mudflats are now affected by algal mats with biomasses of 1–3 kg wet wt. m^{-2}. The links between land-use, nutrients and algal biomass are compelling and we are confident that primary production in this system is subsidised by nutrients from terrestrial sources (Raffaelli 1999).

But what effect does this subsidised macroalgal biomass have on the mudflat invertebrate assemblage? This issue has been explored through a series of manipulative field experiments, where invertebrate assemblages were compared under different levels of algal biomass (Hull 1987; Raffaelli 2000a; Fig. 16.1). In none of these experiments was there any evidence of an increase in macroalgal consumers. In other words, elevation of the resource (algae) by a subsidy (nutrients) did not affect consumer-resource dynamics. Indeed, densities of most invertebrates were dramatically reduced by macroalgal mats, due to the hostile redox environment and physical interference with invertebrate-feeding mechanisms (Raffaelli et al. 1998; Raffaelli 2000a). Only the opportunistic polychaete *Capitella* sp. consistently increased in abundance under macroalgal mats, presumably due to the organically enriched sediments (Fig. 16.1). Thus, whilst enhanced biomass of algae does have effects on invertebrate assemblages at the local scale, these are mainly indirect, through changes in habitat quality rather than trophic dynamics. However, the densities of deposit-feeding invertebrates have increased in other sections of the estuary not affected by macroalgal mats

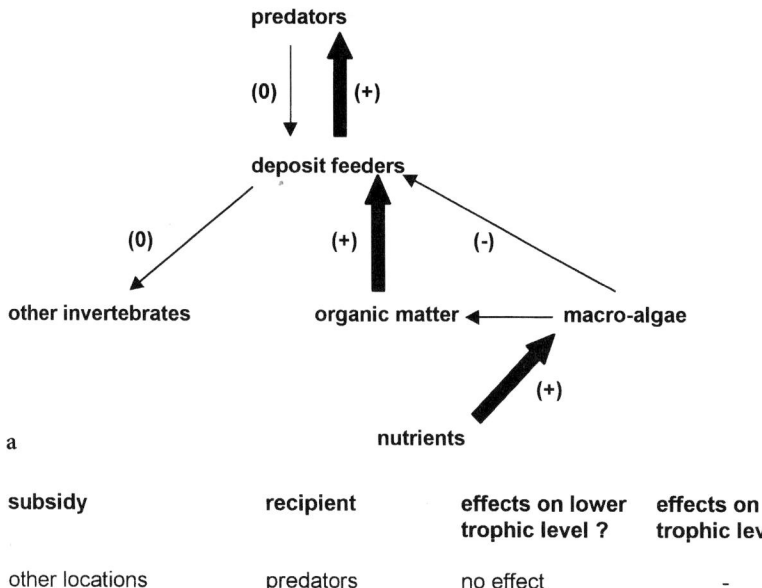

Fig. 16.1. Effects of subsidies on different trophic levels in the Ythan estuary food web (from Riley et al. 2000). **a** Significant effects indicated by the thickness of the *arrows* and **b** summary of effects on different trophic levels

(Raffaelli et al. 1999), almost certainly due to an enhanced subsidy of organic material from mats growing elsewhere in the estuary. Thrush et al. (2000) have addressed similar issues of scale-dependency in the interpretation of field experiments with respect to the dynamics of bivalves.

If there is an enhanced abundance of deposit feeders due to organic subsidies, do these deposit feeders exert top-down effects on other biota? This question has been examined through a series of field experiments where the abundances of the deposit feeders *Corophium volutator*, *Hydrobia ulvae*, *Hediste diversicolor* and *Arenicola marina* were enhanced and the effects on the rest of assemblage assessed (Raffaelli and Hall 1992; Ragnarsson 1996; Limia and Raffaelli 1997). In none of these experiments was there any evidence of significant top-down effects of subsidised deposit feeders on lower trophic levels (Fig. 16.1). In contrast, there is compelling evidence of a bottom-up effect of enhanced deposit-feeder densities on shorebirds, numbers of which have increased in the same areas as those of deposit feeders (Raffaelli 2000a).

16.2.2 Effects of Subsidised Predators

Given the highly mobile nature of many of the predators in the Ythan (shorebirds, fish and epibenthic crustaceans), it could be argued that the abundance of these species is regularly subsidised (tidally or seasonally) by resources located outside of the system. In other words, their abundance within the estuary is maintained at a level higher than can be supported by the invertebrate prey production within the estuary itself. We therefore tested the hypothesis that subsidised predator biomass can depress prey densities through several field experiments where enhanced densities of gobiid fish (*Pomatoschistus minutus* and *P. microps*) and epibenthic crustaceans (*Carcinus maenas* and *Crangon crangon*) were maintained on the mud flat (Jacquet and Raffaelli 1989; Raffaelli et al. 1989). In none of these experiments was there evidence of significant top-down effects of predators on the abundance of their invertebrate prey (Raffaelli and Hall 1992, 1996; Fig. 16.1).

In summary, our research on the role of subsidies in the dynamics of the Ythan estuary food web indicates that there is a significant positive effect of nutrients on macroalgae, which, in turn, has a negative effect on the abundance of invertebrates (and hence their predators) locally, through the production of an organically enriched and hostile physicochemical environment. In contrast, subsidies of organic matter (fuelled ultimately by the nutrient subsidy) may enhance densities of deposit feeders and increase shorebird abundance at the whole estuary scale (Raffaelli 2000a). However, manipulative field experiments provide no evidence to support the idea that top-down effects by these subsidised consumers reduce the abundance of their resources, as postulated by Polis et al. (1997).

16.3 Production and Body-Size Distributions

The association of levels of production with organism body size is well documented for marine benthic systems, as are the reciprocal effects of body size on in situ secondary production (e.g. Gerlach et al. 1985). A feature of the classic Pearson and Rosenberg (1978) model of the response of benthic communities to organic enrichment is the relative predominance of small body-sized species under the most enriched conditions, with larger body-sized species only accommodated under normal inputs or organic matter. The cause-effect relationships in this model are complex, involving size-related life-history traits and variation in the living space made available by the depth of the chemocline, but the empirical pattern of body size is consistent and compelling. However, there have been surprisingly direct experimental tests of the effects of increased system production on body-size distributions in marine benthic assemblages. In this section, we describe two related conceptual approaches – constraint space plots and biomass spectra – that have recently been applied to the Ythan system, and assess their utility for examining the relationship between production and body size. The two approaches are similar in that they both involve plots of body size against abundance, but differ in their representation as a consequence of their different origins in terrestrial and marine ecology respectively.

16.3.1 Constraint Space Plots

Much of the research on the relationship between abundance and body size comes from terrestrial systems (Brown and Maurer 1986; Gaston and Lawton 1988; Lawton 1989, 1990; Cotgreave 1993; Blackburn and Gaston 1994, 1997, 1999), although there have been a few freshwater (Strayer 1994; Cyr et al. 1997a,b) and one rocky shore (Navarette and Menge 1997) study. A potentially useful summary plot of such data from entire communities is the constraint space (Fig. 16.2), described by Brown (1995). This space has three main features: (1) the upper and lower bounds generally decline as average body size increases; (2) for species of very large size there is a region where minimum densities may be independent of body size; and (3) for very small organisms there is a region where minimum densities may be independent of body size and maximum densities may increase with body size (Fig. 16.2).

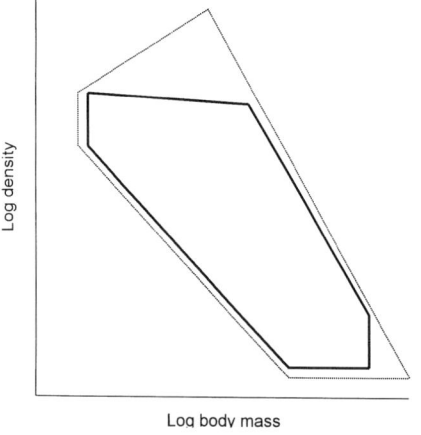

Fig. 16.2. The constraint space hypothesised by Brown (1985; *dotted line*) and that from a real food web, the Ythan estuary (*solid line*). (Leaper and Raffaelli 1999)

Lawton (1990) has proposed that the upper bound of the constraint space will vary with system productivity and there is some evidence for this from freshwater studies (Cyr et al. 1997a). Exploring this proposition in marine systems is difficult, because few have been sufficiently well documented to permit the representation of a constraint space under different production scenarios. One such system is the Ythan food web (Fig. 16.2). In their analysis, Leaper and Raffaelli (1999) plotted the densities of 92 taxa, ranging in body size from a few micrograms to several kilograms (Fig. 16.2). They concluded that, whilst the general form of the regression relationship between body size and density reported for other types of systems was similar for the Ythan, the slope of the main trend line was sensitive to taxonomic resolution (Leaper and Raffaelli 1999). Also, the shape of the Ythan constraint space differed slightly from that proposed by Brown (1995; Fig. 16.2). More importantly, this approach was shown to be insensitive to changes in the body size/abundance plot brought about by whole-system enrichment. The large-scale changes in the ecology of the estuary due to nutrient enrichment (a three-fold increase in nitrogen), described in the preceding section, were not visually obvious as a shift in the position of upper bound (Lawton 1990), because of the use of a log-log plot, whilst the statistical tools available for such comparisons are not sufficiently well developed for complex polygons (Scharf et al. 1998; Leaper and Raffaelli 1999). Thus, whilst the constraint space provides a useful conceptual model for thinking about body size-density relationships, it may have limited application for detecting environmental perturbations, such as enrichment.

16.3.2 Biomass Size Spectra

Plots of the abundance of different sized organisms in sediments (biomass spectra) provide a useful alternative to traditional species-abundance plots for representing benthic assemblages. Biomass spectra have the added advantage that organism size has functional implications for tropho-dynamics (e.g. Gerlach et al. 1985; Sprules and Munawar 1986; Boudreau et al. 1991). Schwinghamer (1981) pioneered the description of body-size spectra from marine sediments (Fig. 16.3), but there have been few subsequent studies, probably due to the large effort involved. From the relatively few studies carried out to date, it seems that benthic biomass spectra are conservative in shape, with a biomass trough at a body size of around 0.5–1 mm, separating meiofaunal and macrofaunal taxa (Schwinghamer 1981, 1985, 1988). Other authors have found evidence for this biomass trough less convincing (Duplesia and Hargrave 1996; Ramsay et al. 1997; Leaper et al. 2000) or only convincing for numbers (i.e. not biomass) spectra (Raffaelli et al. 2000b). If the trough does occur, it may reflect a discontinuity in the physical

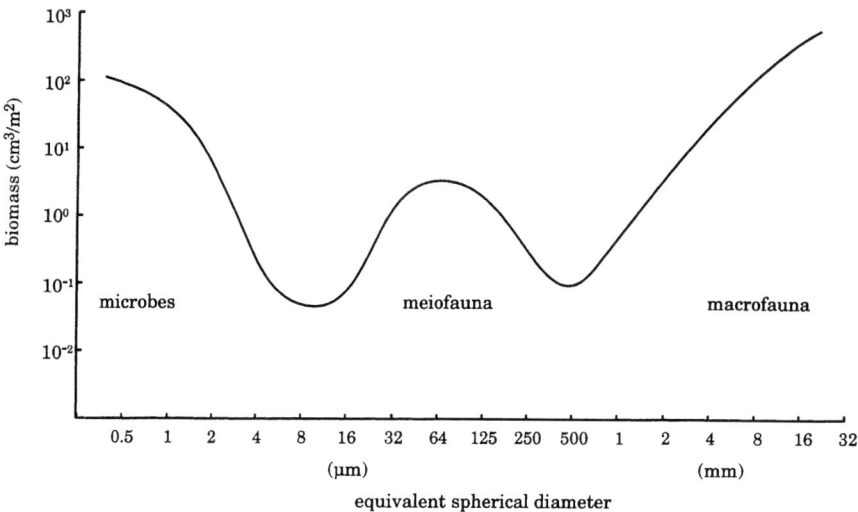

Fig. 16.3. The benthic biomass spectrum. (After Schwinghamer 1981)

architecture of the sediment matrix (Schwinghamer 1981; Leaper et al. 2000) or long-term selection by large predatory meiofauna for macrofauna recruits to settle at large body sizes (Warwick 1989).

We have explored the relationship between production and the shape of the benthic spectrum for areas of mudflat on the Ythan estuary (Raffaelli et al. 2000b). These areas were treated with two levels of enrichment in the form of dried, powdered *Enteromorpha* applied at 10 and 30 g per 225 cm^2. This is equivalent to 150 and 450 g C m^{-2}, an increase in sediment organic carbon of 25 and 75 % respectively. At the end of the experiment (8-week duration) the sediment physicochemistry and the relative abundance of species in the different treatments were consistent with previous enrichment studies on the Ythan (Hull 1987; Raffaelli et al. 1991). That is, a decline in *Corophium volutator*, an increase in *Capitella* sp. and a decrease in sediment redox potential. Responses of spectra to enrichment were assessed by defining the location of the macro- and meiofaunal peaks in the spectrum using a kernel estimation method (see Manly 1996, Leaper et al. 2001; Raffaelli et al. 2000b for details). Our analyses show that there were no significant differences between treatments in the locations of either peaks or troughs, although the locations of the meio-/macrofaunal trough for both enrichment treatments were at a slightly smaller body size than for the control (Fig. 16.4). An obvious feature of the spectrum from the treatment receiving the highest levels of enrichment (30 g) is the lack of individuals in the size classes bordering the trough. This experiment confirmed that the location of peaks and troughs was much less sensitive to enrichment than pelagic spectra (Schwinghamer 1985).

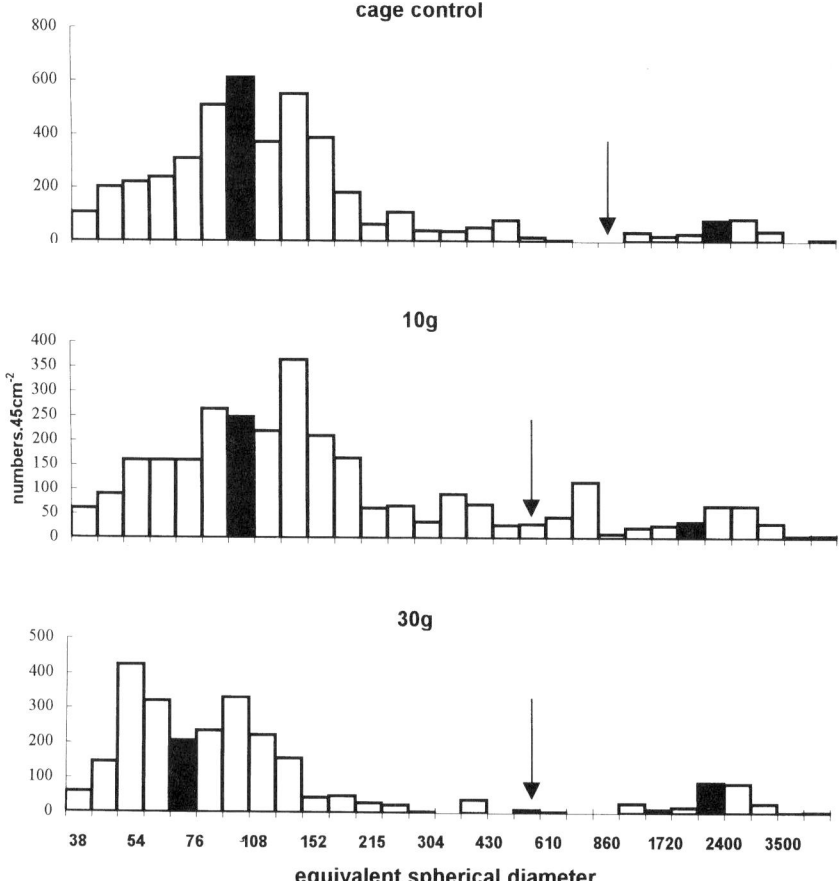

Fig. 16.4. Body-size spectra of invertebrates in an undisturbed area (225 cm^2) of intertidal mudflat, and in similar areas enriched with 10 and 30 g of *Enteromorpha* (see text), Ythan estuary, Scotland. *Solid bars* on spectra show location of meio- and macrofaunal modes and *arrows* indicate locations of troughs, as defined by a kernel estimation procedure. Spectra are the mean of five replicates per treatment. (Raffaelli et al. 2000)

16.4 Production and Biodiversity

Many factors determine species richness (biodiversity), but the supply of useable energy has been increasingly proposed as a mechanism (e.g. Wright et al. 1993; Hall 2000). However, the relationship between energy supply and species richness is probably scale-dependent (Hall 2000). On a global scale, richness increases monotonically with energy supply (Currie 1991), whilst on regional scales (hundreds of kilometres), a hump-shaped relationship exists between diversity and productivity, with the greatest diversity at intermediate levels of productivity (Grime 1973; Rosenzweig and Abramsky 1993). At small scales (metres to kilometres) increases in productivity through the addition of nutrients lead to declines in species richness (DiTommaso and Aarsen 1989; Schindler 1990).

These relationships can be accommodated to some extent within Tilman's (1987) resource heterogeneity hypothesis (RHH). The RHH states that if the landscape is uniform with respect to resource availability, at low levels of resource the average location will sustain few species and productivity will be low. As a mean quality of the landscape increases, spatial variability and diversity of resources also increase, in turn leading to increases in both diversity and productivity. At high levels of resource availability, space and resources become less patchy and diversity is reduced, because when all sites are equally good, competitively superior species dominate. The RHH thus predicts that increasing nutrient supplies will lead to increases in species richness at sites with low initial productivity, but will lead to decreases in species richness at sites where resources are already abundant.

To test these predictions both we (Emmerson 2000) and Hall (2000) manipulated nutrient levels (and hence primary productions) in replicate small-scale artificial habitat units (HUs). In Hall's study, these were placed near the sediment surface in a shallow sub-tidal seagrass bed, Boston Bay, Port Lincoln, South Australia, a region with a particularly low nutrient status, whilst we placed HUs in the Ythan estuary which is characterised by higher nutrient levels (Raffaelli 1999). These two locations thus provide contrasting biogeographies as well as differing in their nutrient status. Although there are some minor differences in protocol between the Australian and the Ythan studies, the experiments are essentially identical and so comparisons are valid. Commercial pan scourers were used to construct artificial HUs into which plant and animals could assemble. Nutrient supply was manipulated by placing 60g of commercial (Osmocote) slow-release fertiliser (nitrate and phosphate) in a fine mesh bag inside each HU (see Hall 2000 for technical specifications regarding Osmocote and nutrient release rates). In the Ythan, HUs were suspended above an intertidal mudflat and left for 20 weeks (May–October). Each week, replicate HUs were collected, macrofauna enumerated, and species richness, species accumulation and Shannon-Weiner indices of diversity calculated. The RHH here predicts a decrease in species richness given the addition of nutrients (additional resources) in the Ythan study. The Ythan has a high nutrient status and hence resources are already widely available. In contrast the RHH predicts an increase in species richness (diversity) and productivity given the addition of nutrients in the Boston Bay study; this area has a low nutrient status and increasing resources here serves to increase the mean quality of the environment.

Analysis of the Ythan pooled data using a two-way ANOVA without replication (treatment and time were considered fixed factors) demonstrated that there was no significant difference in species richness between treatment (Osmocote) and control HUs ($F_{1,18}=2.744$, $P>0.1$, Fig. 16.5c), although more species were identified from enriched HUs than controls during the 20-week period (Fig. 16.5b). However, enriched HUs had a significantly lower diver-sity (Shannon-Weiner measure) than control HUs ($F_{1,18}=4.99$, $P<0.039$, Fig. 16.5a). It appears therefore that the rate of species turnover

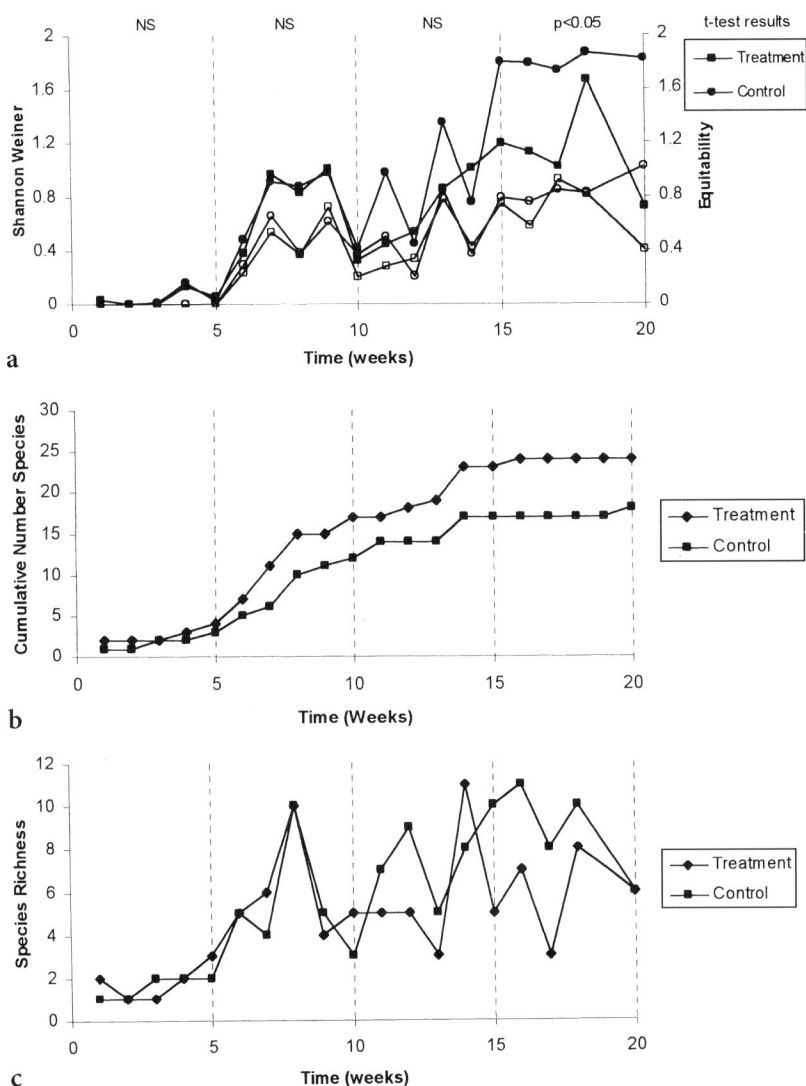

Fig. 16.5. Species diversity (richness and Shannon-Wiener measure) on artificial habitat units (pan scourers), with and without slow-release fertiliser, placed in Ythan estuary, Scotland, for 20 weeks. (Emmerson 2000)

(transient or ephemeral species) was greater in the enriched HUs than in controls.

In South Australia, Hall (2000) demonstrated that algal diversity increased in enriched habitats relative to control areas along with biomass. The total number of faunal species and total number of individuals were also significantly greater in enriched habitats. Thus, Hall's observations are consistent with the predictions of the RHH for a low nutrient site, whereas our own are consistent with respect to the Shannon-Weiner measure of diversity for a high nutrient site.

16.5 Conclusions

Comparisons of ecological processes along environmental gradients and between regions are likely to yield important insights into the key factors organising assemblages on sandy and muddy shores. However, unlike comparisons of the distribution patterns of dominant species, comparisons of processes demand agreed protocols if such comparisons are to be meaningful. The International Biological Programme provided the ideal structure for studies on production and energy flow in marine systems, whereas the lack of agreed protocols for experimental ecologists means that manipulative studies are more idiosyncratic and hence more difficult to compare.

In this chapter, we have discussed several areas where there may be links between production and community structure and dynamics. However, the number of similar studies elsewhere is as yet too few to permit general comparisons between shores. The use of artificial habitat units has merits in this respect, in that they are cheap, standardised and easily deployed. Indeed, other researchers have already used them successfully to explore patterns of marine biodiversity (Gee and Warwick 1996; Kendall et al. 1996). Establishing links between production and dynamics is difficult, but is becoming a pressing issue given the scale of alteration in the fluxes of materials brought about by global climate change – at present we are simply not able to predict the effects of these changes in these fluxes on community structure and dynamics.

References

Baird D, Milne H (1981) Energy flow in the Ythan estuary, Aberdeenshire, Scotland. Estuarine Coast Shelf Sci 12:217–232

Blackburn TM, Gaston KJ (1994) Animal body size distributions: patterns, mechanisms and implications. Trends Evol Ecol 9:471–474
Blackburn TM, Gaston KJ (1997) A critical assessment of the form of the interspecific relationship between abundance and body size in animals. J Anim Ecol 66:233–249
Blackburn TM, Gaston KJ (1999) The relationship between animal abundance and body size: a review of the mechanisms. Adv Ecol Res 28:181–210
Boudreau PR, Dickie LM, Kerr SR (1991) The body size spectrum of biomass as an indicator of ecosystem production. J Theor Biol 152:329–339
Brown AC, McLachlan A (1990) Ecology of sandy shores. Elsevier, Amsterdam, 328 pp
Brown JH (1995) Macroecology. Univ Chicago Press, Chicago
Brown JH, Maurer BA (1986) Body size, ecological dominance and Cope's rule. Nature 324:248–250
Cotgreave P (1993) The relationship between body size and population abundance in animals. Trends Evol Ecol 8:244–248
Currie DJ (1991) Energy and large-scale patterns of animal and plant species richness. Am Nat 137:27–49
Cyr H, Downing JA, Peters RH (1997a) Density – body size relationships in local aquatic communities. Oikos 79:333–346
Cyr H, Peters RH, Downing JA (1997b) Population density and community size structure: comparison of aquatic and terrestrial systems. Oikos 80:139–149
Dahl E (1952) Some aspects of the ecology and zonation of the fauna on sandy beaches. Oikos 4:1–27
DiTomaso A, Aarsen LW (1989) Resource manipulations in natural vegetation: a review. Vegetatio 84:9–29
Duplisea DE, Hargrave BT (1996) Response of meiobenthic size-structure, biomass and respiration to sediment organic enrichment. Hydrobiologia 339:161–170
Emmerson M (2000) Biodiversity and ecosystem function in shallow water marine systems. PhD Thesis, University of Aberdeen, UK
Gaston KJ, Lawton JH (1988) Patterns in the distribution and abundance of insect populations. Nature 331:709–712
Gee JM, Warwick RM (1996) A study of global biodiversity patterns in the marine motile fauna of hard substrata. J Mar Biol Assoc UK 76:177–184
Gerlach SA, Hahrland AE, Schrange M (1985) Size spectra of benthic biomass and metabolism. Mar Ecol Prog Ser 26:161–173
Glenmarec M (1969) La Grande Vasiers: biocenologie. CR Hebd Sceance Acad Sci Paris 268:401–404
Glenmarec M (1973) The benthic communities of the European north-Atlantic. Annu Rev Mar Biol Oceanogr 11:263–289
Gorman ML, Raffaelli D (1993) The Ythan estuary. Biologist 40:10–13
Gray JS (1981) The ecology of marine sediments. Cambridge Univ Press, Cambridge
Grime JP (1973) Control of species density in herbaceous vegetation. J Environ Manage 1:151–167
Hairston NG (1989) Ecological experiments: purpose, design and execution. Cambridge Univ Press, Cambridge
Hall SJ (2000) Biodiversity-productivity relations: an experimental evaluation of the mechanisms. Oecologia (in press)
Hall SJ, Raffaelli D, Thrush S (1994) Patchiness and disturbance in shallow water benthic assemblages. In: Giller P, Hildrew AW, Raffaelli D (eds) Aquatic ecology; scale, pattern and process. Blackwell, Oxford, pp 333–373
Holme N, McIntyre AD (1971) Methods for studying the marine benthos. Blackwell, Oxford

Hull S (1987) The growth of macroalgal mats on the Ythan estuary, with respect to their effects on invertebrate abundance. PhD Thesis, Univ Aberdeen, UK

Hurlbert S (1984) Pseudo replication and the design of ecological field experiments. Ecol Monogr 54:187-211

IBP (1969). Directory of national participation in IBP. Marine Productivity (MP). IBP Central Office

Jaquet N, Raffaelli D (1989) The ecological importance of the sand goby *Pomatoschistus minutus* (Pallas). J Exp Mar Biol Ecol 128:147-156

Jenkins SR, Aberg P, Cervin RA, Coleman RA, Delany S, Della-Santini P, Hawkins SJ, La Croix E, Myers AA, Lindegarth M, Power AM, Roberts MF, Hartnoll RG (1999) Spatial and temporal variation in settlement and recruitment of the intertidal barnacle *Semibalanus balanoides* (L.) (Crustacean : Cirripedia) over a European scale. J Exp Mar Biol Ecol 243:209-226

John DM, Lawson GW (1991) Littoral ecosystems of tropical western Africa. In: Mathieson AC, Nienhuis PH (eds) Intertidal and littoral ecosystems. Elsevier, Amsterdam, pp 297-323

Kendall MA, Widdicombe S, Davey JT, Somerfield PJ, Austen MCV, Warwick RM (1996) The biogeography of islands: preliminary results from a comparative study of the Isles of Scilly and Cornwall. J Mar Biol Assoc UK 76:219-222

Lawton JH (1989) What is the relationship between population density and body size in animals? Oikos 55:429-434

Lawton JH (1990) Species richness and population dynamics of animal assemblages. Patterns in body size: abundance space. Philos Trans R Soc Lond B 330:283-291

Leaper R, Raffaelli D (1999) Defining the abundance body-size constraint space: data from a real web. Ecol Lett 2:191-199

Leaper R, Raffaelli D, Emes C, Manly BFJ (2001) Constraints on body-size distributions: an experimental test of the habitat architecture hypothesis. J Anim Ecol 70:248-259

Limia J, Raffaelli D (1997) Effects of burrowing by the amphipod Corophium volutator on the ecology of intertidal sediments. J Mar Biol Assoc UK 77:408-413

Manly BFJ (1996) Are there clumps in body-size distributions? Ecology 77:81-86

McLachlan A (1990) Dissipative beaches and macrofauna communities on exposed intertidal sands. J Coast Res 86:57-71

Navarette SA, Menge BA (1997) The body size - population density relationship in tropical intertidal communities. J Anim Ecol 66:557-566

Paine RT (1966) Food web complexity and species diversity. Am Nat 100:65-75

Paine RT (1980) Food webs: linkage, interaction strength and community infrastructure. J Anim Ecol 49:667-685

Paine RT, Castillo JC, Cancino J (1985) Perturbation and recovery patterns of starfish dominated assemblages in Chile, New Zealand and Washington State. Am Nat 125: 679-691

Pearson TH, Rosenberg R (1978) Macrobenthic succession in relation to organic enrichment and pollution of the marine environment. Annu Rev Mar Biol Oceanogr 16:229-311

Petersen CGJ (1914) Valuation of the sea. II. The animal communities of the sea bottom and their importance for marine zoogeography. Rep Danish Biol Stn 21:1-44

Petersen CGJ (1915) On the animal communities of the sea bottom in the Skaggera K, the Christiana Fjord and Danish waters. Rep Danish Biol Stn 23:3-28

Petersen CGJ (1918) The sea bottom and its production of fish food. A survey of the work done in connection with the valuation of the Danish waters from 1883-1917. Rep Danish Biol Stn 25:1-62

Petersen CGJ (1924) A brief survey of the animal communities in Danish waters. Am J Sci 7:343–354
Pihl L (1985) Food selection and consumption of mobile epibenthic fauna in shallow marine areas. Mar Ecol Prog Ser 22:169–179
Polis GA, Hurd SD (1996a) Allochthonous input across habitats, subsidized consumers and apparent trophic cascades; examples from the ocean-land interface. In: Polis GA, Winemiller KO (eds) Food webs; integration of patterns and dynamics. Chapman and Hall, London, pp 275–285
Polis GA, Hurd SD (1996b) Linking marine and terrestrial food webs: allochthonous input from the ocean supports high secondary production on small islands and coastal land communities. Am Nat 147:396–423
Polis GA, Winemiller KO (eds) (1996) Food webs; integration of patterns and dynamics. Chapman and Hall, London
Polis GA, Anderson WB, Holt RD (1997) Toward and integration of landscape and foodweb ecology; the dynamics of spatially subsidised food webs. Annu Rev Ecol Syst 28:289–316
Raffaelli D (1999) Nutrient enrichment and trophic organisation in an estuarine food web. Act Oecol 20:449–461
Raffaelli D (2000a) Interactions between macro-algal mats and invertebrates in the Ythan estuary, Aberdeenshire, Scotland. Helgoland Mar Res 54:71–79
Raffaelli D (2000b) Trends in food web research. J Exp Mar Biol Ecol 250:223–232
Raffaelli DG, Hall SJ (1992) Compartments and predation in an estuarine food web. J Anim Ecol 61:551–560
Raffaelli DG, Hall SJ (1996) Assessing the relative importance of trophic links in food webs. In: Polis GA, Winemiller KO (eds) Food webs; integration of patterns and dynamics. Chapman and Hall, London, pp 185–191
Raffaelli D, Hawkins S (1996) Intertidal ecology. Chapman and Hall, London, 356 pp
Raffaelli D, Moller H (2000) Manipulative field experiments in animal ecology: do they promise more than they can deliver? Adv Ecol Res 30:299–338
Raffaelli D, Hull S, Milne H (1989) Long term changes in nutrients, weed mats and shore birds in an estuarine system. Cah Biol Mar 30:259–270
Raffaelli DG, Limia J, Hull S, Pont S (1991) Interactions between invertebrates and macroalgal mats on estuarine mudflats. J Mar Biol Assoc UK 71:899–908
Raffaelli DG, Raven JA, Poole LJ (1998) Ecological impact of green macroalgal blooms. Annu Rev Mar Biol Oceanogr 36:97–125
Raffaelli D, Balls P, Way S, Patterson IJ, Hohman SA, Corp N (1999) Major changes in the ecology of the Ythan estuary, Aberdeenshire: how important are physical factors? Aquat Conserv Mar Freshwater Ecosyst 9:219–236
Raffaelli D, Hall S, Emes C, Manly BFJ (2000) Constraints on body size distributions: an experimental approach using a small-scale system. Oecologia 122:389–398
Ragnarsson S (1996) Successional patterns and biotic interactions in intertidal sediments. PhD Thesis, Univ Aberdeen, UK
Ramsay PM, Rundle SD, Attrill MJ, Uttley MG, Williams PR, Elsemere PS, Abada A (1997) A Rapid method for estimating biomass size spectra of benthic metazoan communities. Can J Fish Aquat Sci 54:1716–1724
Reise K (1985) Tidal flat ecology. An experimental approach to species interactions. Springer, Berlin Heidelberg New York
Riley R, Townsend C, Raffaelli D, Flecker W (2000) Subsidies along the freshwater-estuary continuum: implications for food web dynamics. In: Polis GA (ed) Subsidies in food webs (in press)

Rosenzweig ML, Abramsky Z (1993) How are diversity and productivity related? In: Ricklefs RE, Schluter D (eds) Species diversity in ecological communities. Chicago Univ Press, Chicago, pp 52–65

Salvat B (1964) Les conditions hydrodynamiques interstitielles des sediments meuble intertidaux et la repartition verticale de la faune endogenee. CR Acad Sci Paris 259: 1576–1579

Salvat B (1967) La macrofauna carcinologique endogenee des sediments meubles intertidaux (tanaiduces, isopodes et amphipodes), ethologie, bionomic et cycle biologique. Mem Mus Nat Hist Paris Ser A 45:1–275

Santilices B (1991) Littoral and sublittorla communities of continental Chile. In: Mathieson AC, Nienhuis PH (eds) Intertidal and littoral ecosystems. Elsevier, Amsterdam, pp 347–370

Scharf FS, Juanes F, Sutherland M (1998) Inferring ecological relationships from the edges of scatter diagrams: comparison of regression techniques. Ecology 79:448–460

Schindler DW (1990) Experimental perturbations of whole lakes as tests for hypotheses concerning ecosystem structure and function. Oikos 57:25–41

Schwinghamer P (1981) Characteristic size distributions of integral benthic communities. Can J Fish Aquat Sci 38:1255–1263

Schwinghamer P (1985) Observations on size-structure and pelagic coupling of some shelf and abyssal benthic communities. In: Gibbs PE (ed) Proceedings of the 19th European Marine Biological Symposium. Cambridge Univ Press, Cambridge, pp 347–360

Schwinghamer P (1988) Influence of pollution along a natural gradient and in a mesocosm experiment on biomass-size spectra of benthic communities. Mar Ecol Prog Ser 46:199–206

Sprules GW, Munawar M (1986) Plankton size spectra in relation to ecosystem productivity, size and perturbation. Can J Aquat Sci 43:1789–1794

Stephenson TA, Stephenson A (1949) The universal features of zonation between tide marks on rocky coasts. J Ecol 38:289–305

Stephenson TA, Stephenson A (1972) Life between tide marks on rocky shores. Freeman, San Francisco, 425 pp

Stephenson W, William WT, Cook SG (1971) Computer analysis of Petersen's original data on bottom communities. Ecol Monogr 42:387–415

Strayer DL (1994) Body size and abundance of benthic animals in Mirror Lake, New Hampshire. Freshwater Biol 32:83–90

Thorson G (1957) Bottom communities (sublittoral or shallow shelf). In: Hedgepeth JW (ed) Treatise in marine ecology and paleoecology. I. Ecol Mem Geol Soc Am 67: 461–534

Thorson G (1966) Some factors influencing the recruitment and establishment of marine benthic communities. Neth J Sea Res 3:267–293

Thrush SF, Hewitt JE, Cummings VJ, Green MO, Funnel GA, Wilkinson MR (2000) The generality of field experiments: interactions between local and broad-scale processes. Ecology 81:399–415

Tilman D (1987) Secondary succession and the pattern of plant dominance along experimental nitrogen gradients. Ecol Monogr 57:189–214

Underwood AJ (1981) Techniques of analysis of variance in marine biology and ecology. Annu Rev Mar Biol Oceanogr 19:513–605

Underwood AJ (1997) Experiments in ecology: their logical design and interpretation using analysis of variance. Cambridge Univ Press, Cambridge

Vollenwider R (1969) Methods for assessment of primary productivity in freshwaters. Blackwell, Oxford

Warwick RM (1989) The role of meiofauna in the marine ecosytem: evolutionary considerations. Zool J Linn Soc 96:229-241
Warwick RM, Joint IR, Radford PJ (1975) Secondary production of the benthos in an estuarine mudflat. In: Jeffries RL, Davy AJ (eds) Ecological processes in coastal environments. Blackwell, Oxford, pp 429-450
Worthington EB (1965) The International Biological Programme. Nature 208:225-226
Wright DH, Currie DJ, Maurer BA (1993) Energy supply and patterns of species richness on local and regional scales. In: Ricklefs RE, Schluter D (eds) Species diversity in ecological communities. Chicago Univ Press, Chicago, pp 66-74

Synthesis: Comparative Ecology of Sedimentary Shores

K. Reise

In the biota of sedimentary shores a few dominant species often manage to transform the coastal ecotone. Examples are the reefs or beds of suspension feeders, seagrass beds and algal mats. Dominant bioturbating worms, crabs and sometimes vertebrates may modify topography, hydrodynamics, sediment properties, and alter the habitat structure and nutrient cycling. The rest of the species are relegated to niches positioned relative to these dominants. Trophic pathways are short, usually in three steps from microalgae to invertebrates to fish or birds.

Suspension Feeders

Suspension feeders tend to aggregate in dense patches, beds or reefs on sedimentary shores and dominate the benthic fauna (Dame et al., Chap. 1). These suspension feeders sit smugly in the sediment and are served all their lives by the tidal waters once their pelagic dispersal stages have settled among conspecifics. Currents and waves import food, oxygen and competent larvae and disperse waste and propagules. Suspension feeders take advantage of the shallow coastal conditions where the phytoplankton of the photic zone approaches the bottom and benthic microalgae are stirred up in mixed water masses. By stretching out above the bottom, by being epibenthic or by generating hummocks or reefs, the aggregated suspension feeders themselves enhance turbulent mixing. On a small scale and moderate density, this positive effect of crowding on the food supply may outweigh competition for food, at least along the edges of suspension-feeder patches (Green et al. 1998).

Where the exchange rates of water masses between nearshore and offshore are low, as in estuaries and lagoons, suspension feeders significantly graze down phytoplankton with cascading effects on the pelagic food web. By facilitating the quick remineralization of nutrients, there is also a positive loop from suspension feeders back to the nutrient supply for further primary production. The combination of high grazing pressure with high release rates of nutrients may shift the phytoplankton towards smaller cell sizes. Dense assemblages of suspension feeders are a major agent in the bentho-pelagic coupling of the coastal ecosystem. They also resemble rocky shores by providing settlement sites, crevices and solid surfaces for algae and their grazers. Reefs of mussels, oysters, vermetid gastropods and polychaetes also constitute a transition to reefs of clonal cnidarians. The latter are regarded as hard-bottom structures, and thus are outside the scope of this volume on sedimentary shores.

The occurrence of dense suspension-feeder assemblages may be limited by the supply of planktonic food to a coastal region. Alternatively, new establishment of beds or reefs may depend on the rare coincidence of several factors required to initiate a new dense assemblage. Thus, there may be a spacious capacity of suitable but empty habitat. The recent proliferation of mussel and oyster cultures seems to indicate that many coasts can support more suspension feeders than naturally occur. However, crashes of oyster stocks by disease could be a sign of local over-cultivation.

Over-exploitation of natural oyster beds on sandy bottoms near the island of Sylt in the North Sea prompted Möbius (1877) to sketch the first ecological community concept. He saw the oyster population in the context of its predators, competitors, diseases, and the relevant physical factors, showing that the millions of eggs produced are necessary to sustain the natural stock and are not just a surplus for the benefit of the oyster market. His recommendation for modest and prudent use of this natural resource was in vain. Dense beds of suspension feeders have since then been recognized as habitats to a diverse community of associated species (Commito and Dankers, Chap. 2). In this respect, mussel beds on soft-bottoms from both sides of the North Atlantic are compared (Fig. 1). Barnacles are common epifaunal associates, while oligochaetes with their direct development and tolerance to enriched sediments and low oxygen concentrations dominate the infauna. Field experiments have revealed the significance of grazing snails for the control of algal epigrowth. Algae may enhance suffocation of mussels by accumulating mud. The interaction between the flow regime and the physical structure of mussel aggregates is a key variable for associated species as well as for the performance of the mussels themselves. The complex spatial structure and the dynamics can be quantified by means of fractal geometry.

Sandy beaches are subject to strong physical forcing (Jaramillo and Lastra, Chap. 3). Their biota show a high degree of similarity between continents and climatic zones. At the more exposed beaches, the top contributors to the biomass are very often mobile suspension feeders, composed of anomuran crabs (*Emerita* spp.) and bivalves (*Mesodesma* and *Donax* spp.; Fig. 2). These occur in aggregates with the juveniles more upshore and the adults more

Synthesis: Comparative Ecology of Sedimentary Shores 359

Fig. 1. Mussel bed (*Mytilus edulis*) with barnacle epigrowth and grazing littorinid snails on intertidal sediments in the North Sea

Fig. 2. White ibis feeding on anomuran crabs (*Emerita portoricensis*) and bivalves (*Donax variabilis*) on a sandy beach in the Gulf of Mexico

downshore, and at gentle beach slopes they also perform tidal migrations. Highest abundance and biomass are found at beaches where upwelling waters provide an ample food supply.

In most coastal waters the amount of suspended food is a highly variable commodity and growth phases of suspension feeders are then limited to short seasons. On the other hand, when phytoplankton is abundant, this is a very desirable food source, and some surface deposit feeders find an advantage in switching to suspension feeding (Riisgård and Kamermans, Chap. 4). Various benthic invertebrates of the shore are capable of gathering food directly from the water as well as from the sediment surface. Generally, high flow conditions favour suspension feeding, although there are upper limits, while in calm waters deposit feeding is more profitable.

Some animals may generate their own currents and thus perform suspension feeding even under otherwise low flow conditions. This is done by the ragworm *Nereis diversicolor*, a common polychaete in European estuaries. It lives in a U-shaped burrow and consumes deposits from the sediment surface in the immediate surrounding of its burrow openings. In spring and summer, when phytoplankton concentration is high, the worm remains in the burrow, moves backward and secretes a fine-meshed mucus net, and then pumps water through the funnel-shaped net-bag by undulating movements of its body. After a few minutes the worm moves forward again and swallows the net together with the retained particles. Using this method, the worms are as efficient in suspension feeding as mytilid mussels. They may even deplete their planktonic food source in shallow waters under low flow conditions.

The tellinid clam *Macoma balthica* is less efficient in suspension feeding. It has a narrow inhalant siphon and a rather small gill area for filtering. Nevertheless, this common bivalve of the North Atlantic shores may occasionally be forced to employ suspension feeding. It normally sweeps food particles from the surrounding sediment surface with its long inhalant siphon, while otherwise remaining with its body buried well below the surface. The long siphons are often cropped by fish when the tide is in, and wading birds probe for the whole animal when the tide is out. Thus, there is a trade-off between suspension or deposit feeding and exposure to predators. At times it may be necessary to retract the siphons from the sediment surface when there are too many encounters with siphon nippers. Also, the remaining length of the siphon may be used to stay as deep as possible below the surface, out of reach to flocks of probing birds. In these situations, *M. balthica* switches to suspension feeding. The food supply will be rarely sufficient for long. However, for short periods reduced rations of food may be the preferred evil when predators abound.

Feeding behavior may vary with habitat conditions and individual fates. *M. balthica* in the European Wadden Sea mainly feeds when the tide is in. The related *Tellina opalina* is abundant on tidal flats in the Indo-Pacific. It uses the inhalant siphon to suck in the detritus that accumulates in the troughs between ripples, only when the tide is out (Reise 1987). Presumably, when the tidal flats are submerged, there are too many fish eager to crop the siphons. Comparing feeding behavior of shore animals across distant coasts may indicate what the major food resources are or which kind of predation pressures prevail. Experiments and behavioral studies are essential to determine the causes for switches in feeding modes.

Biogenic Stabilization and Disturbances

Organisms transform their sedimentary environment by stabilizing as well as destabilizing it (Fig. 3). Major stabilizers are the suspension feeders with their hummocks and reefs. Others are the rooted plants that slow down the flow of water. On sheltered shores, mats of microbes bind sediments with their mucus secretions. Some stabilizers may even entail net accretion, causing the sea bottom to rise and the shoreline to prograde seaward. Conversely, many animals destabilize the sediment and enhance erosion, and then occasionally

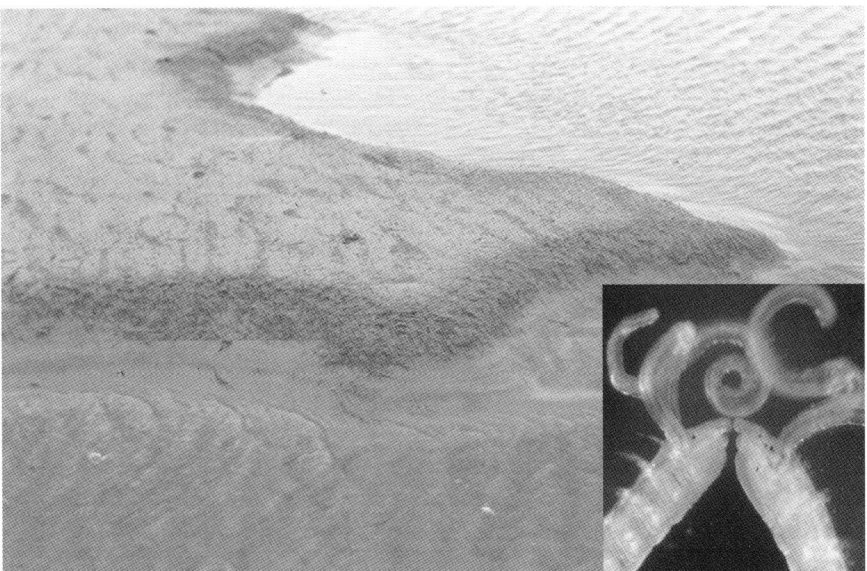

Fig. 3. Tubes of *Pygospio elegans* form a dense mat resisting erosion on a sandy tidal flat in the North Sea. This versatile spionid polychaete can filter by building a mucus net within its tube, and uses its palps to catch plankton or deposits from the sediment surface

cause the shoreline to retreat landwards. Major destabilizers are burrowing and deposit-feeding invertebrates, as well as fish, birds and mammals that dig feeding pits. Direct consumption of stabilizers may also result in a destabilization of the sediment. The effects of stabilizers and destabilizers on sediment layering and composition, erosion and accretion have cascading effects throughout the biotic community and produce a distinct, dynamic patchiness at the sedimentary shore.

Sediments are composed of a continuum of particle sizes where the larger ones are mostly spherical and the smaller ones elongated and flat. While the former behave as independent particles, the latter are attracted to each other by Van der Waals forces and form cohesive particles. This cohesiveness is substantively enhanced by microbes secreting organic material, termed extracellular polymeric substances (Paterson and Hagerthey, Chap. 5). The microbes may also bind larger sand particles to each other. At the surface of shore sediments, it is the microphytobenthos that primarily increases the cohesiveness. Major components are the diatoms which attach to sand grains or move freely on muddy surfaces. Cyanobacteria often bind sediments with their meshwork of filaments. Species diversity of microbes consolidating sediments attains a maximum halfway between clean sand and pure mud.

Hydrodynamics are the major force determining sediment composition and stability, but the microbes counteract and often generate cohesive sediments in spite of a strong flow by trapping and binding small particles, which otherwise would become resuspended. Treating the physical and biotic components separately would inhibit our understanding of the sediment properties; this is a salutary lesion for all soft-sediment ecologists. Sediment stability is also a function of the microphytobenthic species composition. Diatom species differ with respect to their secretion activity and certain species of cyanobacteria may generate conspicuous, consolidated sedimentary structures, including the laminated stromatolites also known from the earliest history of life.

Cycles in the abundances of sediment stabilizers and destabilizers may occur. On both sides of the northern Atlantic, benthic diatom growth proceeds in spring and then comes under grazing pressure by corophiid amphipods in summer. As a consequence, sediment stability first increases and then decreases, and thus accretion gives way to erosion. In late summer, migrant wading birds arrive and feed on the amphipods. This partially releases the diatoms from grazing pressure, and sediment stabilization is achieved again. In winter, light limitation and higher hydrodynamic forces hamper the diatoms, and then the sediment returns to instability. Eroding conditions prevail until the following spring when diatoms again bloom (Daborn et al. 1993; Cadée, Chap. 6).

Sediment reworking by bioturbators is important in the vertical transport of particles. The latter can be quantified as the layer of sediment that is reworked annually. Expressing the magnitude of bioturbation in this way, there is a notable increase from high to low latitudes. In cold to temperate nearshore sediments, lugworms (Arenicolidae) are dominant bioturbators and these rework some 5–40 cm annually, depending on the length of their active season. At warm temperate to tropical shores, large decapod crustaceans take over, particularly callianassid shrimps, which rework some 50 to more than 100 cm annually.

Effects of large invertebrate bioturbators, lugworms and callianassid shrimps on the associated fauna are both positive and negative (Flach and Tamaki, Chap. 7). In the European Wadden Sea, lugworm burrows provide an attractive habitat for some meiofauna, an amphipod and a scale worm species. In southern Japan, callianassid burrowing benefits a mobile isopod near the surface and a snake eel in the burrow. Both of the large deposit feeders decrease sediment stability. This negatively affects various smaller infauna. In the Wadden Sea, tube-building amphipods and spionid polychaetes, in Japan a trochid snail and small polychaetes. Different species of bioturbators from distant tidal flats apparently have very similar effects and the basic processes are the same.

In southern England, erosion occurs at the edge of salt marshes, where in earlier times sediment accretion prevailed with a prograding pioneer vegetation of salt marsh plants. Experiments show that invertebrates, particularly the ragworm *Nereis diversicolor*, play a key role in the sedimentary processes at the upper shore (Hughes, Chap. 8). These worms graze on the microphytobenthos, on filamentous algae and on seeds and seedlings of the vascular plants. This indirectly reduces shear strength of the sediment and may enhance erosion. At level surfaces within the marsh, further net accretion will be impeded. Alongside salt marsh creeks, a positive feedback is initiated by grazing on the sediment binding and accreting algae. Erosion is facilitated and further increased because creeks become wider, are filled with more water, and this increases tidal flow dynamics and hence erosion. More wading birds would help to decrease the invertebrates and this may halt the erosional process. Maintaining salt marshes in front of a dyked land may prove very difficult in the face of sea-level rises.

Not only invertebrates, but also rays, ducks, geese, flamingoes, walrus and Gray whales among others contribute to sediment disturbances (Cadée, Chap. 6). Their feeding pits range from tens of centimeters up to 4 m in length. Gray whales stay with their young in coastal lagoons and hence were among the most susceptible to overkill by early whalers. They have become extinct from the shores of the Atlantic and are presently confined to the eastern North Pacific coasts. These baleen whales sieve benthic fauna out of the sediment. This causes considerable bioturbation and creates a very patchy habitat for the benthos.

Seagrasses and the Benthic Fauna

The seagrasses of the sedimentary shores tend to occur in dense meadows (Den Hartog and Phillips, Chap. 9). These constitute an analogue to the kelp forests of rocky coasts. Seagrass beds are highly productive, delivering leaf material to the detritus pool of the entire coastal ecosystem, attract grazers such as fish, ducks, geese and dugongs, calm down hydrodynamics, and stabilize and modify the sediment where they are rooted. At the lower shore, they may provide a sheltered habitat for a rich community of algae, invertebrates and fish. The composition and diversity of this associated community largely depends on the quality of the surrounding habitats, and disjunct patches of seagrass attract more visitors than large continuous stands.

The center of seagrass diversity resides in the subtidal zone of the tropical Indo-West-Pacific. Plant size, structural complexity, and species richness of the seagrasses tend to decline towards the intertidal and towards the temperate to arctic zones. Intertidal species may be found subtidally when the larger subtidal species fail to occur. In the Baltic Sea, density and biomass of the *Zostera marina* stands decrease with salinity (Baden and Boström, Chap. 10). Seagrass meadows shift from shallow and sheltered sites on a rich substrate under marine conditions to deeper and exposed sites on poor substrates under the most brackish conditions. This niche displacement is apparently caused by competition with fresh water plants in the brackish waters. Seagrass beds occur usually separated from marshes and mangroves, and do not constitute a successional step towards the terrestrial vegetation.

The dynamics of seagrass beds are usually slow, and many questions are still unanswered. How much empty but suitable habitat is there available for seagrasses at the sedimentary shores? Stands persisting over several decades are known as well as spatially shifting beds (Den Hartog and Phillips, Chap. 9). Except for the introduced seagrass species, *Zostera japonica* invading the North American Pacific coast and *Halophila stipulacea* the Mediterranean Sea, no recent expansions of seagrass beds have been recorded. Instead, seagrass beds are in decline all over the world. Although climatic events are often the proximate cause, as with exceptional cloudiness in the northern Atlantic during the 1930s or a cyclone in northern Australia in 1992, the ultimate causes for the decline or the lack of recovery often are anthropogenic. One such factor is deforestation with the resulting erosion and a subsequent increase in turbidity in the coastal waters. Dredging operations and embankments, enhanced eutrophication and industrial waste, and the introduction of competitive algal species, are further factors contributing to the decline of seagrasses.

An ecological comparison at the eastern Atlantic coast from cold to warm temperate to subtropical climate reveals a number of interesting aspects

(Sprung et al., Chap. 11; Michaelis and Wolff, Chap. 12). In the coastal waters along this gradient from north to south, mean temperature, seasonality, insolation and salinity increase. Nutrient and phytoplankton supply to the coastal biota, on the other hand, decreases in the same direction. The seagrass *Zostera noltii* occurs throughout this latitudinal range but increases in areal share from north to south. Vascular plants dominate the primary production in the south. In the Wadden Sea, salt marsh vegetation is marginal. In the Ria Formosa lagoon in Portugal, *Spartina* marshes and stands of shrubby halophytes are extensive. At the Banc d' Arguin in northern Africa, where seagrasses cover most of the tidal flats, there are only a few stands of *Spartina* and stunted mangroves. Due to the arid climate, unvegetated sebkhas extend above mean high tide line. Herbivory on the vascular plants is insignificant in the south, whereas the beds of seagrass and the salt marshes in the Wadden Sea are seasonally grazed by migrant geese and ducks.

In the benthic fauna, species richness increases and biomass decreases from north to south. The decrease in biomass and production is most pronounced in the suspension feeders. The dense beds of cockles and mussels in the Wadden Sea find no analogue in the Ria Formosa and the Banc d'Arguin, presumably because the food supply from the sea decreases from north to south on this eastern Atlantic coast. Connolly and Roughgarden (1998) explain a decreasing coverage at the rocky shore with barnacles and mussels along the North American Pacific coast from north to south with different current regimes. In the north, nearshore circulation promotes larval recruitment, in the south this is inhibited by strong offshore currents. It may be speculated that recruitment to the benthic fauna with pelagic larvae to the sedimentary shores along the eastern Atlantic coast may follow a similar trend with positive effects of the oceanographic regimes in the north and negative ones in the south.

The southern sites also have no equivalent to the large populations of the lugworm (*Arenicola marina*) and the mud snails (*Hydrobia ulvae*) in the north, which primarily feed on the rich microphytobenthos. The provision of food to upper trophic levels by the zoobenthos depends on the nutrient supply to easily digestible unicellular algae in the plankton and in the sediment. Migrant wading birds feed on their flyway on the zoobenthos in all three regions, particularly in the Wadden Sea and the Banc d' Arguin, and must cope with these different food provisions. Predation pressure on the zoobenthos by flood tide visitors has not been quantified sufficiently to allow comparisons along this latitudinal gradient.

The benthic macrofauna on tidal flats in the temperate zone and in the tropics is always strongly dominated by sets of a few species. A main difference seems to be that in the tropics total species richness is higher and there is a greater variety of taxa. Also, within and between regions, there may be a higher variety in the tropics than in the temperate zones. Suspension feeders

Fig. 4. Soldier crabs (*Mictyris longicarpus*) rework the upper sediment when deposit feeding during low tide emergence and affect many other organisms on tidal flats in the Indo-Pacific region

may be more important in the zoobenthos of temperate tidal flats than on most tropical flats. On the other hand, cold temperate regions lack abundant burrowing decapod crustaceans such as fiddler crabs and callianassid shrimps, which often are prominent on tidal flats in the warmer regions. However, for a balanced comparison more studies along latitudinal gradients are needed, and particularly more tropical sites are waiting to be investigated.

Differences between tropical tidal flat faunas, for example between northern Australia and the Pacific coast of Central America, are mainly caused by different types of organisms achieving dominance (Dittmann and Vargas, Chap. 13). Hoards of soldier crabs (*Mictyris longicarpus*) abound on tidal flats from Africa to Australia but have no analogue on east Pacific coasts (Fig. 4). They rework the upper sediment during tidal immersion and profoundly affect the rest of the invertebrate fauna. Thus, in spite of similar physical conditions, evolved differences between biogeographic provinces entail different processes to be important. Conversely, similarities arise when the same taxonomic groups dominate at such distant shores, i.e., fiddler crabs, lingulid brachiopods, capitellid and spionid polychaetes, naticid and nassariid gastropods.

Dynamic Structures and Trophic Supplies

Structural patterns in the benthos of sedimentary shores are highly dynamic. Comparing patterns helps to reveal underlying dynamics. This often requires long-term observations to discover significant correlations between variables, and manipulative field experiments to understand the mechanisms. Patterns of spatial heterogeneity and biodiversity in marine sediments have been explained with the disturbance mosaic model proposed by Johnson (1973), assuming that local disturbances produce patches, and among these the recovery of benthic assemblages may be out of phase. Disturbances are biogenic on a small scale and physical over much larger scales. Anthropogenic sediment disturbances are frequently caused by coastal engineering, changes in land-use or bottom trawl or dredge fisheries.

All these disturbances are very common on sedimentary shores, and an understanding of their role is important. Similar disturbance experiments have been performed on coasts with very different biota, for example in New Zealand and New England (USA) (Thrush and Whitlatch, Chap. 14). Biotic interactions during the course of recovery are highly variable but the provision of habitat structure by larger organisms produces consistent patterns. The differential mobility of benthic organisms is a key variable in the process of recolonization. On a small scale crawling and bed-load transport prevail, on a wider scale drifting of benthic stages with the tidal currents is of importance, and finally planktonic larvae allow for colonization from very distant sources. Because of these different modes of dispersal, the scaling-up from small and short field experiments to the recovery of an entire bay cannot be a linear exercise and is not trivial. Studies on dispersal are an important challenge for ecologists of sedimentary shores because of the complexity of hydrodynamics in these environments and the multiple life stages over which dispersal occurs.

High mobility in prey and predator populations also complicates attempts to answer questions such as whether shorebirds differ in their impact on their bivalve prey between regions (van der Meer et al., Chap. 15). Respective field experiments with cage exclosures have not been sufficiently large or long. Difficulties also arise when consumption/production ratios are calculated by assuming species-specific production rates or assuming data from a single year to be representative. Calculated ratios of shorebird consumption versus zoobenthic production varied along the latitudinal flyway from Europe to Africa without any trend. On average some 30 % of the overall production is consumed by birds. However, what does this mean for the dynamics and the regulation of the prey populations? A data set spanning almost three decades revealed no relation between the death rate of adult bivalves and the numbers of their main shorebird predators, knots and oystercatchers. This lack of correlation still does not imply that shorebirds are unable to exert any selection pressure on the life-history traits of their prey.

Field experiments consistently indicated that small epibenthic predators preying on early bivalve recruits exert a very strong effect on population dynamics. Due to the small size of predator and prey, the biomass consumed is much less than that by the shorebirds that feed on the adult bivalves. Nevertheless, in terms of community structure and dynamics, the former link is more important than the latter. Often it requires a spatio-temporal mismatch between the bivalve recruits and the juvenile epibenthic predators to give the former a chance to reach maturity and to grow up into the prey size spectrum of the birds.

Sedimentary shores are strongly subsidized from the sea and in estuaries also from the land (Raffaelli and Emmerson, Chap. 16). Pelagic and benthic algae thrive where nutrients are plentiful, and then also the zoobenthic fauna may abound. This bottom-up effect may also be a negative one when an excess of nitrogen is supplied. Mats of green algae often quickly cover the entire sediment surface, and then most of the benthic macrofauna either dies or escapes the sulphidic conditions underneath the mats. On a larger scale, however, the green algae constitute an organic enrichment and, as detritus, they benefit those deposit feeders that have not been covered by them. Thus, bottom-up effects are basic for the abundance of benthic invertebrates at sedimentary shores while small predators affect their dynamics.

Increasing nutrient supply is a common corollary of high human population density and intensive agriculture next to sedimentary shores. The effects differ between sites, depending on the hydrodynamic energy and the nutrient supply level. When hydrodynamics are strong, the response potential of the biota to increased resource supply is physically limited. When conditions are sheltered and oligotrophic, an enrichment may increase biotic production and diversity. However, when conditions are already eutrophic, a further enrichment may have no or even negative effects. This has been confirmed by field experiments following the same protocol on a coast in southern Australia with a low and in a Scottish estuary with a high nutrient status. Within artificial habitat units of identical size and structure, diversity of colonizing organisms was higher in units with fertilizers than in unfertilized controls at the oligotrophic site. At the eutrophic site, colonizer diversity was lower in the fertilized habitat units.

General Conclusions

Comparisons show that sedimentary shores constitute ecotones which receive significant organic subsidies from the sea and locally also from the land. The latter is governed by modern land-use practices. Coastal upwelling and nearshore circulations, onshore winds and pronounced seasonality may enhance the seaside supply. The amount of this supply directly accounts for the abundance of suspension feeders, which in turn may affect the pelagic system in lagoons and estuaries. After remineralization of the organic import, the released nutrients support a rich microphytobenthos. This in turn provides food for grazers and deposit feeders.

Suspension feeders tend to dominate the zoobenthos of the more exposed shores, provided the adjacent sea is a productive source region. Grazers and deposit feeders, on the other hand, tend to dominate at the more sheltered shores. Intermediate sedimentary shores with respect to hydrodynamics and sand-mud composition are supposed to be the richest in production, zoomass and diversity of the organisms. Superimposed on these are the gradients in salinity and climate: species richness in the benthic communities generally increases with salinity and mean temperature.

Important are transformations of the sedimentary environment by the organisms at the sheltered shores, and increasingly towards lower latitudes. Sediment-consolidating and sediment-disturbing organisms occur in dynamic patterns of alternating zones perpendicular to the shore and in distinct patches. Biotic interactions often determine whether accretion or erosion prevails. Reef-like structures of suspension feeders, beds of seagrasses and salt marsh plants not only stabilize and accrete sediment but also modify nearshore hydrodynamics. This has cascading effects on habitat structure and the biotic composition. Therefore the worldwide decline in seagrass beds, ultimately caused by anthropogenic habitat deterioration, has far-reaching consequences for the ecology of sedimentary shores.

The mobility of benthic organisms by passive drift or active migrations is a major agent in spatio-temporal dynamics of the biota of sedimentary shores. This mobility has regular tidal, diurnal, lunar and seasonal components, and the recovery after disturbances is governed by differential dispersal. The vagaries, ranges and directions of dispersal need further investigations to include such processes in large-scale comparisons.

Large animals are usually not residents on sedimentary shores but are flood-tide, ebb-tide or seasonal visitors that exploit opportunistically the food resources where these are rich. They mostly export their gains out of the ecotone. Global patterns on the amount of their gains and on their impact on the resident organisms need more attention in future studies. The flood-tide visitors have rarely been quantified sufficiently. The larger these organisms

Fig. 5. Similar surface topographies with fecal mounds are generated by very different deposit feeders: enteropneust worms in the Andaman Sea, callianassid shrimps in southern Chile, lugworms in the North Sea

are, the more they have been affected by humans. The spatio-temporal match between young epibenthic predators and the recruits of their prey are more important for population dynamics than the large predators which consume most of the zoomass.

On a global scale, sandy beaches of exposed coasts are very similar in taxonomic composition and dominant forms of life. Sheltered sand and mud flats exhibit more variety, including regionally unique features, particularly in the tropics. Versatile species may switch their food acquisition and this entails changes up to the ecosystem level. On the other hand, very different forms of life, such as lugworms and callianassid shrimps, may attain very similar roles in feeding, transforming the sediment and affecting other species (Fig. 5). A comparative approach to the biota of sedimentary shores on a worldwide scale may greatly advance our insight and give more power to our judgements and recommendations. To achieve this we need to proceed from studies devoted to the analysis of contingent details on a local scale to an integration of knowledge under a global perspective on the ecology of sedimentary shores.

References

Connolly SR, Roughgarden J (1998) A latitudinal gradient in Northeast Pacific intertidal community structure: evidence for an oceanographically based synthesis of marine community theory. Am Nat 151:311–326

Daborn GR, Amos CL, Berlinsky M, Christian H, Drapeau G, Faas RW, Grant J, Long B, Paterson DM, Perillo GME, Piccolo MC (1993) An ecological 'cascade' effect. Migratory birds affect stability of intertidal sediments. Limnol Oceanogr 38:225–231

Green MO, Hewitt JE, Thrush SF (1998) Seabed drag coefficients over natural beds of horse mussels (*Atrina zelandica*). J Mar Res 56:613–637

Johnson RG (1973) Conceptual models of benthic marine communities. In: Schopf TJM (ed) Model in paleobiology. Freeman and Cooper, San Francisco, pp 148–159

Möbius K (1877) Die Auster und die Austernwirtschaft. Wiegandt, Hempel Parey, Berlin

Reise K (1987) Experimental analysis of processes between species on marine tidal flats. In: Schulze E-D, Zwölfer H (eds) Potentials and limitations of ecosystem analysis. Ecological studies, vol 61. Springer, Berlin Heidelberg New York, pp 391–400

Species Index

Page numbers in *italics* denote figures, photos, or tables.

Abarenicola pacifica   167
Abra tenuis   262, *264*, 265
Acartia tonsa   76
Acesta lopezi   283
Achnanthes longipes   118
Aedicira cf. *belgicae*   262
Alpheus sp.   287
Ampelisca abdita   82
Ampelisca brevicornis   262, *264*
   vadorum   82
Amphibolis antarctica   224
Amphipholis geminata   283
   squamata   81
Amphiura chiajei   81
   filiformis   78, 79, 80, 81
Amphora cymbifera   117
Amygdalum glaberrima   283
Anadara senilis   258, 261, *262*, *263*, 265, 268, 270
Anas penelope   138
Anguilla anguilla   229–230
Angulus tenuis   155
Anser caerulescens   138
Anthuridae sp.1   262, *264*
Apseudidae sp.1   261
Arabella iricolor   262
Arachnoides placenta   287
Arca afra   262
   subglossa   262
Arctica islandica   14
Arenicola marina   75–76, *84,* 130, 134, 138, 140, 150–*152*, 155, 157, 182, 248, 266–268, 270, 288, 342, 365
Argopecten irradians   14
Arcidiea sp.   262
Armandia amakusaensis   164
   secundariopapillata   283

Armeria maritima   243
Arthrocnemum spp.   241
Ascidiella aspersa   14
Asterias rubens   42, *230*
Atrina zelandica   49, 51
Atriplex portulacoides   243
Avicennia africana   260
Axiothella sp.   262

Bacillaria paxillifer   118
Balanus crenatus   14, 25
   perforatus   14
Barantolla sp.   283, 287
Boccardia polybranchia   77
   proboscidea   80
   pugettensis   80
   syrtis   302
Brachidontes rostratus   46
Branchiostoma lanceolatum   76
   senegalense   262
Branta bernicla   138, 141
Bullaria adansoni   262

Calidris canutus   320
Callianassa acanthochirus   132
   californiensis   131, 134–135
   filholi   133, 138
   harmandi   158
   japonica   132, 151, 158, *288*–289
   longiventris   132
   petalura   158
   subterranea   133, 138
Callinectes sp.1   262
   sapidus   136
Cancer pagurus   136

Capitella capitata 48, *155*, *262*, 265
Capitella sp. 341, 346
Caraziella citrona 287
Carcinus sp.1 262
  maenas 53, *157*, 229–230, 251, 326, 343
Cardium echinatum 14
  edule 224
Caulerpa prolifera 200
  taxifolia 207
Celleporella hyalina 76
Ceramium rubrum 200
Cerastoderma edule 13, *14*, 19, 93, *155*,
  181–182, *262*, 266–268, *270*, 320–
  323, *325–328*
  glaucum 224
Cerithium zonatum 283
Chaetopterus variopedatus 15, 76, 88
Chara spp. 217
Chlamys hastata 14
Chorda filum 222
Choromytilus chorus 49
Ciona intestinalis 14, 20, 24, 76
Cirriformia sp. 262
Cirriformia tentaculata 262
Clavatula sp. 262
Coricuma nicoyensis 283
Corophium sp. 287
  arenarium 156
  insidiosum 83, *222*, 224–225, 228
  salmonis 83
  spinicorne 83
  volutator 79, 83, 139, 152, 156–*157*,
    177, 181–186, *342*, 346
Coscinodiscus sp.1 116
Crangon crangon 150, *157*, 230, 266, 343
Crassostrea gigas 14, 19
  virginica 13–*14*, 19, 29–30, 53
Crenella dollfusi 262
Crepidula sp. 262
  fornicata 15, 25
Cychlina grimaldii 262
Cylindrotheca closterium 118
  signata 116
Cymadusa hirsuta 262, 264
Cymatosira belgica 116
Cymbium sp. 262
Cymodocea nodosa 197, 200, 243
  rotundata 199
  serrulata 199
Cyprideis pacifica 283
Dasyatis akajei 163

Dasybranchus lumbricoides 283
Dendropoma irregulare 19
  petraeum 19, 21
Diopatra neapolitana 262, 264
  ornata 287
Donax spp. 61, 358
  hanleyanus 70
  serra 70
  sordidus 70
  variabilis 359
Dosinia sp. 283
  cf. *lupinus* 262
  dunkeri 283
Dreissena polymorpha 50
Drilonereis cf. *monroi* 262
  filum 262
Dugong dugong 138

Ectocarpus confervoides 200
Emerita spp. 61, 358
  agree 66–68
  analoga 62, 64–70
  benedicti 66
  brasiliensis 66
  portoricensis 66, 359
  rathbunae 66
  talpoida 66
Encope stokesi 287
Enhalus acoroides 199
Enhydra lutris 136
Ensis spec. 155
  directus 142
Enteromorpha sp. 180, 346
  radiata 207
Entomoneis paludosa 116
Ericthonius difformis 222, *224–225*, 228
  pugnax 224
Eschrichtius robustus 134
Eteone cf. *foliosa* 262
  longa 266
  siphodonta 262
Euclymene cf. *natalensis* 262
  lüderitziana 262, 264
Eurydice nipponica 164

Festuca rubra 243
Ficopomatus enigmaticus 15, 20, 23–24
  see *Mercierella enigmatica* 23
Ficulina ficus 15

Species Index

Fucus serratus 221
  vesiculosus 221-222
Gammarus spp. 222, 224
Gasterosteus aculeatus 230
Gersemia antarctica 83
Geukensia demissa 14, 42-43
Gibbula cineraria 222
  umbilicalis 262
Gloecapsa sp. 115
Glottidia audebarti 283-284, 287
Glyceria convoluta 262
Gobius niger 229-230
Golfingia sp. 262
Gymnocephalus cernuus 229
Gyrosigma balticum 106
  limosum 116, 118
    (formerly *spencerii*)

Haematopus ostralegus 320
Halichondria panicea 15, 76
Halimione portulacoides 241
Halodule beaudettei 198
    see *H. wrightii*
  emarginata 199
  uninervis 199
  wrightii 198-199, 203
    see *H. beaudettei*
Halophila beccari 199
  ovalis 198-199, 204
  pinifolia 199
  stipulacea 205, 364
Haminea elegans 262
  orbignyana 262
Hantzschia virgata 117
Haploscoloplos elongatus 283
Harmothoe sarsi 166
Harpinia sp. *pectinatus* 262
Haustorius arenarius 265
Hediste diversicolor 342 see *Nereis diversicolor*
Herposiphonia secunda 21
Heteromastus filiformis 48, 138, 155, 262, 264, 266-267
Heterozostera tasmanica 198
Hyale perieri 262
Hyboscolex longiseta 261, 262
Hydrobia spp. 222-224, 228
  ulvae 134, 177, 181, 184, 186, 248, 262, 264, 266-267, 342, 365
Hydroides norvegica 15

Hypsicomus cf. *capensis* 262

Idotea baltica 222, 224
  chelipes 262, 265

Jaera albifrons 222

Labyrinthula macrocystis 206
  zosterae 201, 206
Lanice conchilega 15, 74, 80, 263, 266-267
Larus ridibundus 136
Larus sp. 326
Leprochelia ignota 224
Leptocheirus pilosus 222
Leucothoe richiardi 262
Limulus polyphemus 136
Lingula anatina 283-284, 287
Littorina spp. 222, 224
  littorea 43, 53
Loimia sp. 287
Loripes lacteus 262, 264,
Lumbrinereis spp. 287
  heteropoda 262
Lymnaea spp. 223-224
Lysianassa ceratina 262

Macoma balthica 73. 76, 81-82, 84, 90-95, 140, 154-155, 181-182, 184, 248, 266-268, 270, 320-323, 325-330, 360
  liliana 305
  nasuta 82
Macroclymene monilis 262, 264
Macrophthalmus latreillei 287
Manayunkia aestuarina 77, 79
Marenzelleria viridis 77
Marphysa sanguinea 262, 264
Mediomastus sp. 287
  californiensis 283
Mercenaria mercenaria 14, 18, 19
Merismopedia sp. 115
Mesodesma spp. 61, 358
  mactroides 70
Microcoleus chthonoplastes 115
Microdeutopus gryllotalpa 222, 224, 228
Mictyris longicarpus 287, 289, 366
Modiolus modiolus 14

Musculista senhousia 50
Mya arenaria 155, 266–268, 270
  truncata 135
Mycale sp. 76
Myliobatis californica 136
  tenuicaudatus 136
Myriophyllum spp. 217
Mytilus spp. 224
  californianus 39
  chilensis 49
  edulis 12–13, 15, 19, 39, 41–43, 45–53, 76, 82, 84, 216–217, 223, 228, 263, 359
  galloprovincialis 46–47
  trossulus 217
Myxicola infundibulum 15

Nainereis laevigata 262, 264
Nassarius spp. 262, 287
Natica spp. 287
Navicula cancellata 117
  flanatica 116
  pargemina 116
  pelliculosa 118
  perminuta 118
  phyllepta 116
Nebalia cf. *bipes* 262
Nematonereis unicornis 262
Nephtys hombergi 155, 266–267
Nereis sp.2 262
  caudata 262
  diversicolor 73, 76–77, 84–90, 95, 155, 177, 181, 184–186, 248, 262, 267, 342, 360–361
Nerophis ophidion 230
Nihonotrypaea harmandi 151, 159–165
  japonica 158–162,
Nitzschia epithemioides 116
Nitzschia sp. 115
  sigma 118
  spathulata 117
Notomastus sp. 287
Nucella lapillus 53

Ophiactis lymani 262
Ophiothrix fragilis 14, 25
Oscillatoria spp. 180
  limosa 114
Ostrea edulis 15
Owenia fusiformis 77

Owenia spp. 287

Palaemon adspersus 229–230
  cf. *elegans* 262, 265
Paracaudina sp. 287
Paraonis lyra 262
Paraprionospio pinnata 77, 283
Penaeus cf. *kerathurus* 262
Perca fluviatilis 229–230
Perinereis cultrifera 262, 264
Persicula chudeani 262
Petaloproctus terricola 262, 264
Phacoides adansoni 262
Phaeodactylum tricornutum 115
Phallusia mammillata 14
Phormidium sp. 115
Phragmatopoma lapidosa 20, 22–23
Phragmites australis 217
Phyllochaetopterus limnicolus 75
  prolifica 75, 79
Pinnixia valerii 283
Platichthys flesus 229–230
Polinices spp. 287
Polycirrus aurantiacus 262, 265
Polydora antennata 262
  cornuta 77, 302
Pomatoceros triqueter 15
Pomatoschistus microps 250, 343
  minutus 230, 343
Pontocrates arenarius 265
Posidonia australis 204
  oceanica 197, 203–204, 207
Potamocorbula amurensis 18–19
Potamogeton spp. 217
  pectinatus 197, 221
Prionospio spp. 287
Pseudopolydora kempi 80
  paucibranchiata 164
Puccinellia maritima 243
Pungitius pungitius 230
Pygospio elegans 77, 80, 155, 167, 266–267, 361

Ranunculus spp. 217
Raphoneis minutissima 116
Rhodine sp. 262
Rissoa spp. 222, 224
Ruppia spp. 216
  cirrhosa 197

Species Index

Rutilus rutilus 229–230

Sabella penicillus 15, 23, 76
   spallanzanii 15, 25
Sabellaria alveolata 20, 22
Salicornia spp. 180, 182, *184*
   stricta 243
Salmancina dysterii 15
Sargassum muticum 207
Schizobranchia insignis 15
Scionella lornensis 262
Scirpus spp. 217
   americanus 138
Scolelepis squamata 77, 79, 261–262,
   265–266, 268
Scoloplos sp.1 262, 264
   armiger 155, 266–267
   chevalieri 262
Scopimera inflata 287
Scrobicularia plana 79, 81, 248
Semibalanus balanoides 53, 339
Serpula narconensis 20
   vermicularis 20
Sigambra ocellata 283
Siphonostoma typhle 230 see *Syngnathus typhle*
Sipunculus nudus 283, 287
Solea senegalensis 250
Somateria mollissima 136
Southernia zosterae 225, 228
Spartina spp. 180, *184*
   alterniflora 250
   anglica 177, 243
   maritima 241, 243, 250
Spinachia spinachia 230
Spio setosa 77, 80
Spiochaetopterus sp. 75
   costarum 75, 288
   oculatus 76, 79
Spiophanes bombyx 77
Spirorbis sp. 262
   borealis 15,
Spirulina sp. 115
Styela clava 14
Suaeda vera 241
Syngnathus typhle 229 see *Siphonostoma typhle*
Syringodium filiforme 198
   isoetifolium 199

Tadorna tadorna 136
Tagelus angulatus 262
   plebeius 13
Tanais dulongii 262
Tapes semidiscussata 19
Tellina australis 283
   fabula 82
   opalina 360
   tenuis 82
Terebella lapidaria 262
Tethya crypta 76
Thalassema steinbecki 287
Thalassia hemprichii 199, 201
   testudinum 198, 202–203
Thalassodendron ciliatum 199
Tharyx dorsobranchialis 262
Tharyx sp. 262
   marioni 155
Theodoxus fluviatilis 223–224
   oualaniensis 283
Thioploca spp. 117
Trichobranchus glacialis 262
Trypea australiensis 287–289
Tubifex costatus 267
Tubificoides benedeni 47–48
Turritella torulosa 262
Typosyllis mauretanica 262

Uca spp. 287–288
   tangeri 261–262, 264–265, 270
Umbonium moniliferum 159–165
Urolophus halleri 136
Urothoe elegans 262
   grimaldii 262, 266–267
   poseidonis 166

Venerupis aurea 262
Venus rosalina 262
Vermetus triquetrus 19
Verongia fistularis 76
   gigantea 76
Victoriopisa atlantica 262

Yoldia limatula 134

Zannichellia palustris 216
Zoarces viviparus 229–230

Zostera asiatica 197
  caespitosa 197
  capensis 198
  capricorni 198
  caulescens 197
  japonica 197, 205–206, 365
  marina 141, 173, *196*–198, 200–201, *203*–206, 213–219, 220–231, 243, 364

Zostera mucronata 198
  muellerii 198
  noltii 138, *177,* 197, 204, 206, 242–243, 260, 268, 365
  novozelandica 198

Subject Index

Page numbers in *italics* denote figures, photos, or tables.

adaptation   73, 84, 88
adjustment stability   40
adult-larval interactions   39, 46
adventive algae   207
aggregate   11
Aktuopaläontologie   131
algal blankets   207
– film   201
allochthonous import   338, 340
alternative feeding mechanisms   74
– feeding mode   90
– stable state   177
Andaman Sea   269
annual net production   203
antipredator adaptions   331
Ardbear Lough   *20*
areas (ant)arctic   140
– temperate   140
– tropical   140
Ariake Sound   158
artifical habitat units (HUs)   348–349
assemblages   284
autocatalytic   30

Bacillariophyceae   105 see diatom
baffles   187
Balgzand   319
Banc d'Arguin   255, 257, *271*
Bay of Brest   *20, 28*–29
beach morphodynamics   62
– state index   62
bed roughness   43
– shear stress, critical   179, 188
benthic microalgae   357
benthic-pelagic coupling   11
    see benthopelagic

benthopelagic coupling   6, 358
Bermuda   *19*
biodeposition   51
biodeposits   13
biodiversity   2, 113, 347 see diversity,
    species diversity
biofacies   128
biofilm   180
biogenic reefs   306
– stabilization   118, 361
biogeographic regions   165, 275, 289
biological interactions   70
biomass   202
– size spectra   345
– spectra   344–345
biomixing   89
biostabilization   119
biotic interactions   303
bioturbation   127–*129,* 140, *162,* 164,
    177, 189, 339
bioturbator   166, 362, 363
birds   183–184, 188, 250
blackheaded gulls   136
body-size distribution   343–344
bootstrap procedure   321
bottom-up   39
– effect   339, 342
– forces   251
boundary layer   109
– flow   43
brackish   364
browsing   *246*
brushwood groynes   183
burrowing rates   68
byssus threads   12

Subject Index

C/N ratio   218
callianassid crustaceans   132
Carlingford Lough   19, 27–28
castings   130
Chesapeake Bay   18–19, 28–30
Chilean coast   64
chlorophyll a   81
ciliary mucoid mechanism   76
clams solitary   27
clearance rate   14, 28
– time   17, 29
Clementsgreen Creek   176
coastal squeeze   174, 189
Cockburn Sound   20, 25
cohesive particles   108
– sediment   108, 115, 180, 362
community dynamics   338
structure   46, 199
competition   289
– intraspecific   152
competitive exclusion   289
constraint space   344–345
consumer-resource dynamics   340–341
consumption   317, 322
– production ratio   317, 324
conveyer-belt species   134
Costa Rica   269
current scouring   135
– velocity   27
Cyanobacteria   107, 180

Dean's parameter   61
Delaware Bay   19, 28–29, 30
density dependence   319, 328
– maximum   14
density-governing mechanisms   152
depletion   17
deposit feeder   4, 129, 133, 276, 284, 286, 289, 360, 363, 369–370
– subsurface   73
deposit-feeding   73, 78–79, 83, 246, 362, 366
– behaviour   90, 92
– surface   73, 77, 79, 85, 90
detrital pathway   201
detritivores   214
detritus   220, 245, 248–249
– feeder   73
– cycle   271

diatom   107, 139 see Bacillariophyceae
– migration   116
diffusion model   89
direct interception   80
disturbance   297, 367, 369
diurnal migration   202
diversity   348, 350 see biodiversity, species diversity
– of microphytobenthos   116
ducks   138
dugongs   138

earthworm   130
East Atlantic Flyway   255–256
East Pacific Barrier   275
ecological quality of habitats   202
– roles   286–287, 289
ecosystem engineers   12, 283, 290
– health   309
eddy diffusion   110
edges   42
eelgrass   213 see seagrass
effective fetch   217
Ellis Fjord   20
encrustations   24, 27
energy density   321
– flow   248–249, 338–339
environmental conditions   278
– energy   127
epifauna, mobile   219
epipelon   113
epiphyte   220
– production   203
epiphytic algae   200
epipsammic   114
epipsammon   113
erosion   173, 175, 183, 185–189, 190, 204
estuarine system   158
eustatic SLR   174
eutrophication   5, 364, 368
evolution   29
exclosure experiments   329
exopolymers   139
experimental manipulative approach   338–339
exploitation   5
Extracellular Polymeric Substances (EPS)   108, 111, 118

Subject Index

F/R-value 75, 84
facultative deposit/suspension feeder 90
fecundity 327
feedback loop 18, 30
- positive 18, 30
feeding behaviour 74, 86, 91
- holes 134, 136
- mode 89
- on siphon tips 94
- opportunistic 89
- type, classification 73
field experiments 153
filter feeder 11, 22, 73, 214
- feeding 161
filter-net structure 86
filter-pump effiency 84
filtration capacity 29, 87, 89
- rate 17, 82, 87
fished areas 306
fishing activity, effects 141
flumes 18
food chain dynamics 340
- web 338, 340–341
food-trapping 86
fouling 220, 227
fractal dimension 45
- geometry 44
frame element 199

gardening 134
geese 138–139
generation time 330
geographical distribution 195
global warming 173
gray whales 134, 139
grazer 364, 369
grazing 17–18, 30, 358, 362–363, 365 see grazer
- impact 89–90
- pathway 201
green algae 5, 219, 242, 244, 368
green macroalgae 341
gregarious 23, 29
group living 39
growth 327
- forms 199
- parameters 67
gut absorption effiency 321

habitat destruction 142
- diversities 284
- hydrodynamically isolated 308
- island 46
herbivores 214, *246*
herbivory 177, 189
hierarchy theory 300
high sands 260, 267–268
high-energy environments 128
high-water roosts 320
human activity 4, 5
hummocking 45
hummocks 25, 29
hydrodynamics 301

ice scour 43
ichnology 132
immigrants, limnetic 216
- marine 216
indigenous brackish water species 216
infaunal assemblage 46
- species 13
instantaneous death rate 324–325
interaction 290
- biotic 149
- interspecific 167
- intraspecific 168
- modification 166
- species 286
interface feeders 74
Intermediate Disturbance Hypothesis 114
International Biological Programme (IBP) 338–339, 350
introduced species 5
invasive mussels 50
irrigation 134

Juist Area 255, 258, 269, *271*

Kertinge Nor *20*, 24, *28*
key-species 167
keystone predator 339
Königshafen *19, 28*

Lac de Tunis *20*
laminar flow 109

Langebaan Lagoon   269
leaf age   204
- canopy   214, 227
- fauna   220
life-history traits   331
light-deficit hypothesis   206
Loch Creran   20
long-term change   *162*
low-energy pump   84
lugworm   75, 151

macroalgal mats   341 see green macroalgae
macrobenthic abundances   282
- animals   150
- assemblage   282
- communities   297
macroinfauna   62
    see infaunal assemblage
macrophyte   338
managed realignment   173, *176*, 178, 180, 182–184, 190
- retreat   176
mangrove   265
Marennes-Oléron   *19*, *28*–30
Marina da Gama   *20*, *28*
Mauritanian Coast   255
meiofauna   346
mesocosms   18
meta-analysis   299, 339
metabolic rate   321
micro-atoll   21
microbial mats   267
microphytobenthos   105, 179–181, 241, 244, 362–363, 365, 369
microzooplankton   18
minimal scaling   84
mixed lawns   199
mixed-layer   131
mobility   302
Mont Saint-Michel Bay   *20*
morphodynamic state   338
mortality-summation method   318, 320
mucus filter-bag feeding   75
mud flat   184, *246–247*
- mounds   187
- residual   186
mussel   358–*359*, 365
- bed   18, 39, *246–247*, 266
- density manipulation   49

Narragansett Bay   *19*, *28*
near-bottom flow   51
- phytoplankton depletion   89
New Caledonia   269
New England   298
New Zealand   298
North Inlet   *16*, *19*, *28*
numeral response   328
nursery function   195
- grounds   *159*
nutrient   341
- enrichment   345
- cycle   11, 248
NW-African upwelling area   258

oligochaete   47
omnivory   *246*, 248
Oosterschelde   *19*, *28*
opportunistic responses   302
Orplands   176
oystercatcher   320
oysters   358

palaearctic waders   256
particle-retention effiency   86
patches   42
patchy distribution   68
pattern and process   338
persistence stability   40
Peterson-Thorson communities   337
physical environment   168
phytoplankton   4, 13, 17, 242, 244–245, 357–358, 360, 365
phytoplankton-reduced near-bottom water   89
pioneer vegetation   189, 267
planktonic larvae   308
population abundances   67
- dynamics   160, 319
porosity   110–111
power output   88
predation   42, 89, 288, 361, 365 see predator
predator   157, 214, *246*, 250, 360, 367–368, 371
predator-prey system   329
primary producers   341
production   317, 321–323, 338, 340
- and biodiversity   347

Subject Index

- biomass ratios (P/B ratio)   318, 321
- primary   240–241, 243–244, 270, 341
- secondary   240, 242, 245, 320

pumping costs   88
- rate   82

ragworm   85
ray   136
recovery recent   165
recruitment   45, 156, 320, 326–327, 365
reefs   12–13, 16, 18, 21, 23–24, 27,
regulation   326, 328
residence time   17
resilience   309
Resource Heterogeneity Hypothesis
    (RHH)   348, 350
retention effiency   81
rheotactic response   81
rhizophytic algae   200
Ria de Arosa   19, 28
Rio Formosa   237–239, 245, 249
roughness   29

Sahara Desert   256
salinity   215–216, 219
- gradient   215
- stress   216
saltmarsh   184, 189, 241, 243, 245–247,
    363, 365, 369
- creek   175, 183, 185–187, 189
- erosion   173, 189
- restoration   173
- vegetation   185
San Francisco Bay   29–30
sand flat   160, 246, 247
sandy beach   61, 358, 359 see beach
scale   330
scale-depency   342
scales of mobility   308
scaling-up   298
scavenging   89
seagrass   4–6, 241, 348, 357, 364–365, 369
    see eelgrass
- beds   243, 246, 264–265, 270
- settlement   205
sea-level rise (SLR)   173–174
seasonality   301
sebkha   260, 265, 268–269
sediment accretion   181–182

- non-cohesive   114
- particle size   338
- properties   163
- resuspension   179
- reworking   129
- stability   180
- transport   127
- trapping   51
- water content   337
SE-England   184
self-enhancing   30
Senckenberg am Meer   131
set-back   176
shear stress   179
shelducks   136
shell deposits   266
shellfisheries   141
shifting baselines   309
Shikmona   19
shore crabs   326
shorebirds   342–343
siphon nipping   94
small-scale experiments   304
soft sediment   297
soft-bottom   39
soft-corals   83
South Carolina   16
South San Francisco Bay   19, 28
Southeast Florida   20
Spain   69
spatial and temporal complexity   39
- distribution   150, 284, 289
- heterogeneity   297
- scale   42, 297, 330
- structure   41
- zonation   284–285
species diversity   276, 279, 289, 362
- interactions   286 see interaction
- richness   4, 202, 279, 364–365, 369
stability   40, 118–119
stingray   163
stock-recruitment relationship   327, 329
storm damage   43
strategies   298
stromatolites   113
subsidies   340–343
substrate alteration   195
- extension   195
- protection   195
succession   204, 299
- classical model   299

– recovery end points 299
surface topography 43
survival 324
suspension feeder 4, 6, 24, *26, 28,* 64, 134, 244, 357–358, 360–361, 365, 369
suspension-feeding 11, 23, 27, 40, 73, *79,* 85, 88, 91, *246*
– active 84
– behaviour 92
– costs 85
– mechanism 80
– obligate 82, 84
– passive 76, *79,* 83
switching 73, 91
Sylt *3,* 188
Sylt-Rømø Bay 237–239, *245, 249* see Königshafen
synchronous spawning 29
system clearance time *26, 28*

taxonomic similarity 281
thermoneutrality 321
threshold, critical 306–307
tidal currents 128, 186
– migration 70
– movement 68
time share 89
Tollesbury 174, *176,* 183
Tomioka Bay *160*
top-down 39
– effect 339, 342–343
– forces 251
– processes 228
trace crawling 132
– dwelling 132
– escape 132
– fossils 129
trampling 136
trigger level 89
trochid gastropod 161
trophic-groups 284–285

trophic pathway 357
tropical tidal flats 277, 281, 283, 286, 290
turbulence 29
turbulent flow 109
Turion 221
turnover rate 204

universal zonation pattern 337 see zonation
unmanged realignment site 183

variation latitudinal 139
– seasonal 138, 139
vertical distribution 196
– profiles 45
viscous sub-layer 112

Wadden Sea 19, 28–30, 141, 151, 270, 319
Wallasea Island 176
walrus 135
wasting disease 201
– phenomenon 206
water volume residence time 25–*26, 28*
wave 178
– action 182
– breaks 183, 188
weight-specific filtration rate 82
– pumping rate 82
wind-driven vertical mixing 89
wind-mixing 89

year-to-year patterns 204
Ythan estuary 340–342, *344,* 346–349

zonation 1–2, 166, 196, 284–285, 337
zoned distribution 276
zoomass 2, 4–6, 369, 371

Ecological Studies
Volumes published since 1995

Volume 110
Tropical Montane Cloud Forests (1995)
L.S. Hamilton, J.O. Juvik, and F.N. Scatena (Eds.)

Volume 111
Peatland Forestry. Ecology and Principles (1995)
E. Paavilainen and J. Päivänen

Volume 112
Tropical Forests: Management and Ecology (1995)
A.E. Lugo and C. Lowe (Eds.)

Volume 113
Arctic and Alpine Biodiversity. Patterns, Causes and Ecosystem Consequences (1995)
F.S. Chapin III and C. Körner (Eds.)

Volume 114
Crassulacean Acid Metabolism. Biochemistry, Ecophysiology and Evolution (1996)
K. Winter and J.A.C. Smith (Eds.)

Volume 115
Islands. Biological Diversity and Ecosystem Function (1995)
P.M. Vitousek, L.L. Loope, and H. Adsersen (Eds.)

Volume 116
High Latitude Rainforests and Associated Ecosystems of the West Coast of the Americas: Climate, Hydrology, Ecology and Conservation (1996)
R.G. Lawford, P. Alaback, and E. Fuentes (Eds.)

Volume 117
Global Change and Mediterranean-Type Ecosystems (1995)
J. Moreno and W.C. Oechel (Eds.)

Volume 118
Impact of Air Pollutants on Southern Pine Forests (1996)
S. Fox and R.A. Mickler (Eds.)

Volume 119
Freshwater Ecosystems of Alaska. Ecological Syntheses (1997)
A.M. Milner and M.W. Oswood (Eds.)

Volume 120
Landscape Function and Disturbance in Arctic Tundra (1996)
J.F. Reynolds and J.D. Tenhunen (Eds.)

Volume 121
Biodiversity and Savanna Ecosystem Processes. A Global Perspective (1996)
O.T. Solbrig, E. Medina, and J.F. Silva (Eds.)

Volume 122
Biodiversity and Ecosystem Processes in Tropical Forests (1996)
G.H. Orians, R. Dirzo, and J.H. Cushman (Eds.)

Volume 123
Marine Benthic Vegetation. Recent Changes and the Effects of Eutrophication (1996)
W. Schramm and P.H. Nienhuis (Eds.)

Volume 124
Global Change and Arctic Terrestrial Ecosystems (1997)
W.C. Oechel et al. (Eds.)

Volume 125
Ecology and Conservation of Great Plains Vertebrates (1997)
F.L. Knopf and F.B. Samson (Eds.)

Volume 126
The Central Amazon Floodplain: Ecology of a Pulsing System (1997)
W.J. Junk (Ed.)

Volume 127
Forest Decline and Ozone: A Comparison of Controlled Chamber and Field Experiments (1997)
H. Sandermann, A.R. Wellburn, and R.L. Heath (Eds.)

Volume 128
The Productivity and Sustainability of Southern Forest Ecosystems in a Changing Environment (1998)
R.A. Mickler and S. Fox (Eds.)

Volume 129
Pelagic Nutrient Cycles: Herbivores as Sources and Sinks (1997)
T. Andersen

Volume 130
Vertical Food Web Interactions: Evolutionary Patterns and Driving Forces (1997)
K. Dettner, G. Bauer, and W. Völkl (Eds.)

Volume 131
The Structuring Role of Submerged Macrophytes in Lakes (1998)
E. Jeppesen et al. (Eds.)

Volume 132
Vegetation of the Tropical Pacific Islands (1998)
D. Mueller-Dombois and F.R. Fosberg

Volume 133
Aquatic Humic Substances: Ecology and Biogeochemistry (1998)
D.O. Hessen and L.J. Tranvik (Eds.)

Volume 134
Oxidant Air Pollution Impacts in the Montane Forests of Southern California (1999)
P.R. Miller and J.R. McBride (Eds.)

Volume 135
Predation in Vertebrate Communities: The Białowieża Primeval Forest as a Case Study (1998)
B. Jędrzejewska and W. Jędrzejewski

Volume 136
Landscape Disturbance and Biodiversity in Mediterranean-Type Ecosystems (1998)
P.W. Rundel, G. Montenegro, and F.M. Jaksic (Eds.)

Volume 137
Ecology of Mediterranean Evergreen Oak Forests (1999)
F. Rodà et al. (Eds.)

Volume 138
Fire, Climate Change and Carbon Cycling in the North American Boreal Forest (2000)
E.S. Kasischke and B. Stocks (Eds.)

Volume 139
Responses of Northern U.S. Forests to Environmental Change (2000)
R. Mickler, R.A. Birdsey, and J. Hom (Eds.)

Volume 140
Rainforest Ecosystems of East Kalimantan: El Niño, Drought, Fire and Human Impacts (2000)
E. Guhardja et al. (Eds.)

Volume 141
Activity Patterns in Small Mammals: An Ecological Approach (2000)
S. Halle and N.C. Stenseth (Eds.)

Volume 142
Carbon and Nitrogen Cycling in European Forest Ecosystems (2000)
E.-D. Schulze (Ed.)

Volume 143
Global Climate Change and Human Impacts on Forest Ecosystems: Postglacial Development, Present Situation and Future Trends in Central Europe (2001)
J. Puhe and B. Ulrich

Volume 144
Coastal Marine Ecosystems of Latin America (2001)
U. Seeliger and B. Kjerfve (Eds.)

Volume 145
Ecology and Evolution of the Freshwater Mussels Unionoida (2001)
G. Bauer and K. Wächtler (Eds.)

Volume 146
Inselbergs: Biotic Diversity of Isolated Rock Outcrops in Tropical and Temperate Regions (2000)
S. Porembski and W. Barthlott (Eds.)

Volume 147
Ecosystem Approaches to Landscape Management in Central Europe (2001)
J.D. Tenhunen, R. Lenz, and R. Hantschel (Eds.)

Volume 148
A Systems Analysis of the Baltic Sea (2001)
F.V. Wulff, L.A. Rahm, and P. Larsson (Eds.)

Volume 149
Banded Vegetation Patterning in Arid and Semiarid Environments (2001)
D. Tongway and J. Seghieri (Eds.)

Volume 150
Biological Soil Crusts: Structure, Function, and Management (2001)
J. Belnap and O.L. Lange (Eds.)

Volume 151
Ecological Comparisons of Sedimentary Shores (2001)
K. Reise (Ed.)

Printing (Computer to Film): Saladruck, Berlin
Binding: Stürtz AG, Würzburg